AF847677

Power Systems

Electrical power has been the technological foundation of industrial societies for many years. Although the systems designed to provide and apply electrical energy have reached a high degree of maturity, unforeseen problems are constantly encountered, necessitating the design of more efficient and reliable systems based on novel technologies. The book series Power Systems is aimed at providing detailed, accurate and sound technical information about these new developments in electrical power engineering. It includes topics on power generation, storage and transmission as well as electrical machines. The monographs and advanced textbooks in this series address researchers, lecturers, industrial engineers and senior students in electrical engineering.

Power Systems is indexed in Scopus

More information about this series at https://link.springer.com/bookseries/4622

Vahid Vahidinasab • Behnam Mohammadi-Ivatloo
Editors

Whole Energy Systems

Bridging the Gap via Vector-Coupling Technologies

Springer

Editors
Vahid Vahidinasab
Department of Engineering, School of Science and Technology
Nottingham Trent University
Nottingham, UK

Behnam Mohammadi-Ivatloo
Department of Electrical and Electronics Engineering
Muğla Sıtkı Koçman University
Muğla, Turkey

Faculty of Electrical and Computer Engineering
University of Tabriz
Tabriz, Iran

ISSN 1612-1287 ISSN 1860-4676 (electronic)
Power Systems
ISBN 978-3-030-87652-4 ISBN 978-3-030-87653-1 (eBook)
https://doi.org/10.1007/978-3-030-87653-1

© The Editor(s) (if applicable) and The Author(s), under exclusive license to Springer Nature Switzerland AG 2022
This work is subject to copyright. All rights are solely and exclusively licensed by the Publisher, whether the whole or part of the material is concerned, specifically the rights of translation, reprinting, reuse of illustrations, recitation, broadcasting, reproduction on microfilms or in any other physical way, and transmission or information storage and retrieval, electronic adaptation, computer software, or by similar or dissimilar methodology now known or hereafter developed.
The use of general descriptive names, registered names, trademarks, service marks, etc. in this publication does not imply, even in the absence of a specific statement, that such names are exempt from the relevant protective laws and regulations and therefore free for general use.
The publisher, the authors and the editors are safe to assume that the advice and information in this book are believed to be true and accurate at the date of publication. Neither the publisher nor the authors or the editors give a warranty, expressed or implied, with respect to the material contained herein or for any errors or omissions that may have been made. The publisher remains neutral with regard to jurisdictional claims in published maps and institutional affiliations.

This Springer imprint is published by the registered company Springer Nature Switzerland AG
The registered company address is: Gewerbestrasse 11, 6330 Cham, Switzerland

Preface

Whole Energy System is an approach to face conflicting challenges of security, affordability and sustainability, which is often referred to as the energy trilemma and it is a necessity for shaping the transition to decarbonization targets. This field of study addresses the interactions and inter-dependencies within the energy landscape and its connections with other sectors and systems and Integrating social, economic, technical and environmental issues. Considering the important role of the electricity (power) industry in the future of energy and especially its share in clean energy production, make the power sector a focal point of the whole energy system.

All the stakeholders and active players of this area emphasize not seeing the power sector in isolation and they all believe that the interaction of the electricity sector with other sectors of the economy, including the heat, gas, hydrogen and transport, is of great importance to successfully meet the Trilemma with the least costs to the customers.

This book introduces an overview of the concept of the whole energy system and related vector-coupling technologies for coupling the power sector with others while covering a comprehensive and in-depth analysis of each technology for bridging the gap and meeting energy trilemma.

Chapter 1, entitled "Concept, Definition, Enabling Technologies, and Challenges of Energy Integration in Whole Energy Systems To Create Integrated Energy Systems", the concept and definition of the vector coupling concept in the whole energy systems are discussed in detail. Also, enabling technologies and challenges associated with integrating energy vectors are discussed.

Chapter 2, entitled "Power-to-X for Renewable-Based Hybrid Energy Systems", introduces the Power-to-X approach and its significantly substantial technologies and the limitations and benefits of each. This route has received much attention in recent years to overcome the shortcomings of existing energy systems. A deep understanding of Power-to-X technologies with improving its challenges will contribute to the development and improvement of sustainable energy systems.

Chapter 3, entitled "Whole Energy Systems Evaluation: A Methodological Framework and Case Study", presents a methodological framework for whole energy systems evaluation with underpinning principles, demonstrated through a case study. The framework provides a socio-technical approach for evaluation by combining stakeholders' requirements with the system components and functions through a system-of-systems architecture methodology. The framework application involves three stages: scenario formulation, conceptual modelling and quantitative modelling, in an iterative process of feedback between the stages. The case study demonstrates the framework application and presents evidence on the potential of energy systems integration in the region via vector-coupling technologies, including Combined Heat and Power, Power-to-Gas and Heat Pumps, under different scenarios.

Chapter 4, entitled "Targeting and Design Multigeneration System Through Total Site Integration Approach", starts with explaining the perspective of energy integration in the Total Site approach. Then, the introduction and modelling of thermal network components in the Total Site will be examined. After describing the various components of a thermal network, the method of targeting and coupling heat and power will be examined. After setting a thermal network and targeting power and heat, targeting, coupling, and modelling refrigeration and desalination systems methods will be assessed. Ultimately, the methods of evaluating a Total Site will be introduced, and each of them will be examined separately.

Chapter 5, entitled "Investigating the Effective Methods in Improving the Resilience of Electricity and Gas Systems", demonstrates the importance of resilience in the electricity and gas network's cooperation and investigates the different strategies and methods to increase resilience. For every engineering system (here, gas and electricity systems infrastructure), many definitions of similar essence have been proposed, focusing on the ability to deal with disruptions. Taking the importance of actions prior, during, and afterwards of an adverse event in mind, resilience is defined as a system's ability to anticipate, resist, absorb, respond to, adapt to, and recover from a disturbance.

Chapter 6, entitled "Optimal Placement of Combined Heat and Power (CHP) Systems Considering the Cost of Environmental Pollutants", discusses the importance of environmental issues, environmental pollutants as one of the crucial factors in the long-term planning of various CHP technologies should always be considered. A comprehensive model for the optimal planning and allocation of CHP technologies such as microturbines, fuel cells, and internal combustion engines with environmental pollutants was presented and the results analysed.

Chapter 7, entitled "Optimal Coalition Operation of Interconnected Hybrid Energy Systems Containing Local Energy Conversion Technologies, Renewable Energy Resources, and Energy Storage Systems", investigates the optimal coalition operation of interconnected hybrid energy systems (IHESs) to mitigate operation and CO2 emission costs. The aim is to improve the whole system's optimal performance by different renewable and non-renewable facilities' interactions inside each hybrid energy system.

Chapter 8, entitled "Optimal Co-generation of Electric and Heat Energy Systems Considering Heat Energy Storage Systems and CHP Units", is a co-generation approach between the two of the most important energy system including heat and electricity to analyze the two in a simultaneous operation considering both systems' topologies. The main focus is on heat systems which the authors have considered the whole pipeline of the system. Further, CHP and a heat energy storage system are also considered. The objective, here, is to minimize the total cost of the hybrid system and the result is interesting. Using the mentioned strategy, the authors have gained a cost-saving approach of more than 15% overall.

Chapter 9, entitled "Investigating the Role of Flexibility Options in Multi-vector Energy Systems", introduces the different types of flexibility options like storage systems, bi-directional compressors, and power-to-gas (P2G) systems to cope with the imposed intermittency to the energy system. Specifically, the contribution of these components in mitigating the intermittency and variability of RES is investigated based on previous projects and studies.

Chapter 10, entitled "Impact of Demand Response Programs on the Operation of Power and Gas Systems", is an effort to answer the following questions: (i) What are the types of demand responses in energy systems? (ii) How can the efficiency of the energy system be increased by implementing the demand response correctly? (iii) What is the role of demand response modelling in improving line capacity and reducing costs in energy systems?

Chapter 11, entitled "Two-Stage Stochastic Market Clearing of Energy and Reserve in the Presence of Coupled Fuel Cell-Based Hydrogen Storage System with Renewable Resources", explains how the integration of renewable energy resources, especially wind energy sources, brings technical challenges to the power system. It highlights that the emerging of the hydrogen storage system as a novel facility in energy platforms is one of the viable solutions. Therefore, the fuel cell-based hydrogen storage system technology with wind energy sources could increase the system's flexibility and reliability. Furthermore, providing more operational flexibility has been obtained from taking into account both energy and reserve markets. In line with this issue, presenting a two-stage stochastic network-constrained market-clearing approach to obtain optimal scheduling and provide energy and reserve services could pave the way for the integration of fuel cell-based hydrogen storage system and wind energy sources.

Chapter 12, entitled "Polygeneration Systems in Fossil Fuel Power Plants: The Role of Power-to-X in CO_2 Mitigation", tries to evaluate the concept of polygeneration operation for future power plants with the aim of not only minimizing emission but also sustaining profitability. The current dependence on fossil fuel power generation and its developed infrastructure make fossil fuel still a crucial energy source. However, sustainable power generation is a critical need. The integrated power generation and CO2 utilization process are a power-to-X system for simultaneous power and chemical production. The well-outlined benefits and the flexibility in energy and chemical conversion give this process an encouraging technical and theoretical perspective.

Chapter 13, entitled "The Role of Distributed Multi-vector Energy Assets in Economic Decarbonisation: Early Findings of a UK Demonstrator", outlines the initial findings of a smart local energy system demonstrator that sought to deploy renewable generation and storage, EV charging, heat pumps and a Proton Exchange Membrane H2 electrolyser to be collectively controlled as a Virtual Power Plant.

The greatest challenges were observed in the multi-sectoral integration, optimisation (and modelling) of assets, ownership and management of real-time data, contract formations and achieving sustainable smart local energy system business models and revenue streams.

The provided material in this book fits the needs of advanced undergraduate or graduate modules and courses on energy systems. The book offers an adequate mixture of technology and engineering background and modelling approaches that makes it a suitable reference for the students as well as researchers and engineers in academia and industry who are active in the field.

To conclude, we would like to sincerely thank all of the authors who contributed to this book. Also, the editors would like to extend their deep gratitude to the following reviewers (sorted alphabetically) for their thoughtful comments on all the submitted book chapters including those that were accepted and published in this book: Ali El Hadi Berjawi, Alireza Akbari Dibavar, Amin Mansour-Saatloo, Amir Mirzapour-Kamanaj, Amir Talebi, Arash Moradzadeh, Ayşe Aybike Şeker, Behzad Motallebi Azar, Hadi Rostamzadeh, Mahdi Habibi, Masoud Agabalaye-Rahvar, Mohammad Hemmati, Morteza Nazari-Heris, Seyed Mohsen Hashemi, and Yasin Pezhmani.

The editors, authors, and reviewers of this book have dedicated their time and enthusiasm to creating it to the hope that it will be useful to researchers, graduate students, and practitioners interested in this field.

Nottingham, UK Vahid Vahidinasab

Muğla, Turkey Behnam Mohammadi-Ivatloo

Contents

Contributors

Masoud Agabalaye-Rahvar Faculty of Electrical and Computer Engineering, University of Tabriz, Tabriz, Iran

Adib Allahham School of Engineering, Newcastle University, Newcastle upon Tyne, UK

Hossein Ameli Control and Power Group, Imperial College London, London, UK

Mohammad Taghi Ameli Department of Electrical Engineering, Shahid Beheshti University, Tehran, Iran

Majid Amidpour Energy Systems Department, K. N. Toosi University of Technology, Tehran, Iran

Mohammad Mehdi Amiri Shahid Beheshti University, Tehran, Iran

Amjad Anvari-Moghaddam Department of Energy Technology, Aalborg University, Aalborg, Denmark

Sasan Azad Shahid Beheshti University, Tehran, Iran

Behzad Motallebi Azar Faculty of Electrical Engineering, Sahand University of Technology, Tabriz, Iran

Ali El Hadi Berjawi School of Engineering, Newcastle University, Newcastle upon Tyne, UK

Mohammad Mehdi Davary Department of Electrical Engineering, Shahid Beheshti University, Tehran, Iran

Sahar Davoudi Chemistry & Process Engineering Department, Niroo Research Institute (NRI), Tehran, Iran

Arash Esmaeilzadeh Energy Systems Department, K. N. Toosi University of Technology, Tehran, Iran

Seyed Hamid Reza Hosseini School of Engineering, Newcastle University, Newcastle upon Tyne, UK

Matin Karbasioun Energy Systems Department, K. N. Toosi University of Technology, Tehran, Iran

Kazem Kashefi Development & Optimization of Energy Technologies Research Division, Research Institute of Petroleum Industries (RIPI), Tehran, Iran

Rasool Kazemzadeh Faculty of Electrical Engineering, Sahand University of Technology, Tabriz, Iran

Amirhossein Khalili-Garakani Chemistry & Process Engineering Department, Niroo Research Institute (NRI), Tehran, Iran

Amin Mansour-Saatloo Faculty of Electrical and Computer Engineering, University of Tabriz, Tabriz, Iran

Mohammad Amin Mirazaei Faculty of Electrical and Computer Engineering, University of Tabriz, Tabriz, Iran

Amir Mirzapour-Kamanaj Faculty of Electrical and Computer Engineering, University of Tabriz, Tabriz, Iran

Behnam Mohammadi-Ivatloo Department of Electrical and Electronics Engineering, Muğla Sıtkı Koçman University, Muğla, Turkey
Faculty of Electrical and Computer Engineering, University of Tabriz, Tabriz, Iran

Arman Oshnoei Department of Energy, Aalborg University, Aalborg, Denmark

Charalampos Patsios School of Engineering, Newcastle University, Newcastle upon Tyne, UK

Mohammad Royapoor School of Engineering, Newcastle University, Newcastle upon Tyne, UK

Omid Sadeghian Faculty of Electrical and Computer Engineering, University of Tabriz, Tabriz, Iran

Leila Samiee Development & Optimization of Energy Technologies Research Division, Research Institute of Petroleum Industries (RIPI), Tehran, Iran

Khezr Sanjani Shahid Beheshti University, Tehran, Iran

Vahid Shabazbegian Faculty of Power Engineering, Tehran, Iran

Goran Strbac Department of Electrical and Electronic Engineering, Imperial College London, London, UK

Vahid Vahidinasab Department of Engineering, School of Science and Technology, Nottingham Trent University, Nottingham, UK

Sara Louise Walker School of Engineering, Newcastle University, Newcastle upon Tyne, UK

Kunpeng Wang School of Engineering, Newcastle University, Newcastle upon Tyne, UK

Robin Wardle School of Engineering, Newcastle University, Newcastle upon Tyne, UK

Kazem Zare Faculty of Electrical and Computer Engineering, University of Tabriz, Tabriz, Iran

Chapter 1
Concept, Definition, Enabling Technologies, and Challenges of Energy Integration in Whole Energy Systems To Create Integrated Energy Systems

Omid Sadeghian, Arman Oshnoei, Behnam Mohammadi-Ivatloo, and Vahid Vahidinasab

Status Quo, Challenges, and Outlook
The energy crisis and environmental issues have led to the need for integration of energy systems with the aim of efficient use of energy sources to supply different energy demands. For this aim, optimal switching among different conventional and renewable energy sources should be scheduled to enable different energy systems to efficiently meet the energy demands in every sector of the whole energy system. The vector coupling of energy systems is challenging due to technical and economic barriers. Therefore, the first stage is to answer the question: What is the concept and definition of vector coupling of energy systems and related challenges to achieve whole energy systems? And what enabling technologies are required to integrate different energy systems?

O. Sadeghian (✉)
Faculty of Electrical and Computer Engineering, University of Tabriz, Tabriz, Iran
e-mail: omidsadeghian@tabrizu.ac.ir

A. Oshnoei
Department of Energy, Aalborg University, Aalborg, Denmark

B. Mohammadi-Ivatloo
Department of Electrical and Electronics Engineering, Muğla Sıtkı Koçman University, Muğla, Turkey

Faculty of Electrical and Computer Engineering, University of Tabriz, Tabriz, Iran
e-mail: bmohammadi@tabrizu.ac.ir

V. Vahidinasab
Department of Engineering, School of Science and Technology, Nottingham Trent University, Nottingham, UK

© The Author(s), under exclusive license to Springer Nature Switzerland AG 2022
V. Vahidinasab, B. Mohammadi-Ivatloo (eds.), *Whole Energy Systems*, Power Systems, https://doi.org/10.1007/978-3-030-87653-1_1

1.1 Introduction

Nowadays, vector coupling of energy systems, i.e., integration of different energy systems to achieve comprehensive energy-efficient systems, is ongoing [38]. The energy crisis and air pollution issues [69] and also restraining the uncertainty and intermittency of renewable energy sources in a high penetration [11] are the main reasons for the transition from conventional single-carrier energy systems toward integrated (or coupled) energy systems [48]. Attaining energy sustainability can also be accelerated by the integration of energy systems [47]. Furthermore, the uninterrupted and reliable energy supply is an advantage of integrated energy systems [71]. By vector coupling of energy systems, in addition to improving the energy efficiency, the system flexibility is also improved [26]. This results in a reduction in load interruption in case of failure in the conventional generators [65]. Figure 1.1 illustrates the Sankey diagram of the energy exchange in an integrated energy system.

Multi-energy systems are mainly based on synergy among different energy carriers such as electricity, gas, heat, and hydrogen carriers [59]. In such systems, there are degrees of freedom for both the supply and demand sides [79], where the much energy-efficient way to meet the load is optimal scheduling of the energy sources [34]. The vector coupling in energy systems is accomplished with different goals such as improving energy saving, operational costs, reliability, or carbon emission. Due to the abovementioned advantages, the planning and operation of integrated energy systems instead of single-carrier energy systems have been increasing in recent years [18].

Integration of several multi-carrier energy systems is affordable via coupling technologies such as combined heat and power (CHP) [66]; combined cooling, heating, and power (CCHP) [36]; heat pump [77]; electric boiler [9]; and gas boiler [88] systems. The characteristics of the coupling points among the multi-energy systems for improving the operation and efficiency of the overall system can be

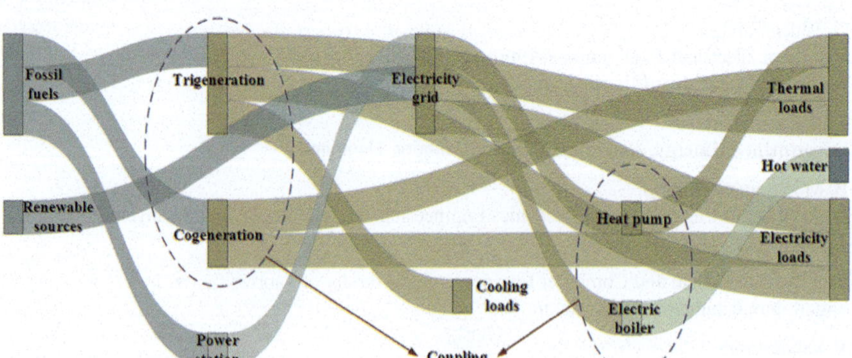

Fig. 1.1 Sankey diagram of some components in integrated energy systems

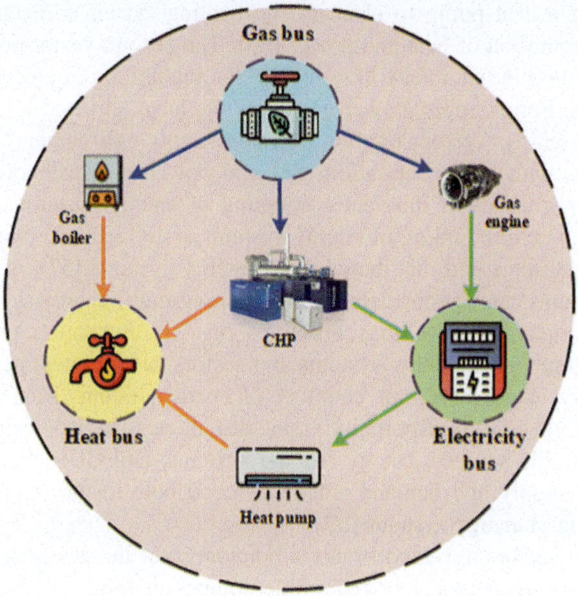

Fig. 1.2 The role of energy conversion units in the coupling of energy systems

studied from the viewpoint of dynamic and static features. In Fig. 1.2, a schematic diagram of the energy conversion among gas, electric, and heat networks using coupling components (i.e., energy conversion systems) is presented.

A key concept for energy integration is the distributed generation concept since a large amount of energy losses has occurred in the generation, transmission, and distribution parts of the electricity systems (respectively, generation, distribution, and transmission), which requires “on-site” and “near-site” power generation to overcome [14]. One important type of distributed generation units is renewable energy sources, which are critical to meet energy crisis and decarbonization [4]. Among renewable energy sources, variable renewable sources, including solar photovoltaic and wind turbines, are the most promising ones due to having no operation cost [86]. However, renewable sources are largely dependent on environmental conditions and exhibit intermittent and uncertain behavior [67]. Integrating these generation units can, therefore, compromise the system’s security if effective energy management approaches are not used in the case of the high penetration of such renewable energy systems [73]. Energy conversion [76], energy-vector shifting among resources [13], utilizing energy storage systems [7], and demand flexibility [54] are available solutions to meet the surplus energy of variable renewable sources, for a case when the conventional generators are not able to accommodate the uncertainty of variable renewable sources.

There are two major kinds of vector coupling in the energy sector, including end-use sector coupling and cross-sector coupling [26]. Examples for the first one are using a heat pump to electrify the heating system [52] and using CHP units for cogeneration of heat and power [66]. The second vector-coupling type is energy integration in main networks, such as the integration of electricity, gas, and heat networks. For example, power-gas networks have attracted a lot of attention in recent years [45]. Vector coupling between supply and demand in renewable-based systems with high penetration [58] and power-to-X technologies [29] are some other approaches of the vector coupling of energy systems. From another point of view, vector coupling of energy systems is divided into two categories of multi-energy systems and distributed multi-energy systems [57]. Vector coupling of energy systems can be considered for integrating multi-energy systems on every scale, such as individual dwellings, buildings, district, city, region, or country level [42]. A wide number of energy systems and sectors can be integrated to attain the mentioned advantages of vector coupling of energy systems such as residential, commercial, industrial, transportation, agricultural, etc. The increasing penetration level of renewable sources, energy storage systems, and CHP energy units have evolved both supply and demand sides and contribute to the implementation of sector coupling of energy systems [53].

The energy vectors imported from the boundaries of the whole system could be electricity from conventional or renewable sources, natural gas, waste heat from the industrial processes, etc. [22]. The grid parameters influence on performance and efficiency of integrated energy systems, for instance, the parameters of the electricity, heat, and gas networks [37] and parameters of conversion components (like CHP, heat pump, energy storage systems, and diesel generators). On the other hand, the output of integrated energy systems can be the nodes' voltage, active/reactive power, and power losses (for electrical grids); heat power, supply and return temperatures, mass flow rates, and losses (for district heat networks); and nodal pressures, gas flow rates, and losses (for gas networks) [37]. In Fig. 1.3, the schematic diagram of an integrated energy system is given.

In integrated systems, in low-demand hours, when the RES's generation is additional, the surplus electrical energy is adopted to generate hydrogen from water, in which the produced hydrogen is taken into account as renewable energy. The surplus energy could be used in transportation, gas grid, and even chemical industry to generate high-temperature thermal energy such as hydrogen usage in the steam reforming process. In addition, hydrogen usage in the aviation and marine sectors is an appropriate choice due to the problems of using electricity. However, more flexible consumption and energy storage systems can also be employed to meet the surplus generation of variable renewable sources (i.e., solar photovoltaic system and wind turbine) [82].

Despite the complexity of integrated multi-energy systems, they have received significant attention in terms of research and practical aspects. There are a lot of cases implemented in practice, such as a Euro project in [27], the University of Parma Campus in Northern Italy [89], the CCHP systems located in eastern Tehran [3], and the decentralized micro-CHP systems in the UK [44]. China's energy policy with social capital supports the construction of integrated multi-energy systems, which are increasingly developing [38]. Furthermore, in [17], the multi-vector

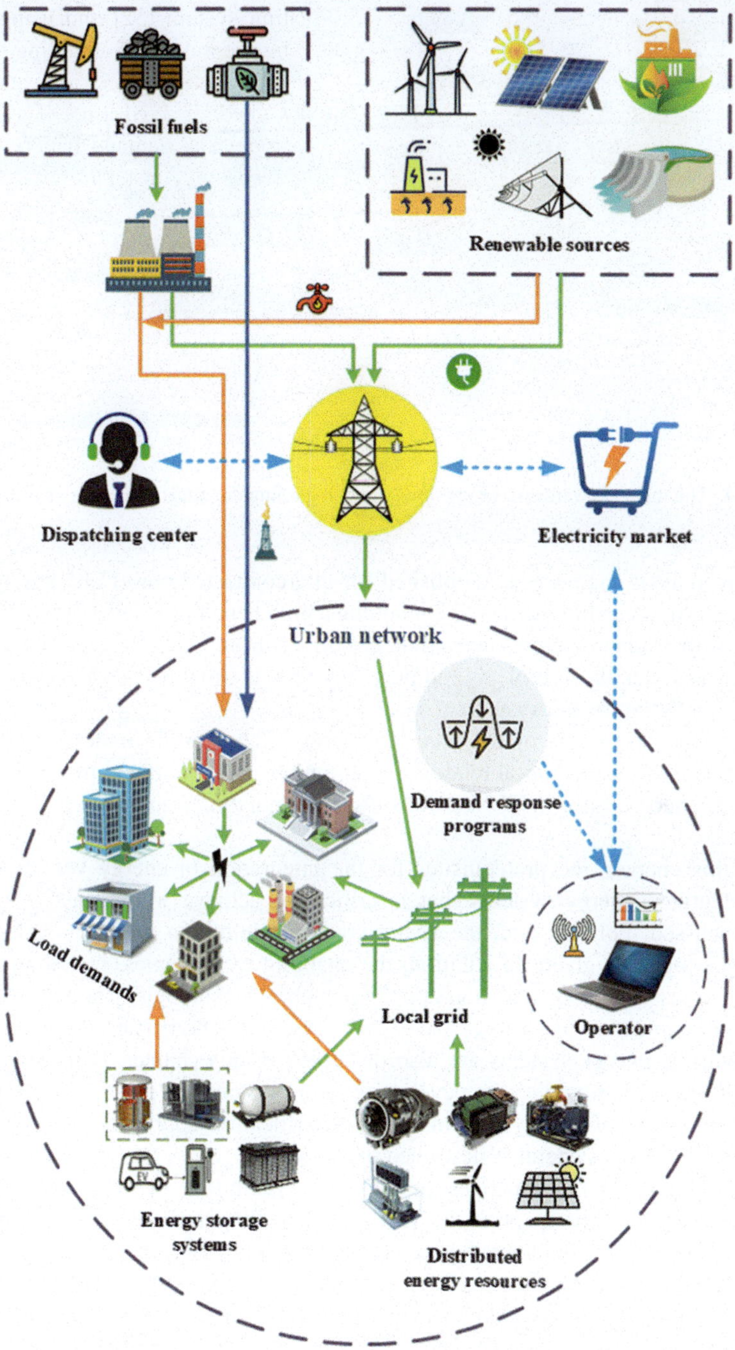

Fig. 1.3 Schematic diagram of an integrated energy system

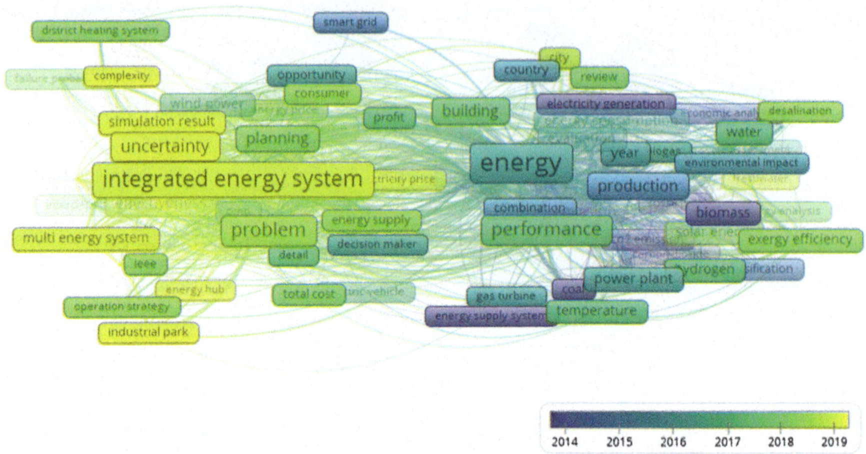

Fig. 1.4 The main keywords of the previous works regarding the integrated energy systems

energy analysis has been accomplished for interconnected power and gas systems of Great Britain and Ireland, both containing a high penetration level of wind power, whereas the Ireland systems depend on gas import from Great Britain. The analysis has revealed that the hybrid gas and power system of Great Britain is more resilient compared to the Ireland system.

The main keywords of the previous works related to the understudy context associated with the temporal relationship among them are outlined in Fig. 1.4. This graph has been created in VOSviewer software using the "Web of Science" research tool.

In this chapter, the prerequisites for the integration of energy vectors to create integrated energy systems that is critical to achieve whole energy systems are discussed. For this aim, the vector-coupling definition related to integrating energy systems is given. In addition, the enabling technologies, including energy technologies (coupling and non-coupling energy technologies) and information and communication technology (ICT), are reviewed. The barriers and challenges for coupling energy systems are also discussed in this chapter. Furthermore, the role of demand flexibility in developing such systems is given. Moreover, the decarbonization of energy systems with electrification approaches as a goal for integrating energy systems is described in this study.

The chapter is continued as follows: The definition of vector coupling in multi-energy systems is given in Sect. 1.2. Enabling technologies for coupling energy systems are enumerated in the next section. Section 1.4 discusses the demand-side management programs. The electrification options and their role in the decarbonization of the whole energy system are outlined in Secti. 1.5. The next section relates to the challenges and barriers for integrating energy systems. Finally, the paper is concluded in Sect. 1.7.

1.2 Definition

Integration of all the energy carriers has been stated as the definition of the vector coupling of energy systems [62]. In addition, sector coupling has been defined as converting the surplus power to other forms of energy to be used efficiently in high-demand hours or used in other applications such as the industrial sector [55]. This is while the waste or excess heat of power stations or distributed generation units and the waste heat from industrial processes is not generally considered as a vector-coupling action in energy systems and only considered as multi-energy systems [62]. Some distinctions between multi-carrier energy systems and integrated energy systems can be found in the literature [62]. Therefore, there is not a widely accepted standard for the vector coupling of energy systems.

In this chapter, the vector coupling of energy systems is defined as follows:

> Vector coupling of energy systems is integration/combination of different fuels (such as fossil fuels, coal, natural gas, hydrogen, biogas, and biomass), different energy systems (such as electricity, gas, hydrogen, and district heat networks), different energy sectors (such as residential, transport, commercial, industrial, and agricultural) or energy systems (such as energy hubs, microgrids, virtual power plants, and building) to coordinate the centralized and distributed energy supply systems (such as power stations, renewable sources, diesel generators, CHP plants, and fuel cells) and energy storage systems (such as battery, thermal, cooling, hydrogen, and compressed air energy storage systems) with every purpose (such as improving the environmental performance, economic operation, energy efficiency, flexibility, and reliability/availability) via suitable coupling technologies (such as CHP, heat pumps, electric and gas boilers, chillers, microturbines, and fuel cell) to supply different energy demands (like electrical, heating, lighting, cooling and every combination of them), by optimal interaction/linkage and switching among energy sources in order to handling a considerable share of renewable sources and better managing sudden failures.

1.3 Enabling Technologies

Vector coupling of the current energy systems will be affordable by respective enabling technologies [62]. These technologies include energy technologies as well as information and communication technology (ICT), which are discussed. These technologies are categorized in Fig. 1.5.

1.3.1 Energy Technologies

The energy technologies that are required for coupling energy systems are divided into two categories, including coupling elements and non-coupling elements. Coupling components locate in inter-network locations [37]. For instance, the CHP systems are located among gas, electrical, and district heat networks, or electric heat pumps are located among electricity and heat networks [28]. On the other

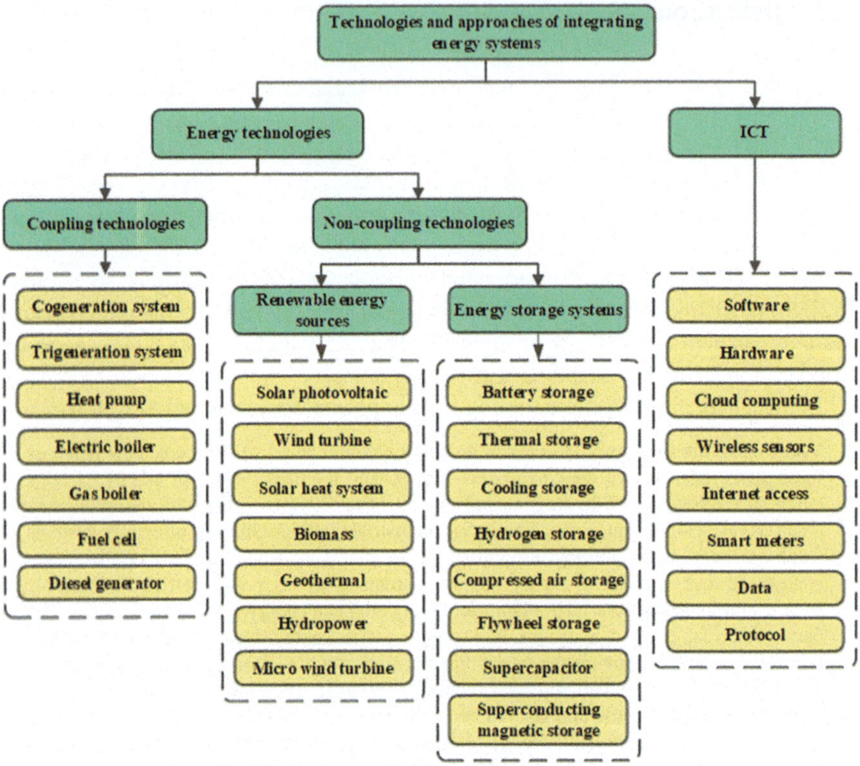

Fig. 1.5 Required technologies for integrating energy systems

hand, the non-coupling components deal with one type of energy. For instance, the battery storage is located in the electricity network, or thermal storage is situated in the district heat network. In the following, the enabling energy technologies and approaches required for vector coupling of energy systems are discussed.

1.3.1.1 Coupling Energy Technologies (Energy-Converting Systems)

The vector coupling of different energy systems needs combination and management of multi-energy together. Coupling components (or conversion systems) relate to different energy carriers to efficiently use the energy sources. The most known conversion systems to couple energy systems are CHP, CCHP, heat pump, boiler, diesel generator, and fuel cell. Among them, CHP plants and heat pumps are seen as the key components for integrating energy systems.

The most noticeable technology for integrated multi-energy systems is the CHP system, which is capable to efficiently generate heat and power, simultaneously [13]. Mainly, CHP systems are gas-fired and coal-fired systems. A CHP system could be a gas turbine with a corresponding efficiency and heat-to-power ratio [37]. The CHP units are mainly integrated with an auxiliary boiler, which the fuel for such boiler is supplied from various sources like electricity. In addition, gas- and hydrogen-based fuel cells are also seen as promising cogeneration systems for the implementation of whole energy systems. Fuel cells are new but nonrenewable energy systems that, in addition to environment-friendly generation, can be fueled by hydrogen or different gases, which is another advantage of such generation systems. Cogeneration systems have various advantages containing 30% decarbonization, efficiency over 80%, increasing supply security, 30% saving on energy bills, and loss reduction in both the distribution and transmission levels [90].

CCHP systems or so-called trigeneration systems are another coupling facility to generate cooling, heat, and power, simultaneously. CCHP is mainly a developed form of CHP. Accordingly, CHP systems can be a trigeneration system when they are associated with heat-to-cooling equipment, which are absorption/adsorption machines [43]. By using CHP and CCHP systems, 30–50% energy saving is achieved [35] in comparison to separate production of heat and power, which have about 30% energy efficiency [14]. In other words, using cogeneration and trigeneration systems, 75–80% of the input fuel is converted into useful energy [14]. These systems mainly contain gas engines and gas turbines [35].

Heat pumps (generally means electric heat pumps) are another technology effective on vector coupling of energy systems used to transfer the heat from a colder source (such as the external environment of building or water) to a hotter sink (such as indoor air) for heating aims [39]. Based on their application, the heat pumps are divided into three types containing water-source [56], air-source [80], and ground-source [25] heat pumps. Heat pumps can also operate as chillers for cooling aims. However, chillers can operate in cogeneration mode by producing cooling and heating through using a heat recovery condenser integrated with the chiller. There is also another type of heat pump called gas-fired heat pumps, which are located between gas and heat networks.

Based on the literature [37], a system with CHP systems and heat pumps has the lowest carbon emission, whereas the same system with gas boilers has the highest carbon emission. The most effectiveness of the presence of CHP or heat pump on the total emission depends on the emission of the electricity grid. In addition, a system with heat pumps has the least energy cost, whereas the same system with gas boilers results in the highest energy cost. The most effectiveness of the presence of CHP or heat pump on the total energy cost depends on the energy cost of the electricity grid. By using a combination of technologies such as CHP systems and heat pumps, the balance between techno-economic and carbon emission aspects might be in place.

Boilers as steam-generating systems are other technologies to enable the energy systems to be efficiently integrated [72]. Such conversion systems are responsible for a significant portion of energy use in the world. Heating water or steam is the result of using boilers for heating aims by consuming natural gas, oil, coal, electricity, etc. [75]. The energy efficiency of boilers is about 75–90% [6]. In industrial processes, the generated heat is used by steam turbines to generate electricity.

The microturbine (or so-called micro-gas turbine), as another coupling component, is a high-speed and small gas turbine. Such systems generate electricity by consuming different kinds of fuels. The electrical efficiency of a microturbine is about 20–30% and its heat-to-power ratio is about 1.3–2 [61].

1.3.1.2 Non-coupling Energy Technologies

Non-coupling components play critical roles in coupled energy systems. These components include renewable sources and energy storage systems. As mentioned, increasing the penetration/share of renewable sources is the main reason for integrating energy systems, particularly the variable renewable sources, including solar systems and wind turbines, because of their zero generation cost. However, other renewable sources such as geothermal, biomass, hydro energy, and tidal energy are developing. Apart from grid connection of renewable sources, the surplus power of a decentralized photovoltaic system can directly electrify transport or indirectly supply a heat pump to supply the heating loads. Hydrogen or methane may also be stored by using the surplus electricity of renewable sources or fed into the natural gas network [90].

Energy storage systems are critical facilities for vector coupling of energy systems to increase the degree of freedom in such systems. Using such systems can significantly reduce wind and photovoltaic curtailment and also operation costs [68]. There are different types of energy storage systems for storing energy. Electricity storage has different types such as battery storage, electric vehicles, supercapacitor, flywheel, pumped storage, and compressed air energy storage [46]. The other types of energy storage systems include heat storage, cold water storage, and hydrogen storage tank. There is also another energy storage system called seasonal energy storage systems, which are able to meet the seasonal intermittency of renewable sources. Such systems can play a backup role in the case of system failure. Hybrid energy storage systems have also been focused in recent years. These storage systems are the combination of two or more storage technologies to utilize the advantage of different storage technology such as the combination of high-power and high-energy battery storage technologies [70].

1.3.2 Information and Communication Technologies

Information and communication technology (ICT) is another enabling technology for integrating energy systems. ICT is a diverse set of electronic technologies and related approaches used to store, create, manipulate, receive, share, exchange, manage, and transmit information from one place to another by using gadgets such as cell phones, the wireless network, and the Internet. Through ICT, the message delivery is accomplished faster, more convenient, and easy to access. ICT tools include software, hardware, services, and communications. Smart meters, controllers, wireless technologies, and the Internet of Things technology are examples of ICT. Internet of Things technology as a novel ICT form (from 2009) is a developed form of the radio-frequency identification technology (from the 1980s) and the wireless sensor networks (from the 1990s) [16]. Although the ICT concept was first born in the 1940s, still there is not a uniform definition for this concept [85].

This technology is essential for linking and coordinating energy systems, especially in the presence of variable renewable energy sources. The optimal switching among energy sources to efficiently save energy consumption considering the comfort level of energy customers is not possible without ICT equipment and related services. This technology provides intelligent monitoring and management systems for energy saving as well as energy efficiency by automation of energy management [5]. The role of ICT is critical in the intelligent management of connecting modern energy networks, in which the coordination of distributed energy resources for enabling collaborative storage and demand response scenarios should be accomplished by this technology for increasing the share of renewable energy resources [31]. The management of grid stability and power quality issues in integrated energy systems are guaranteed by ICT [81]. The ICT contributed to accumulating information from every point of an integrated energy system for demand forecasting, control, protection, and optimal operation of the grid. This technology has increased the requirement for DC power since the equipment of such technology utilized DC power to accomplish related tasks [84]. Energy trading in regional markets is also possible with ICT infrastructures and equipment [60]. Self-healing, energy loss minimization, and emission control are some other applications of ICT in power systems [41]. Through the ICT, the observability of the energy grid is significantly enhanced by real-time monitoring. The ICT has an important role in the automation of load management with switching among energy sources based on the customers' behavior. For instance, the comfort requirement in a smart building includes indoor air quality, visual comfort, and thermal comfort [2].

1.4 Importance of Demand-Side Management in Developing Integrated Energy Systems

In addition to using technologies, demand-side management is also effective in developing integrated energy systems [74]. Energy efficiency and demand flexibility as the roles of demand-side management are essential for integrating energy systems. This is because the aim of developing integrated systems is energy conservation as well as enhancing the share of renewable energy sources in supplying the energy demands. Accordingly, the demand-side management concept is divided into energy efficiency and demand response programs. Energy efficiency means using less energy for a typical task and covers several subjects, including the use of new energy-efficient technologies instead of conventional ones, energy

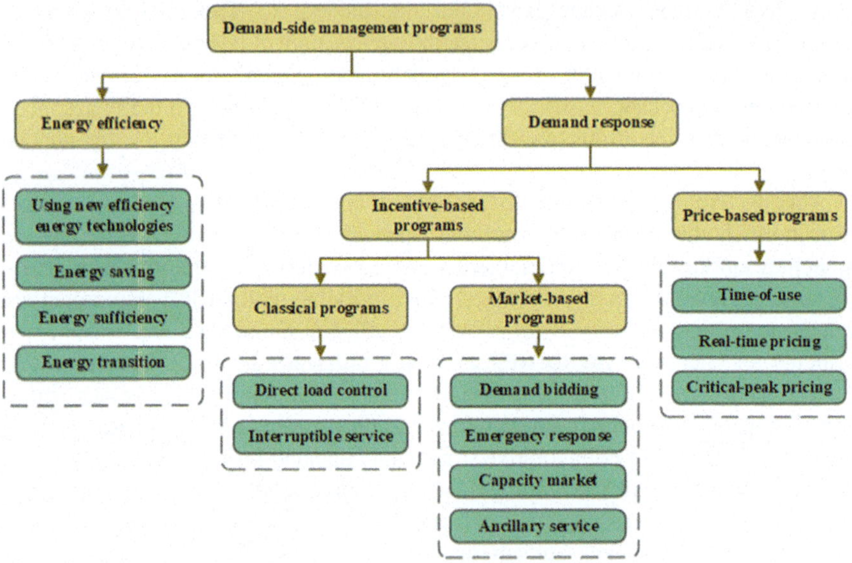

Fig. 1.6 Demand-side management approaches and related subcategories

saving, and energy transition, whereas demand-side response programs enable the responsive loads to participate in demand flexibility programs. Demand response programs in integrated energy systems have been focused on in research studies [54]. For effective implementation of such programs, load forecasting for different load kinds of demands [49] is an indispensable action to manage different load types for optimal participation in demand response programs. However, in the existing research studies regarding load forecasting, a few of them are consistent with integrated energy systems [38]. The demand-side management approaches and related subcategories are listed in Fig. 1.6.

1.5 Role of Electrification in Decarbonizing Energy Systems

Coal and oil are responsible for 80% of carbon emission, and therefore, using renewable sources and electrification of the transport sector are the main steps toward decarbonization [4], which is one of the major aims of integrating energy systems [1]. Among the energy sectors, more than two-thirds of all energy is used in cities, whereas urbanization is increasing that accelerates the role of cities in carbonization [40]. Decarbonization is not limited to a specific sector of energy systems and is related to the whole energy system. By efficient use of energy sources, the higher efficiency of integrated energy systems can be achieved, which results in lower fuel use and carbon emission [8]. Like renewable sources, the fuel

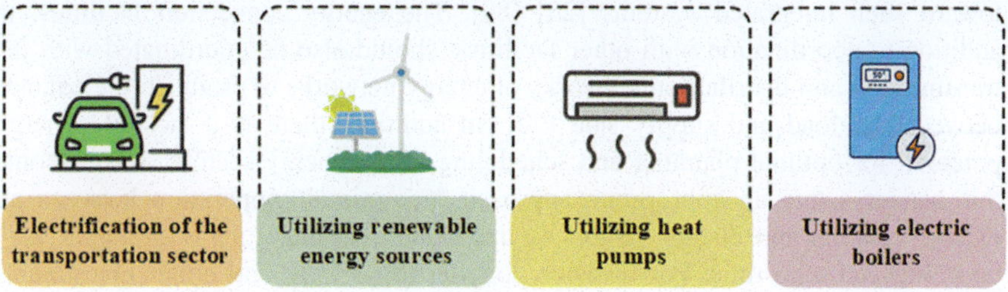

Fig. 1.7 Electrification options to decarbonize integrated energy systems

cell systems generate zero/negligible greenhouse gas emissions and are seen as a clean combined energy system since it enables direct electrification by electrolytes [62]. Energy storage systems are other effective facilities for decarbonization by reducing the peak load to reduce the power generation of the high-emission conventional generators or diesel-based and gas-based distributed generators [33]. Electric heat pumps, electric boilers, solar heat, and geothermal are also effective in decarbonization by heat generation without carbon emission. In such units, the electricity directly converts to thermal energy for space heating or hot water, whereas some other distributed generators such as gas boilers convert fossil fuel to thermal energy by producing carbon dioxide. Figure 1.7 depicts the main actions for electrification in integrated energy systems [78].

As mentioned, electrification is an action to move toward low-carbon technologies [13]. Electrification needs additional electricity to meet the new electricity demands. The required electricity is mainly supplied via renewable sources, biomass gas, and nuclear facilities. Among energy sectors, the heat, industry, and transport sectors are the most fossil-based sectors [64]. However, electrification of heat and transport is more expected due to their impact on optimizing energy use in low-demand time periods that lead to improvement of the energy efficiency [32]. The electrification for fossil-based heating systems is recognized as an appropriate option to produce low-temperature heat. This is while there is not a commercially feasible option for electrification of high-temperature applications. Electrification for many industries seems difficult as they need high-temperature and high-rate heat processes [10]. Renewable sources, specially biomass-based renewable energy sources, are mentioned as the only solution for electrification of some processes in literature [87].

1.6 Challenges and Barriers

Despite the advantages of vector coupling of energy systems, there are a lot of challenges and barriers against the coupling of energy vectors [51]. Optimal design and management of coupled multi-energy systems is a great challenge to move toward such integrated systems [22] [89]. The energy conversion facilities, in addition to coordination with other facilities, should also be coordinated with the transmission and distribution networks of energy networks to establish the balance between the load and supply side [22]. In addition, there is a lack of synergy potential for optimal planning and scheduling of the energy sectors. To overcome this barrier, using an appropriate approach for optimal cooperation between all energy systems and energy sectors is inevitable. The investment strategies may be influenced with time. For instance, long-term gas CHP and boiler investments decrease with increasing strategic behavior, which leads to an increase in gas price 2.6 times more than the case of gas price without strategic behavior [24]. Moreover, the combination of higher gas and CO2 prices increases power prices and fosters renewable investments [24]. For another challenge, the transition of classical single-carrier systems toward distributed multi-energy systems introduces complex physical and commercial interactions between different energy vectors. Optimal sizing of coupling elements is another challenge for moving toward such systems.

Integrating energy systems leads to complex systems, in which the planning and scheduling problem of such systems needs specific approaches [4]. Available computer tools to simulate and analyze the integrated multi-energy systems combined with renewable sources have been discussed in [15]. Most of the available tools that can be used for individual energy systems are not capable of being used for integrated energy systems [37]. To solve the resulted model in such systems, agent-based models or model coupling are required [30]. A lot of approaches for optimal management and operation of integrated energy systems have been introduced in the literature [83]. Load flow in integrated energy systems requires hybrid load flow methods. It has been stated in the previous studies that no systematic framework has, so far, been developed for techno-economic analysis of complex and distributed coupled energy systems [21]. For every integrated energy system, a coordinated scheduling model is required to optimally use all the components for demands supply [90]. Temporal and spatial resolutions are the major specific challenges for modeling the energy infrastructure [30], which need a fundamental understanding of the modeling, dynamics, and interdependency among the systems [90].

One other barrier is the need to further innovation in technologies. For instance, a combination of renewable sources with energy storage systems is not still competitive with other existing solutions due to the high price of battery energy storage systems. For another sample, converting electricity in low-demand hours into hydrogen and methane as usable gases will be 20% more expensive compared to fossil-based fuels in 2050. Therefore, the cost efficiency of transforming power to other forms of energy such as hydrogen is important for developing integrated energy systems. Since the existing technologies vary in each region, a comprehensive study regarding the understudy region is required before the energy integration.

Energy is produced, stored, and transported over distances in one of the three basic forms containing thermal, electric, and chemical. For instance, the energy in the form of natural gas is transmitted to be used in power stations to generate electricity or used in buildings to generate heat by boilers or gas-based CHP systems [22]. The case of Denmark shows that district heating and the gas network have much more capacity than the electricity grid, both in terms of distribution and storage [22], but this capacity is not available elsewhere. Reinforcement and extension of energy networks may be required for vector coupling of energy systems. Development of distributed generation systems and electrification of heating systems influence line flows in electricity systems. In addition, there is a need for investment in heat networks, which have not been developed compared to gas and electricity grids. District heating systems are uncompetitive with existing decentralized heating systems. However, it is expected that the building space heating demand is 20% lower in 2050 compared to 2016, due to improvement in building energy efficiency measures in the future [4]. Although electricity networks are more developed compared to the other energy networks, grid expansion to maximize the usage of renewable sources' generation is an alternative for developing integrated systems. The authors in [63] have revealed that the second solution only costs 30% of an electrolyzer having the same capacity. However, the hydrogen can be sold without re-electrification in a competitive environment, and transmission expansion can also be economically beneficial. For the most efficient way for the supply of demand in integrated energy systems, the best network topology/configuration should be adopted. The energy losses and cost of such networks are effective factors in the adoption and construction of multi-energy-based network configuration [12]. A new concept of transmission network has been introduced for integrated energy systems as combined transmission or interconnector concept with the aim of transmitting electrical, thermal, and chemical energy in one underground device [19]. This layout seems to be a hollow conductor with the capability of carrying gas inside. The advantage of using this layout is efficiency improvement by storing the heat generated by conductors in the carried gas. The efficiency of such transmission lines is increased when the stored heat in the gas is used at the end of the link [20]. The integration of energy systems may lead to congestion in the capacity of the individual energy systems. For instance, using the CHP units increased the used capacity of gas networks in Germany [23]. However, distributed generation units in integrated energy systems can solve the congestion according to their location in the network.

Another barrier in integrating the energy systems is regarding the market condition, which can restrict the development of technologies with small scales such as electric-to-gas technologies like hydrogen generation from surplus electricity. For another example, low-scale biomass in distributed areas is not expected to be exploited. Another challenge is the possibility that low electricity prices may lead to direct electrification instead of power-to-gas and power-to-heat transformations in low-price time intervals [62].

Uncertainty is another challenge in such systems. The uncertainty sources may be load demand, generation of renewable energy sources, fuel (or energy) price, etc. One important uncertainty source is the power generation of variable renewable sources. The transformation mechanisms are associated with uncertainty in the long term. For instance, global warming can change the demand from heating to cooling. To overcome this challenge, using low-temperature district heat systems can enable the system for cogeneration of heating and cooling [62]. Another challenge for integrated energy systems is that a threatening factor to one energy system may endanger the security of all the integrated energy systems by rising mismatch issues between demand and supply sides.

The current limitation in using ICT equipment and approaches is another challenge of developing integrated energy systems. A limited number of companies currently exist that have restricted this development. However, nowadays, new emerging and promising ICT approaches such as the Internet of Things technology are increasingly developing [50].

1.7 Conclusions

In this chapter, the integration of the energy systems, which is ongoing with the aim of improving energy efficiency, climate protection, and also enhancing reliability in supplying energy demands, was discussed. Firstly, the concept and definition of vector coupling in whole energy systems was discussed. In addition, the coupling technologies, including energy (coupling and non-coupling energy technologies) and the information and communication technology (ICT), were enumerated in this chapter. Furthermore, the challenges and barriers related to integrated energy systems and available solutions to overcome them were explained. Demand-side management and decarbonization as the main steps toward energy integration were also outlined. The conducted study showed that there is still a need for a lot of actions to couple the energy systems in order to achieve the whole energy system.

Acknowledgements This publication was partially supported by award NPRP12S-0125-190013 from the QNRF-Qatar National Research Fund, a member of The Qatar Foundation. The information and views set out in this publication are those of the authors and do not necessarily reflect the official opinion of the QNRF.

References

1. Abeysekera, M., & Wu, J. (2015). Method for simultaneous power flow analysis in coupled multi-vector energy networks. *Energy Procedia, 75*, 1165–1171.
2. Aduda, K. O., Zeiler, W., Boxem, G., & Labeodan, T. (2014). On defining information and communication technology requirements and associated challenges for 'energy and comfort active' buildings. *Procedia Computer Science, 32*, 979–984.
3. Ameri, M., & Besharati, Z. (2016). *Optimal design and operation of district heating and cooling networks with CCHP systems in a residential complex* (Vol. 110). Elsevier B.V.
4. Arabzadeh, V., Mikkola, J., Jasiūnas, J., & Lund, P. D. (2020). Deep decarbonization of urban energy systems through renewable energy and sector-coupling flexibility strategies. *Journal of Environmental Management, 260*(January).
5. Asare-Bediako, B., Ribeiro, P. F., & Kling, W. L. (2012). Integrated energy optimization with smart home energy management systems. In *IEEE PES innovative smart grid technologies conference Europe* (pp. 1–8).

6. Barma, M. C., Saidur, R., Rahman, S. M. A., Allouhi, A., Akash, B. A., & Sait, S. M. (2017). A review on boilers energy use, energy savings, and emissions reductions. *Renewable and Sustainable Energy Reviews, 79*, 970–983.
7. Bartolini, A., Carducci, F., Muñoz, C. B., & Comodi, G. (2020). Energy storage and multi energy systems in local energy communities with high renewable energy penetration. *Renewable Energy, 159*, 595–609.
8. Brown, M. A., & Li, Y. (2019). Carbon pricing and energy efficiency: Pathways to deep decarbonization of the US electric sector. *Energy Efficiency, 12*(2), 463–481.
9. Bühler, F., Zühlsdorf, B., Van Nguyen, T., & Elmegaard, B. (2019). A comparative assessment of electrification strategies for industrial sites: Case of milk powder production. *Applied Energy, 250*(April), 1383–1401.
10. Bühler, F., Holm, F. M., & Elmegaard, B. (2019). Potentials for the electrification of industrial processes in Denmark. In *ECOS 2019 - proceedings of the 32nd international conference on efficiency, cost, optimization, simulation and environmental impact of energy systems* (pp. 2137–2152).
11. Canale, L., Rita, A., Fazio, D., Russo, M., & Frattolillo, A. (2021). An overview on functional integration of hybrid renewable energy systems in Multi-Energy Buildings. *Energies, 14*(4), 1078.
12. Carradore, L., & Turri, R. (2009). Modeling and simulation of multi-vector energy systems. In *2009 IEEE Bucharest PowerTech: Innovative ideas toward the electrical grid of the future.*
13. Chicco, G., Riaz, S., Mazza, A., & Mancarella, P. (2020). Flexibility from distributed multienergy systems. *Proceedings of the IEEE, 108*(9), 1496–1517.
14. Cho, H., Smith, A. D., & Mago, P. (2014). Combined cooling, heating and power: A review of performance improvement and optimization. *Applied Energy, 136*, 168–185.
15. Connolly, D., Lund, H., Mathiesen, B. V., & Leahy, M. (2010). A review of computer tools for analysing the integration of renewable energy into various energy systems. *Applied Energy, 87*(4), 1059–1082.
16. Da Xu, L., He, W., & Li, S. (2014). Internet of things in industries: A survey. *IEEE Transactions on Industrial Informatics, 10*(4), 2233–2243.
17. Devlin, J., Li, K., Higgins, P., & Foley, A. (2017). A multi vector energy analysis for interconnected power and gas systems. *Applied Energy, 192*, 315–328.
18. Dolatabadi, A., Mohammadi-Ivatloo, B., Abapour, M., & Tohidi, S. (2017). Optimal stochastic design of wind integrated energy hub. *IEEE Transactions on Industrial Informatics, 13*(5).
19. Favre-Perrod, P., Kienzle, F., & Andersson, G. (2010). Modeling and design of future multi-energy generation and transmission systems. *European Transactions on Electrical Power, 20*(8), 994–1008.
20. Geidl, M., Koeppel, G., Favre-Perrod, P., Klöckl, B., Andersson, G., & Fröhlich, K. (2007). Energy hubs for the future. *IEEE Power & Energy Magazine, 5*(1), 24–30.
21. Good, N., Martínez Ceseña, E. A., Zhang, L., & Mancarella, P. (2016). Techno-economic and business case assessment of low carbon technologies in distributed multi-energy systems. *Applied Energy, 167*, 158–172.
22. Guelpa, E., Bischi, A., Verda, V., Chertkov, M., & Lund, H. (2019). Towards future infrastructures for sustainable multi-energy systems: A review. *Energy, 184*, 2–21.
23. Hauser, P., Heidari, S., Weber, C., & Möst, D. (2019). Does increasing natural gas demand in the power sector pose a threat of congestion to the German gas grid? A model-coupling approach. *Energies, 12*(11), 1–22.
24. Heidari, S. (2020). How strategic behavior of natural gas exporters can affect the sectors of electricity, heating, and emission trading during the European energy transition. *Energies, 13*(19).
25. Huang, J., Fan, J., & Furbo, S. (2020). Demonstration and optimization of a solar district heating system with ground source heat pumps. *Solar Energy, 202*, 171–189.
26. Ilo, A. (2020). Use cases in sector coupling as part of the link -based holistic architecture to increase the grid flexibility. In *Proceedings of the CIRED 2020 Berlin workshop, Berlin, Germany* (pp. 22–23).

27. Jacobs, T., et al. (2016). Increasing the utilization ratio of photovoltaic energy by network hybridization. In *Proceedings - 2016 IEEE international conferences on big data and cloud computing, BDCloud 2016, social computing and networking, SocialCom 2016 and sustainable computing and communications, SustainCom 2016* (pp. 445–452).
28. Khezri, R., Oshnoei, A., Hagh, M. T., & Muyeen, S. M. (2018). Coordination of heat pumps, electric vehicles and AGC for efficient LFC in a smart hybrid power system via SCA-based optimized FOPID controllers. *Energies, 11*(2).
29. Koj, J. C., Wulf, C., & Zapp, P. (2019). Environmental impacts of power-to-X systems - a review of technological and methodological choices in life cycle assessments. *Renewable and Sustainable Energy Reviews, 112*, 865–879.
30. Kriechbaum, L., Scheiber, G., & Kienberger, T. (2018). Grid-based multi-energy systems-modelling, assessment, open source modelling frameworks and challenges. *Energy, Sustainability and Society, 8*(1).
31. Lazzaro, M., et al. (2019., no. 731268). Smart ICT framework for the intelligent management of different modern energy systems. In *Proceedings - 2019 IEEE international conference on environment and electrical engineering and 2019 IEEE industrial and commercial power systems Europe, EEEIC/I and CPS Europe 2019* (pp. 1–6).
32. Lechtenböhmer, S., Nilsson, L. J., Åhman, M., & Schneider, C. (2016). Decarbonising the energy intensive basic materials industry through electrification – Implications for future EU electricity demand. *Energy, 115*, 1623–1631.
33. Lee, H., Jung, S., Cho, Y., Yoon, D., & Jang, G. (2013). Peak power reduction and energy efficiency improvement with the superconducting flywheel energy storage in electric railway system. *Physica C: Superconductivity and its Applications, 494*, 246–249.
34. Li, B., & Roche, R. (2020). Optimal scheduling of multiple multi-energy supply microgrids considering future prediction impacts based on model predictive control. *Energy, 197*, 117180.
35. Li, H., Fu, L., Geng, K., & Jiang, Y. (2006). Energy utilization evaluation of CCHP systems. *Energy and Buildings, 38*(3), 253–257.
36. Li, G., et al. (2017). Optimal dispatch strategy for integrated energy systems with CCHP and wind power. *Applied Energy, 192*, 408–419.
37. Liu, X., & Mancarella, P. (2016). Modelling, assessment and Sankey diagrams of integrated electricity-heat-gas networks in multi-vector district energy systems. *Applied Energy, 167*, 336–352.
38. Liu, D., Wang, L., Qin, G., & Liu, M. (2020). Power load demand forecasting model and method based on multi-energy coupling. *Applied Sciences, 10*(2), 1–24.
39. Lo Basso, G., de Santoli, L., Paiolo, R., & Losi, C. (2021). The potential role of trans-critical CO2 heat pumps within a solar cooling system for building services: The hybridised system energy analysis by a dynamic simulation model. *Renewable Energy, 164*, 472–490.
40. Lund, P. D., Mikkola, J., & Ypyä, J. (2015). Smart energy system design for large clean power schemes in urban areas. *Journal of Cleaner Production, 103*, 437–445.
41. Mallikarjuna, M., Khanam, A., & Poonia, A. (2018). Smart grids: A perspective on the integration and encapsulation of power energy systems with ICT systems-new research directions and challenges. In *Proceedings of the 2017 international conference on smart Technology for Smart Nation, SmartTechCon 2017* (pp. 648–654).
42. Mancarella, P. (2014). MES (multi-energy systems): An overview of concepts and evaluation models. *Energy, 65*, 1–17.
43. Mancarella, P., & Chicco, G. (2013). Real-time demand response from energy shifting in distributed multi-generation. *IEEE Transactions on Smart Grid, 4*(4), 1928–1938.
44. Merkel, E., McKenna, R., & Fichtner, W. (2015). Optimisation of the capacity and the dispatch of decentralised micro-CHP systems: A case study for the UK. *Applied Energy, 140*, 120–134.
45. Mirzaei, M. A., Nazari-Heris, M., Mohammadi-Ivatloo, B., Zare, K., Marzband, M., & Anvari-Moghaddam, A. (2020). A novel hybrid framework for co-optimization of power and natural gas networks integrated with emerging technologies. *IEEE Systems Journal, 14*(3), 3598–3608.

46. Mirzaei, M. A., et al. (2020). Integrated energy hub system based on power-to-gas and compressed air energy storage technologies in the presence of multiple shiftable loads. *IET Generation Transmission and Distribution, 14*(13), 2510–2519.
47. Mohammadi, M., Noorollahi, Y., Mohammadi-ivatloo, B., & Yousefi, H. (2017). Energy hub: From a model to a concept – A review. *Renewable and Sustainable Energy Reviews, 80.*
48. Mohammadi, M., Noorollahi, Y., Mohammadi-ivatloo, B., Hosseinzadeh, M., Yousefi, H., & Khorasani, S. T. (2018). Optimal management of energy hubs and smart energy hubs – A review. *Renewable and Sustainable Energy Reviews, 89.*
49. Moradzadeh, A., Mansour-Saatloo, A., Mohammadi-Ivatloo, B., & Anvari-Moghaddam, A. (2020). Performance evaluation of two machine learning techniques in heating and cooling loads forecasting of residential buildings. *Applied Sciences, 10*(11), 3829.
50. Motlagh, N. H., Mohammadrezaei, M., Hunt, J., & Zakeri, B. (2020). Internet of things (IoT) and the energy sector. *Energies, 13*(2), 1–27.
51. Nazari-heris, M., Jabari, F., Mohammadi-ivatloo, B., Asadi, S., & Habibnezhad, M. (2020). An updated review on multi-carrier energy systems with electricity, gas, and water energy sources. *Journal of Cleaner Production*, 123136.
52. Neirotti, F., Noussan, M., & Simonetti, M. (2020). Towards the electrification of buildings heating - real heat pumps electricity mixes based on high resolution operational profiles. *Energy, 195*, 116974.
53. Ni, L., et al. (2018). Optimal operation of electricity, natural gas and heat systems considering integrated demand responses and diversified storage devices. *Journal of Modern Power Systems and Clean Energy, 6*(3), 423–437.
54. Niu, J., Tian, Z., Zhu, J., & Yue, L. (2020). Implementation of a price-driven demand response in a distributed energy system with multi-energy flexibility measures. *Energy Conversion and Management, 208*(February), 112575.
55. Noussan, M. (2018). Performance based approach for electricity generation in smart grids. *Applied Energy, 220*(December 2017), 231–241.
56. Oshnoei, A., Khezri, R., & Muyeen, S. M. (2019). Model predictive-based secondary frequency control considering heat pump water heaters. *Energies, 12*(3).
57. Pan, Z. J., & Zhang, Y. (2016). A novel centralized charging station planning strategy considering urban power network structure strength. *Electric Power Systems Research, 136*, 100–109.
58. Pandžić, H. (2018). Optimal battery energy storage investment in buildings. *Energy and Buildings, 175*, 189–198.
59. Petkov, I., & Gabrielli, P. (2020). Power-to-hydrogen as seasonal energy storage: An uncertainty analysis for optimal design of low-carbon multi-energy systems. *Applied Energy, 274*(April), 115197.
60. Pouttu, A., et al. (2017). P2P model for distributed energy trading, grid control and ICT for local smart grids. In *EuCNC 2017 - European conference on networks and communications.*
61. Qiu, J., Zhao, J., Yang, H., Wang, D., & Dong, Z. Y. (2018). Planning of solar photovoltaics, battery energy storage system and gas micro turbine for coupled micro energy grids. *Applied Energy, 219*, 361–369.
62. Ramsebner, J., Haas, R., Ajanovic, A., & Wietschel, M. (2021). The sector coupling concept: A critical review. *Wiley Interdisciplinary Reviews: Energy and Environment*, (January), 1–27.
63. Robinius, M., et al. (2018). Power-to-gas: Electrolyzers as an alternative to network expansion – An example from a distribution system operator. *Applied Energy, 210*, 182–197.
64. Ruhnau, O., Bannik, S., Otten, S., Praktiknjo, A., & Robinius, M. (2019). Direct or indirect electrification? A review of heat generation and road transport decarbonisation scenarios for Germany 2050. *Energy, 166*, 989–999.
65. Sadeghian, O., Mohammadpour Shotorbani, A., & Mohammadi-Ivatloo, B. (2019). Risk-based stochastic short-term maintenance scheduling of GenCos in an oligopolistic electricity market considering the long-term plan. *Electric Power Systems Research, 175*, 105908.
66. Sadeghian, O., Moradzadeh, A., Mohammadi-Ivatloo, B., Abapour, M., & Marquez, F. P. G. (2020). Generation units maintenance in combined heat and power integrated systems using the mixed integer quadratic programming approach. *Energies, 13*(11), 2840.

67. Sadeghian, O., Oshnoei, A., Kheradmandi, M., Khezri, R., & Mohammadi-Ivatloo, B. (2020). A robust data clustering method for probabilistic load flow in wind integrated radial distribution networks. *International Journal of Electrical Power & Energy Systems, 115*, 105392.
68. Sadeghian, O., Oshnoei, A., Khezri, R., & Muyeen, S. M. (2020). Risk-constrained stochastic optimal allocation of energy storage system in virtual power plants. *Journal of Energy Storage, 31*, 101732.
69. Sadeghian, O., et al. (2021). A comprehensive review on energy saving options and saving potential in low voltage electricity distribution networks: Building and public lighting. *Sustainable Cities and Society, 72*, 103064.
70. Serpi, A., Porru, M., & Damiano, A. (2017). An optimal power and energy management by hybrid energy storage systems in microgrids. *Energies, 10*(11).
71. Shariatkhah, M. H., Haghifam, M. R., Chicco, G., & Parsa-Moghaddam, M. (2016). Adequacy modeling and evaluation of multi-carrier energy systems to supply energy services from different infrastructures. *Energy, 109*, 1095–1106.
72. Sinha, R., Bak-Jensen, B., Pillai, J. R., & Zareipour, H. (2019). Flexibility from electric boiler and thermal storage for multi energy system interaction. *Energies, 13*(1), 1–21.
73. Sinsel, S. R., Riemke, R. L., & Hoffmann, V. H. (2020). Challenges and solution technologies for the integration of variable renewable energy sources—A review. *Renewable Energy, 145*, 2271–2285.
74. Strbac, G. (2008). Demand side management: Benefits and challenges. *Energy Policy, 36*(12), 4419–4426.
75. Syrodoy, S. V., Kostoreva, J. A., Kostoreva, A. A., & Asadullina, L. I. (2020). Ignition of wood and coal particle mixtures in conditions of steam and water boiler furnaces. *Journal of the Energy Institute, 93*(2), 443–449.
76. Tabar, V. S., & Abbasi, V. (2019). Energy management in microgrid with considering high penetration of renewable resources and surplus power generation problem. *Energy, 189*.
77. Thygesen, R., & Karlsson, B. (2013). Economic and energy analysis of three solar assisted heat pump systems in near zero energy buildings. *Energy and Buildings, 66*, 77–87.
78. Vahidinasab, V., Ardalan, C., Mohammadi-Ivatloo, B., Giaouris, D., & Walker, S. L. (2021). Active building as an energy system: Concept, challenges, and outlook. *IEEE Access, 9*, 58009–58024.
79. Van Beuzekom, I., Gibescu, M., & Slootweg, J. G. (2015). A review of multi-energy system planning and optimization tools for sustainable urban development. In *2015 IEEE Eindhoven PowerTech, PowerTech 2015* (pp. 1–7).
80. Vivian, J., Prataviera, E., Cunsolo, F., & Pau, M. (2020). Demand side management of a pool of air source heat pumps for space heating and domestic hot water production in a residential district. *Energy Conversion and Management, 225*, 113457.
81. Vrba, P., et al. (2014). A review of agent and service-oriented concepts applied to intelligent energy systems. *IEEE Transactions on Industrial Informatics, 10*(3), 1890–1903.
82. Wang, A., Liu, J., & Wang, W. (2018). Flexibility-based improved model of multi-energy hubs using linear weighted sum algorithm. *Journal of Renewable and Sustainable Energy, 10*(1).
83. Wei, C., Xu, X., Zhang, Y., Li, X., & Bai, X. (2019). A survey on optimal control and operation of integrated energy systems. *Complexity, 2019*.
84. Wikstrom, R. (2017). Innovative energy management to utilize energy efficient solutions in the ICT infrastructure. In *INTELEC, international telecommunications energy conference (proceedings)* (Vol. 2017, pp. 570–573).
85. Yang, T. (2018). ICT technologies standards and protocols for active distribution network. In *Smart power distribution systems: Control, communication, and optimization* (pp. 205–230). Elsevier.
86. Yang, J., Dong, Z. Y., Wen, F., Chen, G., & Qiao, Y. (2020). A decentralized distribution market mechanism considering renewable generation units with zero marginal costs. *IEEE Transactions on Smart Grid, 11*(2), 1724–1736.

87. Zailan, R., Lim, J. S., Manan, Z. A., Alwi, S. R. W., Mohammadi-ivatloo, B., & Jamaluddin, K. (2021). Malaysia scenario of biomass supply chain-cogeneration system and optimization modeling development: A review. *Renewable and Sustainable Energy Reviews, 148*, 111289.
88. Zare Oskouei, M., Mirzaei, M. A., Mohammadi-Ivatloo, B., Shafiee, M., Marzband, M., & Anvari-Moghaddam, A. (2021). A hybrid robust-stochastic approach to evaluate the profit of a multi-energy retailer in tri-layer energy markets. *Energy, 214*.
89. Zatti, M., Gabba, M., Rossi, M., Morini, M., Gambarotta, A., & Martelli, E. (2018). Towards the optimal design and operation of multi-energy systems: The 'efficity' project. *Environmental Engineering and Management Journal, 17*(10), 2409–2419.
90. Zhu, F., Fu, J., Zhao, P., & Xie, D. (2020). Robust energy hub optimization with cross-vector demand response. *International Transactions on Electrical Energy Systems, 30*(10), 1–14.

Chapter 2
Power-to-X for Renewable-Based Hybrid Energy Systems

Sahar Davoudi, Amirhosein Khalili-Garakani, and Kazem Kashefi

2.1 Introduction

Reduced reserves [1], and also economic and political issues generated by the unequal distribution of fossil fuels in the world [2], have led to the development of sustainable energy systems. Furthermore, the combustion of fossil fuels and consequent greenhouse gas (GHG) emissions, such as carbon dioxide (CO_2), is a much more important and worrying issue [3]. Since, according to reports [4], CO_2 net emissions in the entire energy segment must be close to zero to avoid global mean temperature from rising above $1.5^{\circ}C$. That is why CO_2-free power generation is needed [5]. As a result, climate change and global warming are also major reasons for developing new, clean, and renewable energy production systems like wind turbines (WT) and photovoltaic panels (PV) [6, 7].

Shortcomings and problems in large-scale systems of renewable electricity generation have led to the insecurity and inflexibility of these energy systems [8]. One of these barriers is the mismatch between supply and electrical demand, which leads to fluctuations and intermittency of renewable energy sources (RES) [9]. Grid congestions occur due to the shutdown of wind and solar generators. As a result, energy cannot be fed into the transmission grid. This energy is called surplus energy. Therefore, when the production rate is high, the transmission and distribution of renewable energy to consumers face challenges [10]. Electrical energy storage (EES) can be a good solution to overcome these challenges [11]. This system stores

S. Davoudi · A. Khalili-Garakani (✉)
Chemistry & Process Engineering Department, Niroo Research Institute (NRI), Tehran, Iran
e-mail: Sdavoodi@nri.ac.ir; Akhalili@nri.ac.ir

K. Kashefi
Development & Optimization of Energy Technologies Research Division, Research Institute of Petroleum Industries (RIPI), Tehran, Iran
e-mail: Kashefik@ripi.ir

© The Author(s), under exclusive license to Springer Nature Switzerland AG 2022

V. Vahidinasab, B. Mohammadi-Ivatloo (eds.), *Whole Energy Systems*, Power Systems, https://doi.org/10.1007/978-3-030-87653-1_2

surplus energy and makes it available when needed [12]. There are different types of EES technologies depending on the form of energy. In EES technologies, electrical energy is converted into another storable and transportable form of energy carrier [13].

Another major obstacle to the use of renewable energy is the fuel necessities of the whole transportation segment, in fact, the electrification of passenger jets, heavy working machines, and so on. One solution to this problem is to electrify them indirectly by the application of hydrogen or to use hydrocarbons as fuel. Chemical production is also a barrier. The feedstock in the production of plastics and nitrogen-based fertilizers are oil and gas. Therefore, the chemical industry must use renewable sources to produce chemicals [14].

The power-to-X (PtX) concepts have been developed in recent years. It is presented as an effective and efficient strategy to increase the flexibility and application of the RES [15]. The PtX method allows renewable energy to become a potential alternative to fossil fuels and reduces GHG emissions [16]. PtX technology, as a new and potential approach for chemical renewable energy storage [14, 17], is a state-of-the-art transformation technology to convert renewable electricity (power) into various chemical products (X): gases, liquids, and chemicals [16].

In PtX, excess renewable energies are used to produce energy carriers, i.e., green, gaseous, or liquid fuels and chemicals. In other words, the PtX is a sustainable platform for storing surplus electrons from the electricity network and creating a decarbonization route to produce hydrogen, methane, CO, syngas, formic acid, oxy-hydrocarbons, hydrogen peroxide (H_2O_2), and ammonia [2, 12]. The use of this technology and its products reduces the need for fossil fuels and helps to use renewable energy in sectors such as agriculture, transportation, and manufacturing [7].

Today, there is a strong interest in carbon dioxide capture and utilization (CCU) technology, due to global climate change. The concept of the circular economy, which refers to the use of CO_2 emissions to produce valuable products, is a promising new perspective. If the CCU technology integrates with PtX technologies, the CO_2 cycle closes and prevents CO_2 from being released directly into the atmosphere. The development of PtX technology is much needed to use the recovered CO_2 emissions as feedstock for industrial production processes. Flue gases from power plants and industrial plants are a rich source of CO_2 and potential candidates for carbon dioxide capture. Combustion of fossil fuels such as coal, oil, and natural gas in boilers and furnaces to generate power, the exhaust gas from industrial plants such as cement production, biomass or biogas plants, and waste incineration plants emit CO_2. Ambient air is also a source of CO_2 that can be directly utilized to produce materials and fuels [3, 18]. From the production of fuel using waste CO_2 emissions and renewable energy, it can mention the production of kerosene and other heavy fuels. They can be used for indirect electrification of the aviation and heavy-duty transportation areas [14].

Since the potential sources of waste CO_2 emissions are enormous, the production of CO_2-based products for environmental protection must be developed. Using CO_2

emissions shortly could be a great opportunity to produce carbon-neutral fuels as an alternative to fossil carbon sources.

There is promoting interest in PtX technologies since they have many positive effects on the environment [3]. This technology includes various methods that decrease the volume of atmospheric CO_2 by using electricity in two ways. One is that it uses atmospheric CO_2 directly to produce chemicals, and the other is to reduce CO_2 by using hydrogen. The development and application of these technologies, which are based on renewable energy, can have a tremendous impact on the environment. Therefore, it is necessary to carefully examine and identify the various processes and technologies of this method and investigate the problems and opportunities of existing routes.

To develop future renewable energy systems, the PtX chain has attracted significant interest during the last decade, and this technology has gained more and more popularity. A variety of pathways exist in the PtX chain, which denote the methods of transforming renewable energy into different energy carriers such as gas or liquid fuels and chemicals.

The elements of the future energy system, which are considered as PtX pathways, can be divided into the following categories:

- Power-to-gas (PtG).
- Power-to-liquid (PtL) [10].

The scheme of the process chain of the PtX concept is shown in Fig. 2.1. There is a strong need to store energy and produce green fuels without carbon emissions and alternatives to fossil fuels.

Therefore, studying these methods and identifying the PtX chain and these routes is essential and could be key to provide a more feasible future energy supply.

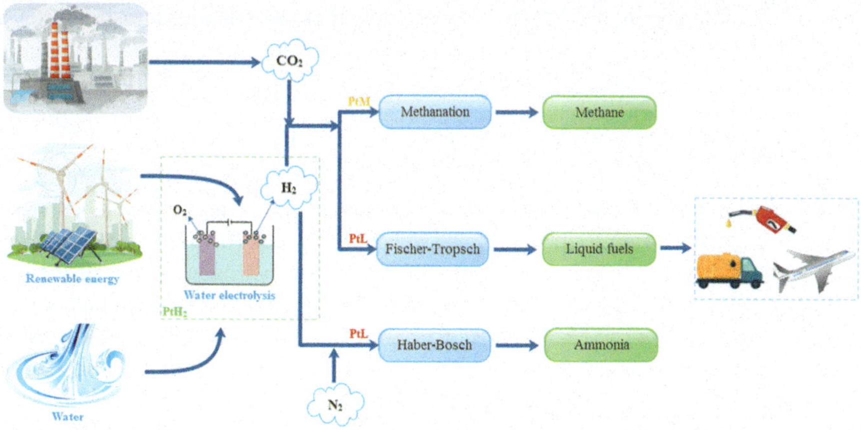

Fig. 2.1 Schematic of the process chain of power-to-X (PtX) technology

In the following, each of these segments will be discussed separately, and their characteristics will be investigated to better understand and implement these infrastructures.

2.2 Power-to-Gas (PtG)

The power-to-gas route, also known as PtG and P2G, is a new energy storage system that was first introduced by Koji et al. in 1994 [9]. The technological chain of PtG refers to the conversion of excess renewable energy into clean gaseous substances such as hydrogen or methane. The main purpose of this concept is to store electrical energy in the form of gas, which is an excellent energy carrier for storage. Because without wasting energy content, it is easy to store for a long time and in large quantities. The production of renewable fuels for transportation and the production of chemicals is also one of the goals of PtG. The main phase of this route is the production of hydrogen by the process of electrolysis of water. This hydrogen has many applications as a carrier of clean energy. It can be stored and used directly as a final and alternative energy carrier in the transportation segment or as raw material for the chemical production. Also, hydrogen can be converted to methane, synthesis gas, liquid fuels like diesel, and chemicals using waste carbon dioxide. It can also be utilized to create electrical energy in a fuel cell [19, 20].

The combination of water electrolysis, which is operated by renewable electricity, with the methanation process, forms the PtG section of the PtX route. PtG technology is a key segment of the future energy storage systems. Because it has many advantages over other storage processes and provides new possibilities for energy transfer, in addition to long-term storage, it creates capacity transfer between energy networks. PtG systems are currently being developed, but more research is needed to implement them industrially and apply them to existing infrastructure.

2.2.1 Power-to-Hydrogen (PtH$_2$)

Hydrogen (H_2) is the clean and sustainable energy carrier and is produced from water without GHG emissions [21]. Its density as a gas (0.0899 kg/Nm^3) is 15 times lower than that of air. The flammable range of hydrogen fuel is in the air, from 4 to 75 vol. %, and in oxygen, from 4 to 95 vol. % [22]. It is also fuel with a high energy density (140 MJ/kg) [23]. Currently, a total of about 500 billion cubic meters (b m^3) of hydrogen is produced per year worldwide [24], that about 96% of this value of fossil fuels [25, 26]. This is especially done through a well-established technology called steam methane reforming (SMR), which is currently the main method to produce hydrogen on an industrial scale. In this mature process, hydrogen production is achieved through the reaction of fossil fuel with water vapor at a temperature in the range of 700 to 1100 ° C and a pressure in the range of 3 to

25 bar (1 bar = 14.5 psi) in the presence of a metal-based catalyst (usually nickel) [27, 28]. However, the hydrogen produced by this method is less pure, and it is associated with the production of a high concentration of harmful GHGs [29, 30]. Steam methane reforming is described by Eq. (2.1) [20]:

$$CH_4 + H_2O \rightleftharpoons CO + 3H_2 \tag{2.1}$$

In PtX, hydrogen is the simplest carrier of energy. Power-to-hydrogen (PtH_2) is part of the general concept of PtX, which deals with the production of renewable hydrogen chemical fuels [13]. Renewable hydrogen production is a good way to effectively decarbonize. Electrolysis is at the core of PtX technologies as well as the main technology for converting power into this type of energy carrier. Electrolysis is an electrochemical process in which, using renewable electricity, water or seawater split into its elementary components, i.e., hydrogen and oxygen. It can also be split by electrolysis, NO_x into ammonia and carbon dioxide into carbon monoxide, syngas, and formic acid [12].

Renewable hydrogen as the main product of the water electrolysis process can be used to produce many hydrocarbon products in processes such as methanation, hydrogenation, and Fischer-Tropsch synthesis (FTS). Hydrogen can also be used for ammonia production in the Haber-Bosch process [31–33]. To store energy, hydrogen produced through electrolysis and CO_2 can be used to produce fuels and chemicals such as synthetic natural gas, methanol, polymers, and so on [14].

Oxygen as a by-product of the water electrolysis process can be used in many industrial and civil applications, in order to the economic profitability of PtX applications [34]. These include the steel industry, pharmaceutical applications, as the oxidant in combustion, and so on [35].

The coupling between electrolysis and geothermal plants has been suggested in a study by Baccioli et al. [34]. They proposed a solution that would make it possible to use CO_2 emitted from geothermal plants, as well as hydrogen and oxygen from an electrolysis plant, to produce liquefied synthetic natural gas (LSNG) and liquefied oxygen (LOx). Geothermal power plants operate more than 8000 hours a year [34], and over 12 GW [36] of geothermal systems are existing around the world. Therefore, large amounts of carbon dioxide are extracted from the fluids used in the plants, which are rich sources for the production of SNG. Liquefaction of these chemicals makes them easy to store and transport from remote areas, and they can be used in various industries and applications [34].

Currently, storage and cost challenges have led to the limitation of hydrogen production through electrolysis [28].

Hydrogen production through electrolysis is affected by the availability of water. The volume of water theoretically required to produce 1 kilogram of hydrogen is 8 liters. But usually, for reasons such as losses in the water purification process, this amount increases to 15 liters [37]. Lack of freshwater can be a problem. The use of seawater and wastewater can solve this crisis. It is also very effective in reducing the cost of the hydrogen production through electrolysis. Due to the presence of many impurities in seawater, its direct use can cause problems for electrodes and membranes. Therefore, a combination of purification and electrolysis processes can be used to solve this problem. Further research is needed to develop this area [38]. But some commercial systems can electrolyze the wastewater to produce hydrogen [39]. The water crisis for hydrogen production by electrolysis can be solved by studying and examining these technologies.

2.2.1.1 Water Electrolysis Technologies

Sustainable hydrogen can be produced through water electrolysis technology using renewable sources without emitting GHG. That is why this technology has attracted a lot of attention today [40, 41]. At present, due to economic problems, only about 4% of hydrogen is gained by water electrolysis [42]. However, to make more use of renewable energy, which is an alternative to fossil fuels, and with the development of electrolysis technology, this amount is expected to increase.

Water electrolysis is a critical section in the development of the PtX technology. The well-known electrolysis process was discovered by Troostwijk et al. in 1789. In this process, water is decomposed into hydrogen and oxygen inside an electrolysis cell using an electric current and electrocatalyst (Eq. (2.2)) [43]:

$$H_2O \rightarrow H_2 + 1/2O_2 \tag{2.2}$$

$\Delta H^0{}_{298,15} = -285.8\ kJ/mol$ [44].

In the cell, the electrodes, i.e., anode and cathode, are disconnected by a diaphragm or membrane with high electrical resistance and ionic conductivity. The electrodes are immersed in an electrolyte to increase ionic conductivity. The role of the electrolyte in the electrolysis process is to enable ion exchange and to separate the two reactions that occur at the electrodes. Depending on the type of electrolyte, materials, and operating conditions used in this process, there are different types of electrolysis technologies [45]. Three major and conventional technologies of the water electrolysis process are alkaline electrolysis (AEL), polymer electrolyte membrane (PEM), and solid oxide electrolysis (SOE).

The main characteristics of the fundamental technologies for water electrolysis are provided in Table 2.1, and the schematic of their operating principle is shown in Fig. 2.2. However, a more detailed description of these technologies is mentioned below.

2.3 Alkaline Electrolysis (AEL)

Alkaline electrolysis is the developed knowledge in hydrogen production and is the most widely used cost-effective electrolytic technology worldwide [45, 46]. Therefore, it has been used to produce hydrogen on a large scale in the early twentieth century [47]. Audi e-gas, the largest power-to-gas plant around the world,

Table 2.1 Characteristic of the three main electrolysis technologies, AEL, PEM, and SOE [2, 7, 20, 28, 43]

	AEL	PEM	SOE
Electrolyte	Aqueous alkaline solution (NaOH or 20–40 wt% KOH)	Solid polymer membrane, e.g., fumapem, Nafion™ thickness ~ 20–300 μm	Yttria (Y2O3)-stabilized zirconia (ZrO2) ceramic (YSZ)
Cell temperature	40–90 $^{\circ}C$	20–150 $^{\circ}C$	600–1000 $^{\circ}C$
Cell pressure	1–200 bar	1–350 bar	<25 bar
Cell voltage	1.8–2.4 V	1.8–2.2 V	0.7–1.5 V
Efficiency range	54–85%	52–79%	Up to 100%
Cell area	<4 m^2	<0.3 m^2	<0.01 m^2
Production rate	1–760 Nm^3/h	1–100 Nm^3/h	<40 Nm^3/h
Advantages	• Most mature technology. • Cost-effective. • Possibility of use in the large plant. • High durability. • Non-noble catalysts. • Use of mature stack components.	• Almost mature. • No corrosive material. • High hydrogen purity. • Flexible integration with other systems. • Rapid system response. • More efficient at a high current density. • Good partial load range.	• High efficiency. • Waste heat recovery possible. • Less power consumption. • Possible application of heat sources such as nuclear or geothermal power. • Non-noble catalysts.
Disadvantages	• Low current density. • Low partial load range. • Short lifetime. • Hydrogen purity less than other technologies. • High maintenance costs due to the application of corrosive electrolyte solution.	• High cost due to fast degradation of membranes and application of expensive catalyst.	• Not mature. • Short lifetime. • Low durability due to the application of brittle ceramics.

also used AEL to produce hydrogen [48]. Hydrogen production by AEL electrolyzer is also used in the PtG plant under the development BioCat Project [49].

Alkaline electrolysis uses a liquid alkaline electrolyte, usually a 20–40 wt% aquatic KOH or NaOH solution [35, 50]. An asbestos diaphragm is used to disconnect the electrodes, which are usually made of nickel [51]. These electrodes are immersed in the electrolyte solution. At the cathode, electrons, current from the anode to the cathode, combine with two molecules of water to form a hydrogen

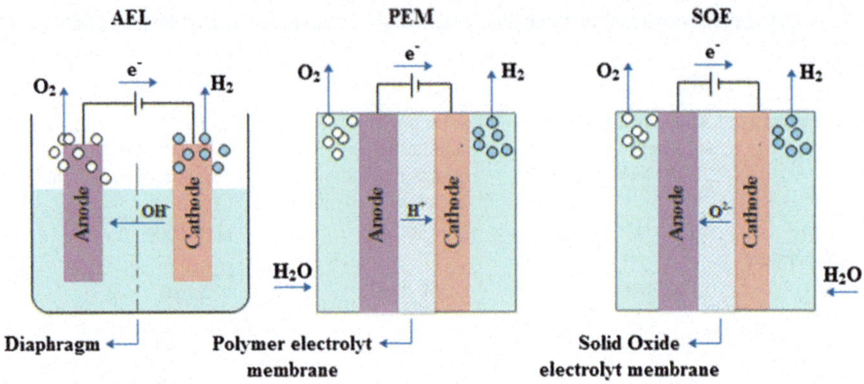

Cathode:	$2H_2O + 2e^- \rightarrow H_2 + 2OH^-$	$2H^+ + 2e^- \rightarrow H_2$	$H_2O + 2e^- \rightarrow H_2 + O^{2-}$
Anode:	$2OH^- \rightarrow 1/2O_2 + H_2O + 2e^-$	$H_2O \rightarrow 1/2O_2 + 2H^+ + 2e^-$	$O^{2-} \rightarrow 1/2O_2 + 2e^-$
Total:	$H_2O \rightarrow H_2 + 1/2O_2$	$H_2O \rightarrow H_2 + 1/2O_2$	$H_2O \rightarrow H_2 + 1/2O_2$

Fig. 2.2 Schematic of the operating principle of the main technologies for water electrolysis

molecule. OH^- ions are transferred through the diaphragm from the cathode to the anode. At the anode, these ions are oxidized into oxygen and water [52]. The relevant partial electrochemical reaction at the cathode and anode is given by [16]:

$$2H_2O + 2e^- \rightarrow H_2 + 2OH^- \text{Cathode} \tag{2.3}$$

$$2OH^- \rightarrow 1/2O_2 + H_2O + 2e^- \text{Anode} \tag{2.4}$$

The purity range of 99.5–99.9% of the produced hydrogen can be obtained with the AEL technology. The purity range of the produced O_2 is also 99–99.8%, which can be increased to 99.999% using a deoxidizer [16]. Despite all the advantages of AEL technology, it also has its drawbacks. Among them, we can mention limited current density and dynamic operation, etc., which lead to a reduction in process productivity and increase the cost of hydrogen production [28, 53]. Researchers have recently conducted research on the use of novel diaphragms as well as electrodes to increase the AEL electrolysis efficiency and issues such as cost and health risks [54]. Adding slits and holes on the surface of the electrode and thus physical modification of the surface, as well as the use of alloys and noble metals, such as platinum (Pt) and ruthenium (Ru), can improve the activity of the electrode and reduce ohmic losses [52].

2.4 Polymer Electrolyte Membrane (PEM) Electrolysis

To overcome the shortcomings of the AEL electrolysis system, a promising technology called PEM was developed [55]. In this system, a proton exchange membrane or a polymer electrolyte membrane (including Nafion®, fumapem®) is used to place between two half-cells [28]. Water is electrochemically split into oxygen (O_2), protons (H^+), and electrons (e^-). Protons are moved from the anode to the cathode. Electrons also flow to the cathode through an external direct current (DC). H^+ and e^- then combine to produce hydrogen (H_2) [56].

PEM technology is a promising option for the PTX chain, as it is better coupled with intermittent systems than other systems [49]. The advantages of PEM electrolysis include compact system design, high operating current density compared to the AEL method, high proton conductivity, high operational pressure capability (equal or differential across the electrolyte), etc. [23]. The capability to operate at very great current densities in a PEM electrolyzer reduces the overall cost of operation [28]. Electrochemical compression of hydrogen can be provided in an electrolyzer under high-pressure operation, which results in less energy being used to store hydrogen [57].

Despite the advantages of PEM electrolysis under high operating pressure, this system also has its drawbacks. Among them is the need for thicker membranes at pressures greater than 100 bar, which reduces the penetration of gas [58]. Another problem with this type of electrolysis system is that the highly corrosion-resistant materials should be used that are stable in harsh conditions. Also, due to the low pH conditions and the high overvoltages, its components and materials are resistant to these conditions [28].

Catalytic activity and long-term stability are two key parameters in evaluating the performance and quality of electrocatalysts in the PEM water electrolysis and water splitting in general. To develop electrocatalysts in hydrogen (HER) and oxygen evolution reactions (OER), studies were performed on the use of nanostructured electrocatalysts [23]. However, in addition to the noble and expensive metal-based electrocatalysts used, such as platinum-group metals (PGM), e.g., Pt, iridium (Ir), and Ru, other components, including current collectors and separator plates, must also be improved [28]. Hence, the PEM electrolysis process is more expensive than the AEL system. If this process is to be used to produce hydrogen on a large scale, costs must be reduced. So, researchers are looking for ways to use more suitable and low-cost electrode materials [23].

To replace expensive precious metals, the use of non-precious-metal electrocatalysts was proposed, which can be prepared through transition-metal materials such as iron and nickel phosphides, cobalt, iron, molybdenum and tungsten sulfides, nickel-molybdenum alloys, etc. It was also proposed that costs could be decreased by decreasing the precious-metal loading [59]. Various synthesis methods for the development of cost-effective, high-quality alternative electrocatalysts are being studied by researchers to reduce noble metal contents and significantly reduce capital costs. The production of mixed oxides was proposed by adding cheap oxides such as a tin oxide (SnO_2), niobium oxide (Nb_2O_5), tantalum oxide (Ta_2O_5), or titanium oxide (TiO_2) to precious-metal-based electrocatalysts such as IrO_2 and RuO_2 while preserving properties [60].

Therefore, to commercialize and implement this technology on a large scale, it is necessary to remove some limitations, including the replacement of expensive catalysts and components of this system with cheap ones, while maintaining quality.

2.5 Solid Oxide Electrolysis (SOE)

One of the best options for achieving a hydrogen economy and efficient hydrogen production is solid oxide electrolysis (SOE) technology that operates at high operating temperatures (600–1000$^{\circ}C$). This thermal energy comes from an external energy source [45]. The efficiency of high-temperature SOE water electrolysis technology is more than 90% [61] since the acceleration of the reaction kinetics occurs due to high heat [45].

High temperatures in SOE systems have advantages. Among them, it can be mentioned that the cell will need low voltage, which leads to a decrease in electricity demand and low energy consumption and increases electrical efficiency [49]. This process can also be integrated with exothermic systems (such as methanol production) that use the heat released by these processes in the SEO system. SEO can also be used to produce syngas and consequently liquid fuels [16].

However, it also has challenges such as the need to use heat-resistant materials and lots of space. This technology also uses water vapor, so a re-treatment process is needed to separate the water vapor from the hydrogen to increase the purity of the hydrogen [28]. Therefore, since SOEC technology has many capabilities and is very attractive and promising, it should continue to be developed due to problems.

2.5.1 *Power-to-Methane (PtM)*

The term power-to-methane (PtM) refers to the sustainable production of synthetic or substitute natural gas (SNG). It is a sector of the PtG system, which uses carbon dioxide and renewable energy to produce methane (CH_4) [62]. Methane is one of the most important energy carriers that has a higher volumetric energy content than hydrogen [63]. In the PtM method, also called the "Sabatier process," CH_4 and H_2O are produced by the reaction of hydrogen and CO_2 (Eq. (2.5)). CO_2 is electrochemically reduced to CH_4 in the methanation, so it is a promising technology with great potential for reducing GHG:

$$CO_2 + 4H_2 \rightleftharpoons CH_4 + 2H_2O \tag{2.5}$$

$\Delta H^0{}_{298} = -165.0\ kJ/mol$ [62].

There are two different pathways for producing methane in PtG technology: catalytic and biological methanation.

2.5.1.1 Catalytic Methanation

One most common and flexible route for renewable PtM is the thermochemical catalytic methanation of CO_2. This process operates at temperatures around 300 to $550^{\circ}C$ and high pressure [49]. Mostly nickel-based catalysts are used in the catalytic methanation process because of their cost-effectiveness, suitable activity, and high methane selectivity [62]. However, other catalysts based on different metals such as Ru, Co, and Pd were also examined, and Ru-based catalysts showed the highest activity [43]. The researchers examined the use of copper electrodes to improve process performance. According to studies, nanoparticles supported on glassy carbon performed better than copper foils and increased the efficiency of the system [64].

The Sabatier reaction is highly exergonic. Therefore, temperature control is very important in reactor design. Researches on different types of reactors have shown that fixed-bed, fluidized-bed, multichannel, three-phase, and structured reactors are the most widely used for the catalytic methanation process [65]. However, usually, fixed-bed reactors are utilized that operate at a pressure of 20–25 bar [66]. One of the problems with this route is the low solubility of CO_2 in water, as well as the development of catalysts to increase the methane selectivity [67]. Measures have been taken in these areas. For example, Kaneco et al. [68] examined the use of sodium salts, such as NaCl, $NaNO_3$, NaH_2PO_4, NaBr, $NaHCO_3$, NaF, Na_2SO_4, NaSCNm, and $NaClO_4$ to improve the solubility of carbon dioxide in methanol. They used methanol because it is a better solvent for carbon dioxide than H_2O. Therefore, sodium salts increased the solubility of carbon dioxide in methanol, and $NaClO_4$ obtained the highest efficiency and produced the highest amount of methane. The process flow diagram of the PtM part of a PtX plant is shown in Fig. 2.3.

To produce methane, hydrogen, and carbon dioxide, gases are used as inlet flows, which reach the desired pressure, and then heated, before entering the reactor. Inside the fixed-bed reactor, a Sabatier reaction occurs and methane and water are produced. The outflow from the reactor includes CH_4 and H_2O as the main products, unreacted components (H_2, CO_2), and CO as the by-products. To separate water from this gas mixture, the outflow from the reactor enters a condenser. It then enters a fractionating column to separate the methane. The efficiency of the whole PtM section is highly dependent on the efficiency of this process. The output of the column is reached by the compressor at a certain inlet pressure and enters the reactor to react unreacted materials [43].

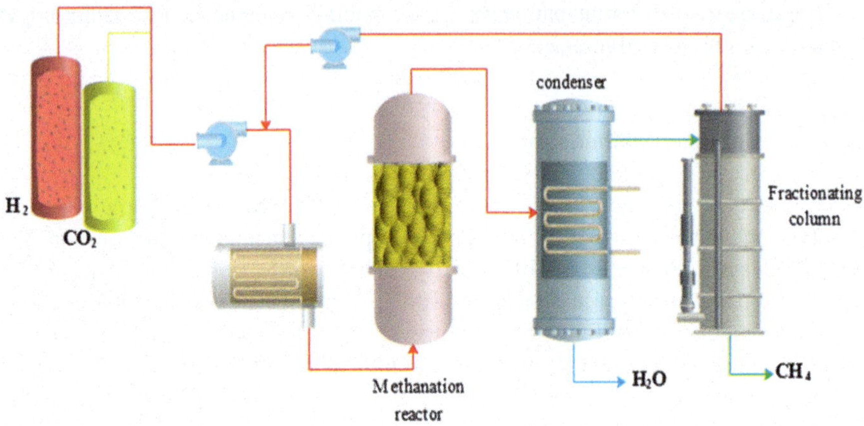

Fig. 2.3 Schematic diagram of the PtM part of a PtX plant

2.5.1.2 Biological Methanation

Biological methanation is obtained through organic matter. In it, autotrophic hydrogenotrophic methanogens (the main class of methanogens) act as autocatalysts and produce methane through metabolic processes [69]. This route of methanation is a cost-effective process due to the simplicity of reactor design (trickle-bed or stirred tank reactor [70]) and the use of low temperature (below $70^{\circ} C$) and pressure compared to catalytic methanation [71]. Also, the biogas process can be combined with the biological methanation system. Since the biogas produced by this method has a high calorific value, it reduces the cost of improving biogas quality [72]. Therefore, biological methanation can be implemented in all biogas plants. Despite the advantages mentioned, this process also faces challenges, such as a low reaction rate, which makes this process still in the pilot phase of the plant [73].

There are two pathways for biological methanation: in situ and ex situ methanation. In situ methanation is to add H_2 directly to a biogas digester as a rich source of carbon dioxide [70]. Ex situ methanation takes place through the parallel injection of carbon dioxide and hydrogen into an adjacent external reactor [72, 74].

2.6 Power-to-Liquid (PtL)

Transport systems on the road, aviation, and navigation sectors play a key role in emitting GHG, by consuming fossil fuels such as diesel and petrol. To solve this fundamental problem, constructive solutions need to be put on the agenda. Electric cars and biofuels have been proposed to replace fossil fuels and reduce GHG emissions, but they are still new technologies and need to be developed [75]. The goal of the power-to-liquid (PtL) way is to use renewable electricity, carbon dioxide, and hydrogen to produce promising substitute liquid fuels, including ethanol, diesel for transportation, and the production of chemicals such as methanol. Liquid fuels are produced by PtL technologies, using the two main processes of modified Fischer-Tropsch synthesis and methanol synthesis [3, 7].

FTS is a highly exothermal process [76] in which liquid hydrocarbons are synthesized using H_2 and CO_2. These hydrocarbons have the highest volumetric energy density among all energy carriers [77] and can be upgraded to fuels, for example, gasoline (C_5-C_{11} hydrocarbons), diesel, and jet fuel using hydrocracking, isomerization, or distillation upgrading processes [78]. State of the art of modified FTS pathway is a two-step process consisting of the reduction of carbon dioxide by the reverse water-gas-shift (RWGS) reaction ((Eq. (2.6)) at temperatures above $800^{\circ} C$ and then CO hydrogenation to liquid hydrocarbons by FTS ((Eq. (2.7)) [79]:

$$CO_2 + H_2 \rightarrow CO + H_2O \tag{2.6}$$

$$CO + 2H_2 \rightarrow -(CH_2)- + H_2O \tag{2.7}$$

A multi-tubular fixed-bed reactor is commonly used for the FTS reaction [77]. Various catalysts are used, including ruthenium, iron, and cobalt-based catalysts [10]. Syngas usually contains CO_2 as a by-product due to the thermodynamic limitations in the RWGS reactor and the resulting incomplete conversion of CO_2. The high contents of CO_2 can affect the FTS reaction kinetics and conversion of CO. Cobalt-catalyzed fixed-bed reactor for FTS is recommended, as it produces no or very low amounts of CO_2 [77].

Production of methanol through direct hydrogenation of CO_2 is one of the simplest methods to produce alternative liquid hydrocarbon fuels such as olefins, dimethyl ether, biodiesel, diesel, kerosene, etc. Moreover, methanol (CH_3OH) is a promising alternative to fossil fuels that can be stored easily and cost-effectively [75]. The catalytic hydrogenation of CO_2 to produce methanol is shown by Eq. 2.8:

$$CO_2 + 3H_2 \rightleftharpoons CH_3OH + H_2O \tag{2.8}$$

$\Delta H^0{}_{298} = -49.2\ kJ/mol$ [7].

To investigate the possibility of expanding and developing power-to-methanol technology on a large scale, research was conducted [80, 81]. To implement this key pathway in industrial projects, evaluation of economic performance potentials and further studies is required.

Power-to-ammonia (PtA) technology produces ammonia (NH_3) as a potential alternative and green fuel to renewable hydrogen, which helps reduce GHG emissions. For this reason, it has attracted a lot of attention in renewable energy systems. In this immature technology, NH_3 is produced under high pressure and temperature through the Haber-Bosch process. The benefits of using ammonia as a sustainable fuel include low storage costs and no CO_2 emissions, and disadvantages include toxicity, which is not a big problem in energy systems [82].

Other potential carriers of liquid hydrogen include formic acid (HCOOH) and formaldehyde (CH_2O), which are synthesized using renewable electricity and CO_2 through various processes such as hydrogenation of CO_2 [7].

2.7 Status Quo, Challenges, and Outlook

This chapter discusses power-to-X technology and its great potential for overcoming the limitations of renewable electricity generation. It is hoped to help develop sustainable energy systems and fill research shortcomings in this area. Power-to-X (PtX) pathway has been developed in recent years to reduce greenhouse gas emissions, especially CO_2. This approach is designed to produce sustainable carbon-neutral fuels as well as to close the CO_2 cycle.

Key technologies for PtX fall into two general categories: power-to-gas (PtG) and power-to-liquid (PtL), each of which involves a variety of conversion processes. These promising processes generate sustainable energy carriers, directly or indirectly, using renewable power. The goals of the PtX chains are to develop renewable energy infrastructures with different applications of power-to-hydrogen, power-to-methane, power-to-methanol, power-to-ammonia, and so on. Integrating the PtX chains with carbon capture processes can be a promising way to reduce environmental impact. Therefore, these approaches have received much attention in recent years.

Despite all the potential benefits that PtX technology has for modifying energy systems and thus the favorable impact it has on the environment, as well as the progress that has been made in recent years, it also has its limitations. Successful implementation of the PtX methods requires addressing these challenges. The main problems are high costs of capital and production using these methods. In addition to the economic problems, technical problems such as the low efficiency of some of these processes have also prevented scaling up and application of these approaches in the existing energy systems. Therefore, technological advancement using more research is needed to widely implement this approach. Challenges need to be further explored to reduce costs as well as significantly improve efficiency.

References

1. Van de Krol, R., & Grätzel, M. (2012). *Photoelectrochemical hydrogen production*. Springer.
2. Khalilpour, K. R. (2019). Interconnected electricity and natural gas supply chains: The roles of power to gas and gas to power. In *Polygeneration with Polystorage for chemical and energy hubs* (pp. 133–155). Elsevier.
3. Koj, J. C., Wulf, C., & Zapp, P. (2019). Environmental impacts of power-to-X systems-a review of technological and methodological choices in life cycle assessments. *Renewable and Sustainable Energy Reviews, 112*, 865–879.
4. Fuss, S., et al. (2014). Betting on negative emissions. *Nature Climate Change, 4*(10), 850–853.

5. Pleßmann, G., Erdmann, M., Hlusiak, M., & Breyer, C. (2014). Global energy storage demand for a 100% renewable electricity supply. *Energy Procedia, 46*(0), 22–31.
6. Bosman, A., & Van Daal, H. (1970). Small-polaron versus band conduction in some transition-metal oxides. *Advances in Physics, 19*(77), 1–117.
7. de Rego Vasconcelos, B., & Lavoie, J.-M. (2019). Recent advances in power-to-X technology for the production of fuels and chemicals. *Frontiers in Chemistry, 7*, 392.
8. Buffo, G., Ferrero, D., Santarelli, M., & Lanzini, A. (2019). Reversible solid oxide cell (ReSOC) as flexible polygeneration plant integrated with CO2 capture and reuse. *E3S Web of Conferences, 113*, 02009. EDP Sciences.
9. Bailera, M., Lisbona, P., Romeo, L. M., & Espatolero, S. (2017). Power to gas projects review: Lab, pilot and demo plants for storing renewable energy and CO2. *Renewable and Sustainable Energy Reviews, 69*, 292–312.
10. Schnuelle, C., Thoeming, J., Wassermann, T., Thier, P., von Gleich, A., & Goessling-Reisemann, S. (2019). Socio-technical-economic assessment of power-to-X: Potentials and limitations for an integration into the German energy system. *Energy Research & Social Science, 51*, 187–197.
11. Storage, E. E. (2011). International Electrotechnical Commission (IEC). *White Paper, Geneva, Switzerland*.
12. Daiyan, R., MacGill, I., & Amal, R. (2020). *Opportunities and challenges for renewable power-to-X*. ACS Publications.
13. Buffo, G., Marocco, P., Ferrero, D., Lanzini, A., & Santarelli, M. (2019). Power-to-X and power-to-power routes. In *Solar hydrogen production* (pp. 529–557). Elsevier.
14. Vázquez, F. V., et al. (2018). Power-to-X technology using renewable electricity and carbon dioxide from ambient air: SOLETAIR proof-of-concept and improved process concept. *Journal of CO2 Utilization, 28*, 235–246.
15. Gür, T. M. (2018). Review of electrical energy storage technologies, materials and systems: Challenges and prospects for large-scale grid storage. *Energy & Environmental Science, 11*(10), 2696–2767.
16. Buttler, A., & Spliethoff, H. (2018). Current status of water electrolysis for energy storage, grid balancing and sector coupling via power-to-gas and power-to-liquids: A review. *Renewable and Sustainable Energy Reviews, 82*, 2440–2454.
17. Sternberg, A., & Bardow, A. (2015). Power-to-what?–environmental assessment of energy storage systems. *Energy & Environmental Science, 8*(2), 389–400.
18. Naims, H. (2016). Economics of carbon dioxide capture and utilization—A supply and demand perspective. *Environmental Science and Pollution Research, 23*(22), 22226–22241.
19. Wulf, C., Linssen, J., & Zapp, P. (2018). Power-to-gas—Concepts, demonstration, and prospects. In *Hydrogen supply chains* (pp. 309–345). Elsevier.
20. Lewandowska-Bernat, A., & Desideri, U. (2017). Opportunities of power-to-gas technology. *Energy Procedia, 105*, 4569–4574.
21. Kumar, S. S., & Himabindu, V. (2019). Hydrogen production by PEM water electrolysis–a review. *Materials Science for Energy Technologies, 2*(3), 442–454.
22. Rajeshwar, K., McConnell, R., & Licht, S. (2008). *Solar hydrogen generation*. Springer.
23. Chi, J., & Yu, H. (2018). Water electrolysis based on renewable energy for hydrogen production. *Chinese Journal of Catalysis, 39*(3), 390–394.
24. Acar, C., & Dincer, I. (2014). Comparative assessment of hydrogen production methods from renewable and non-renewable sources. *International Journal of Hydrogen Energy, 39*(1), 1–12.
25. Lee, B., et al. (2018). Economic feasibility studies of high pressure PEM water electrolysis for distributed H2 refueling stations. *Energy Conversion and Management, 162*, 139–144.
26. Borgschulte, A. (2016). The hydrogen grand challenge. *Frontiers in Energy Research, 4*, 11.
27. Ferreira-Aparicio, P., Benito, M., & Sanz, J. (2005). New trends in reforming technologies: From hydrogen industrial plants to multifuel microreformers. *Catalysis Reviews, 47*(4), 491–588.
28. Carmo, M., Fritz, D. L., Mergel, J., & Stolten, D. (2013). A comprehensive review on PEM water electrolysis. *International Journal of Hydrogen Energy, 38*(12), 4901–4934.

29. Rashid, M. M., Al Mesfer, M. K., Naseem, H., & Danish, M. (2015). Hydrogen production by water electrolysis: A review of alkaline water electrolysis, PEM water electrolysis and high temperature water electrolysis. *International Journal of Engineering and Advanced Technology, 4*(3), 2249–8958.
30. Damyanova, S., Pawelec, B., Arishtirova, K., & Fierro, J. (2012). Ni-based catalysts for reforming of methane with CO2. *International Journal of Hydrogen Energy, 37*(21), 15966–15975.
31. Iida, S., & Sakata, K. (2019). Hydrogen technologies and developments in Japan. *Clean Energy, 3*(2), 105–113.
32. Bruce, S., et al. (2018). *National hydrogen roadmap*. CSIRO.
33. FCH, J. (2019). *Hydrogen roadmap Europe: A sustainable pathway for the European energy transition*. Bietlot.
34. Baccioli, A., et al. (2020). Cost effective power-to-X plant using carbon dioxide from a geothermal plant to increase renewable energy penetration. *Energy Conversion and Management, 226*, 113494.
35. Lehner, M., Tichler, R., Steinmüller, H., & Koppe, M. (2014). *Power-to-gas: Technology and business models*. Springer.
36. Dhabi, A. (2017). *Geothermal power: Technology brief*. IRENA.
37. Tractebel, H. (2017). Study of early business cases for H2 in energy Storage and more broadly power to H2 applications. In *Fuel Cells and Hydrogen Joint Undertaking (FCH JU): Brussels, Belgium*.
38. Dresp, S. R., Dionigi, F., Klingenhof, M., & Strasser, P. (2019). Direct electrolytic splitting of seawater: Opportunities and challenges. *ACS Energy Letters, 4*(4), 933–942.
39. Paudel, S., Kang, Y., Yoo, Y.-S., & Seo, G. T. (2015). Hydrogen production in the anaerobic treatment of domestic-grade synthetic wastewater. *Sustainability, 7*(12), 16260–16272.
40. Leaver, J. D., Gillingham, K. T., & Leaver, L. H. (2009). Assessment of primary impacts of a hydrogen economy in New Zealand using UniSyD. *International Journal of Hydrogen Energy, 34*(7), 2855–2865.
41. Balat, M. (2008). Potential importance of hydrogen as a future solution to environmental and transportation problems. *International Journal of Hydrogen Energy, 33*(15), 4013–4029.
42. Cipriani, G., et al. (2014). Perspective on hydrogen energy carrier and its automotive applications. *International Journal of Hydrogen Energy, 39*(16), 8482–8494.
43. Hermesmann, M., Grübel, K., Scherotzki, L., & Müller, T. Promising pathways: The geographic and energetic potential of power-to-x technologies based on regeneratively obtained hydrogen. *Renewable and Sustainable Energy Reviews, 138*, 110644.
44. Reich, G., & Reppich, M. (2013). *Regenerative Energietechnik*. Springer.
45. Ursua, A., Gandia, L. M., & Sanchis, P. (2011). Hydrogen production from water electrolysis: Current status and future trends. *Proceedings of the IEEE, 100*(2), 410–426.
46. Holladay, J. D., Hu, J., King, D. L., & Wang, Y. (2009). An overview of hydrogen production technologies. *Catalysis Today, 139*(4), 244–260.
47. Yan, X. L., & Hino, R. (2016). *Nuclear hydrogen production handbook*. CRC press.
48. Lambert, M. (2018). *Power-to-Gas: linking electricity and gas in a decarbonising world*. Online OIES, available at: https://www.oxfordenergy.org/wpcms/wp-content/uploads/2018/10/Power-to-Gas-Linking-Electricity-and-Gas-in-a-Decarbonising-World-Insight-39.pdf. Accessed 9 Oct 2019.
49. Götz, M., et al. (2016). Renewable power-to-gas: A technological and economic review. *Renewable Energy, 85*, 1371–1390.
50. Steinmüller, H., Tichler, R., & Reiter, G. (2014). Power-to-gas–eine Systemanalyse. In *Markt- und Technologiescouting und–analyse. Energieinstitut an der Johannes Kepler Universität Linz, TU Wien, MU Leoben, JKU Linz*.
51. Shiva Kumar, S., Ramakrishna, S., Srinivasulu Reddy, D., Bhagawan, D., & Himabindu, V. (2017). Synthesis of polysulfone and zirconium oxide coated asbestos composite separators for alkaline water electrolysis. *Journal of Chemical Engineering & Process Technology, 3*(1035), 1–1035.

52. Zeng, K., & Zhang, D. (2010). Recent progress in alkaline water electrolysis for hydrogen production and applications. *Progress in Energy and Combustion Science, 36*(3), 307–326.
53. Schmidt, O., Gambhir, A., Staffell, I., Hawkes, A., Nelson, J., & Few, S. (2017). Future cost and performance of water electrolysis: An expert elicitation study. *International Journal of Hydrogen Energy, 42*(52), 30470–30492.
54. Coutanceau, C., Baranton, S., & Audichon, T. (2018). Hydrogen production from water electrolysis. *Hydrogen Electrochemical Production*, 17–62.
55. Russell, J., Nuttall, L., & Fickett, A. (1973). Hydrogen generation by solid polymer electrolyte water electrolysis. *American Chemical Society and Division of Fuel Chemistry Preprints, 18*, 24–40.
56. Martinson, C. A., Van Schoor, G., Uren, K., & Bessarabov, D. (2014). Characterisation of a PEM electrolyser using the current interrupt method. *International Journal of Hydrogen Energy, 39*(36), 20865–20878.
57. Grigor'ev, S., Khaliullin, M., Kuleshov, N., & Fateev, V. (2001). Electrolysis of water in a system with a solid polymer electrolyte at elevated pressure. *Russian Journal of Electrochemistry, 37*(8), 819–822.
58. Millet, P., Ngameni, R., Grigoriev, S., & Fateev, V. (2011). Scientific and engineering issues related to PEM technology: Water electrolysers, fuel cells and unitized regenerative systems. *International Journal of Hydrogen Energy, 36*(6), 4156–4163.
59. Giovanni, C. D., et al. (2016). Low-cost nanostructured iron sulfide electrocatalysts for PEM water electrolysis. *ACS Catalysis, 6*(4), 2626–2631.
60. Datta, M. K., et al. (2013). High performance robust F-doped tin oxide based oxygen evolution electro-catalysts for PEM based water electrolysis. *Journal of Materials Chemistry A, 1*(12), 4026–4037.
61. Brisse, A., Schefold, J., & Zahid, M. (2008). High temperature water electrolysis in solid oxide cells. *International Journal of Hydrogen Energy, 33*(20), 5375–5382.
62. Huynh, H. L., & Yu, Z. (2020). CO2 Methanation on Hydrotalcite-derived catalysts and structured reactors: A review. *Energy Technology, 8*(5), 1901475.
63. De Saint Jean, M., Baurens, P., Bouallou, C., & Couturier, K. (2015). Economic assessment of a power-to-substitute-natural-gas process including high-temperature steam electrolysis. *International Journal of Hydrogen Energy, 40*(20), 6487–6500.
64. Manthiram, K., Beberwyck, B. J., & Alivisatos, A. P. (2014). Enhanced electrochemical methanation of carbon dioxide with a dispersible nanoscale copper catalyst. *Journal of the American Chemical Society, 136*(38), 13319–13325.
65. Thema, M., Bauer, F., & Sterner, M. (2019). Power-to-gas: Electrolysis and methanation status review. *Renewable and Sustainable Energy Reviews, 112*, 775–787.
66. Lewandowska-Bernat, A., & Desideri, U. (2018). Opportunities of power-to-gas technology in different energy systems architectures. *Applied Energy, 228*, 57–67.
67. Machado, A. S. R., Nunes, A. V., & da Ponte, M. N. (2018). Carbon dioxide utilization—Electrochemical reduction to fuels and synthesis of polycarbonates. *The Journal of Supercritical Fluids, 134*, 150–156.
68. Kaneco, S., Katsumata, H., Suzuki, T., & Ohta, K. (2006). Electrochemical reduction of CO2 to methane at the cu electrode in methanol with sodium supporting salts and its comparison with other alkaline salts. *Energy & Fuels, 20*(1), 409–414.
69. Taubner, R.-S., Schleper, C., Firneis, M. G., & Rittmann, S. K.-M. (2015). Assessing the ecophysiology of methanogens in the context of recent astrobiological and planetological studies. *Life, 5*(4), 1652–1686.
70. Rittmann, S., Seifert, A., & Herwig, C. (2015). Essential prerequisites for successful bioprocess development of biological CH4 production from CO2 and H2. *Critical Reviews in Biotechnology, 35*(2), 141–151.
71. Van Dael, M., et al. (2018). Techno-economic assessment of a microbial power-to-gas plant–case study in Belgium. *Applied Energy, 215*, 416–425.
72. Lecker, B., Illi, L., Lemmer, A., & Oechsner, H. (2017). Biological hydrogen methanation–a review. *Bioresource Technology, 245*, 1220–1228.

73. Seifert, A., Rittmann, S., Bernacchi, S., & Herwig, C. (2013). Method for assessing the impact of emission gasses on physiology and productivity in biological methanogenesis. *Bioresource Technology, 136*, 747–751.
74. Voelklein, M., Rusmanis, D., & Murphy, J. (2019). Biological methanation: Strategies for in-situ and ex-situ upgrading in anaerobic digestion. *Applied Energy, 235*, 1061–1071.
75. Schmidt, P., Batteiger, V., Roth, A., Weindorf, W., & Raksha, T. (2018). Power-to-liquids as renewable fuel option for aviation: A review. *Chemie Ingenieur Technik, 90*(1–2), 127–140.
76. Kasten, P., Blanck, R., Loreck, C., & Hacker, F. (2013). Strombasierte Kraftstoffe im Vergleich-Stand heute und die Langfristperspektive. *Öko-Institut Working Paper, 1*, 2013.
77. Kaiser, P., Pöhlmann, F., & Jess, A. (2014). Intrinsic and effective kinetics of cobalt-catalyzed Fischer-Tropsch synthesis in view of a power-to-liquid process based on renewable energy. *Chemical Engineering & Technology, 37*(6), 964–972.
78. Schmidt, P., Weindorf, W., Roth, A., Batteiger, V., & Riegel, F. (2016). *Power-to-liquids: Potentials and perspectives for the future supply of renewable aviation fuel*. German Environment Agency.
79. Panzone, C., Philippe, R., Chappaz, A., Fongarland, P., & Bengaouer, A. (2020). Power-to-liquid catalytic CO2 valorization into fuels and chemicals: Focus on the Fischer-Tropsch route. *Journal of CO2 Utilization, 38*, 314–347.
80. Hank, C., et al. (2018). Economics & carbon dioxide avoidance cost of methanol production based on renewable hydrogen and recycled carbon dioxide–power-to-methanol. *Sustainable Energy & Fuels, 2*(6), 1244–1261.
81. Bellotti, D., Rivarolo, M., Magistri, L., & Massardo, A. (2017). Feasibility study of methanol production plant from hydrogen and captured carbon dioxide. *Journal of CO2 Utilization, 21*, 132–138.
82. Ikäheimo, J., Kiviluoma, J., Weiss, R., & Holttinen, H. (2018). Power-to-ammonia in future north European 100% renewable power and heat system. *International Journal of Hydrogen Energy, 43*(36), 17295–17308.

Chapter 3
Whole Energy Systems Evaluation: A Methodological Framework and Case Study

Ali El Hadi Berjawi, Adib Allahham, Sara Louise Walker, Charalampos Patsios, and Seyed Hamid Reza Hosseini

3.1 Introduction

The transition to a net-zero carbon society requires a deep transformation of the energy system. The transformation is expected to affect not only the electricity system but also other energy systems including gas, heat and transport, due to the need to decarbonise those sectors as well as exploiting flexibility measures across them to manage future uncertainties [1]. The energy transition is therefore anticipated to bring major changes to the energy system architecture, with changes in the planning and operations paradigm, the market structure and the regulatory framework [2]. This calls for a whole energy systems (WES) approach that spans multiple energy vectors (electricity, gas, heat, transport), different levels of the energy supply chain from generation to end use through infrastructure and markets and the various stakeholders involved from different domains (social, technical, economic, environmental, political, etc.).

Originating from the WES approach is the concept of energy systems integration (ESI), which aims to connect energy systems physically and virtually across infrastructure and markets and exploit synergies among them [3]. ESI is enabled by vector-coupling technologies such as combined heat and power (CHP), power-to-gas (P2G), heat pumps (HPs) and electric vehicles (EVs) [4]. Hence, the WES approach has been applied for the integrated energy systems planning and operation [5], but gaps have been identified with its application for evaluation of those systems. This includes the need for evaluation methodologies that incorporate a WES approach to provide evidence on the effectiveness of ESI while capturing the interdependencies and emergent complexity of integrated energy systems [6–8].

A. E. H. Berjawi (✉) · A. Allahham · S. L. Walker · C. Patsios · S. H. R. Hosseini
School of Engineering, Newcastle University, Newcastle upon Tyne, UK
e-mail: a.e.h.berjawi2@ncl.ac.uk; adib.allahham@ncl.ac.uk; sara.walker@ncl.ac.uk; haris.patsios@ncl.ac.uk; hamid.hosseini@ncl.ac.uk

© The Author(s), under exclusive license to Springer Nature Switzerland AG 2022

V. Vahidinasab, B. Mohammadi-Ivatloo (eds.), *Whole Energy Systems*, Power Systems, https://doi.org/10.1007/978-3-030-87653-1_3

A novel methodological framework that exhibits the WES evaluation principles has been developed by the authors to address these gaps [9], which is first summarised in this chapter. Building on this work, this chapter presents a practical approach for the implementation of the framework demonstrated by a case study, with a focus on the interaction and feedback between the methodological stages making up the framework. The stages are, namely, the scenario formulation, conceptual modelling and quantitative modelling.

The evaluation principles are identified in correspondence to the WES approach definition being multidimensional, multivectoral and systemic. Additionally, the evaluation should be futuristic, systematic and applicable. The framework provides a socio-technical approach for evaluation by incorporating stakeholders' requirements and the energy system technical components and functions through a system-of-systems architecture methodology. This is done through abstraction and decomposing the integrated energy system under study to facilitate its evaluation using a set of indicators. Indicators are deduced through systems analysis using model-based systems engineering techniques.

The evaluation framework is demonstrated through a case study of a local energy system in the North of Tyne region in the UK, analysing scenarios with different system configurations under varying energy supply and demand conditions. The analysis is supported by an optimisation model for integrated energy networks operation. The case study demonstrates the framework application and presents evidence on the value of energy systems integration in the region via vector-coupling technologies, including CHP, P2G and HPs under different scenarios. Figure 3.1 shows a graphical summary of the work presented in this chapter.

Section 3.2 presents a brief discussion around WES analysis and evaluation. Section 3.3 provides an overview of the WES evaluation principles and a description of the evaluation framework design and implementation. Section 3.4 presents the case study to demonstrate the evaluation framework in practice. Section 3.5 discusses the contributions and limitations of this chapter, and Sect. 3.6 summarises the chapter and concludes with future work.

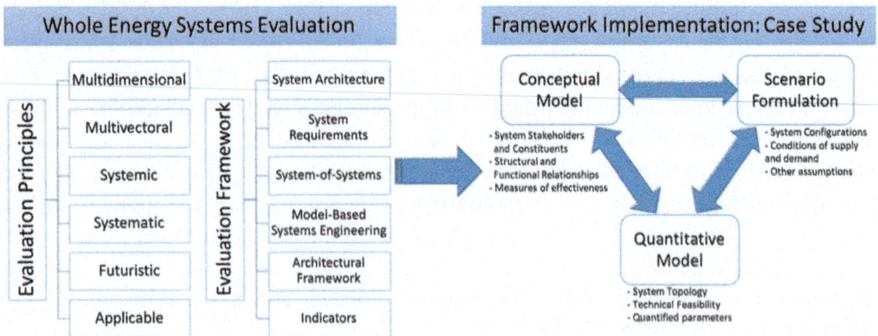

Fig. 3.1 Graphical summary of this chapter

3.2 Status Quo, Challenges and Outlook

Previous work has been presented in recent years to explore ESI in terms of the overall concept and potential benefits [6–8, 10–12], modelling for planning and operation of integrated energy systems [5, 13, 14], enabling technologies [4], flexibility provision [1, 15, 16] and economics and policy [3, 17]. This is a selection of papers that summarise the state of the art in different areas of ESI research, which is based on a WES approach to the modelling and analysis of multivector energy networks.

In terms of evaluation of ESI, previous reviews on ESI have identified gaps around evaluation and suggested research recommendations. Specifically, while ESI provides an opportunity to improve the system performance in terms of the energy trilemma, there is still a need for (i) more quantified evidence to validate this claim and support decision-making in this direction, (ii) new tools and metrics to identify the full range of benefits of ESI under different situations and (iii) comprehensive assessment methodologies to capture the interdependencies across energy systems and the emergent complexity of the whole system [6–8, 11]. Evaluation has been since considered in some studies yet not in a holistic manner, that is, focusing on particular technologies [18, 19] or looking at a limited scope of indicators [20, 21]. Therefore, the literature still lacks methodologies that address the identified gaps in the evaluation of ESI as a pathway to achieve the energy transition objectives. In this regard, a WES approach for evaluation is needed to evaluate the overall system benefits and drawbacks, identify the interdependencies among the different energy systems and adapt to future changes.

A multidisciplinary approach is required to investigate ESI and to achieve the energy transition to a net-zero carbon society [3, 4, 6, 8]. One approach for multidisciplinary research is based on systems thinking that addresses complexity in systems with multiple and interacting factors [22]. Systems thinking is a broad set of principles spanning diverse fields of physical and social sciences, engineering and management useful for considering interrelationships between system elements and their effect on the wider system behaviour [23]. A number of tools and techniques use a systems thinking approach to solve complex problems such as those developed in systems engineering, which itself is a discipline that integrates multiple disciplines to enable the realisation of successful systems across its lifecycle [23]. Given the complexity of the energy system, a systems thinking approach is valuable to understand system change and reflect the interactions between its heterogeneous elements [24].

Going one step beyond is the notion of whole systems thinking, which extends the systems thinking to look at interrelationships within and between systems due to the increasing prospects of systems integration [24]. Accordingly, the concept of SoS has emerged to study large-scale interdisciplinary problems that span multiple, distributed systems [23]. Systems thinking also provides an interdisciplinary approach for the evaluation of systems [25]. This can be used to inform planning and decision-making while ensuring that the evaluation is not focused on a single aspect of the system [23]. In this chapter, a WES approach for evaluation is considered as it is closely related to the concept of ESI. Moreover, concepts and methods from systems engineering are used, mainly the SoS and system architecture, to develop a methodological framework for ESI evaluation that is demonstrated through a case study of a regional energy system.

3.3 Whole Energy Systems Evaluation

This section describes the proposed WES approach to evaluate integrated energy systems. The WES evaluation principles are first identified before describing the developed evaluation framework.

3.3.1 Evaluation Principles

The WES approach for evaluation is defined through a number of principles. Three principles correspond directly to the WES definition, and three other principles are related to the nature of the evaluation framework in the scope of the energy transition. The six principles are identified and elaborated further in [9].

The six principles are important for the evaluation of future integrated energy system towards achieving the energy transition objectives, where:

- A multidimensional evaluation is essential to investigate trade-offs and synergies between the different perspectives and objectives required by multiple stakeholders.
- A multivectoral evaluation is necessary to account for interactions and interdependencies between the coupled energy vectors.
- A systemic evaluation is needed to cover the energy system supply chain from generation through networks and markets to end use, in order to capture emerging properties, such as flexibility and resilience, at different system levels.
- The evaluation should be futuristic as well to accommodate major future changes to the energy system structure and function, which would alter the way it is evaluated.
- The evaluation must be systematic with regard to the procedures for the deduction and interpretation of the evaluation criteria and indicators, which is relevant for the transparency, validity and replicability of the evaluation in different contexts.
- It is important for the evaluation framework to be applicable as a useful means to provide support for decision-making in practice.

3.3.2 Evaluation Framework

To address the identified gaps, an evaluation framework that exhibits the six principles is developed. The framework design and implementation procedure are described below.

3.3.2.1 Framework Design

The framework is based on a system-of-systems architecture methodology which is developed using concepts and methods from systems engineering for the purpose of evaluation. Those are summarised as follows:

- System architecture, which outlines the principles and guidelines governing the system structure, functions, the relationships between its components and with its environment and how the system will meet its requirements [26].
- System requirements, which refer to the functions and capabilities that the system needs to fulfil or acquire and are mainly related to the needs of stakeholders [27].
- System-of-systems (SoS), which is defined as integration of independent systems that act jointly towards a common goal, through synergies, to collectively offer emergent functionality that cannot be provided by constituent systems alone, as a modelling approach [28].
- Model-based systems engineering (MBSE) is the formalised application of modelling to support system design, architecture, analysis and evaluation [29].
- Architectural framework is a structured practice specifying the system views required to describe a system architecture [30].

Figure 3.2 shows the concepts and methods in relation to the evaluation principles.

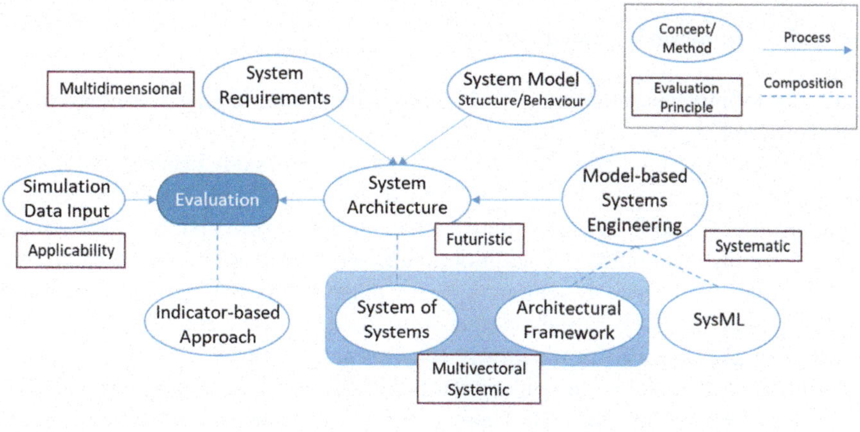

Fig. 3.2 The proposed conceptual framework for whole energy systems evaluation

The main aim of adopting those concepts and methods is to develop a conceptual system model of the energy system, whereby the system structure, function and requirements are identified and combined into a system architecture description. This allows for a socio-technical evaluation by assessing the system performance against system requirements. System performance is realised from actual system datasets or data from quantitative models, while system requirements typically correspond to stakeholders' needs and objectives. This is carried out by tracing the system requirements to the relevant functionalities delivered by the system and the capabilities it acquires. This will facilitate the deduction of appropriate evaluation criteria and indicators that represent the level of satisfaction of the system requirements. This delivers on the evaluation principles identified for being multidimensional and futuristic.

Using a SoS modelling approach allows the evaluation to be multivectoral and systemic, where the energy system is modelled at different system levels and decomposed into different constituent systems (CSs). This will highlight the interactions, interdependencies and emergent behaviour between different energy vectors (electricity, gas, heat) and system components (supply, networks, storage, demand).

MBSE is the method used to develop conceptual models that represent the system architecture, including its structure, functions and requirements. MBSE is supported by the Systems Modelling Language (SysML), which is a graphical modelling language for designing and analysing complex systems [31]. An architectural framework is needed to systematically guide the system modelling in order to capture different perspectives and viewpoints critical for the analysis. The architectural framework reflects the designated system architecture such as the SoS.

The evaluation is conducted using an indicator-based approach, where indicators are the final means for the evaluation. The applicability principle of the evaluation mainly depends on the availability and suitability of the data used for the indicator quantification.

3.3.2.2 Framework Implementation

In order to implement the framework, three methodological stages are considered as shown in Fig. 3.3.

The scenario formulation describes the system under study bearing in mind the scope and objective of the evaluation. This includes specifying the system configurations (combination of energy networks and technologies) and the conditions of energy supply and demand, in addition to any other assumptions to the conditions surrounding the energy system such as for policies, markets, demographics and behavioural changes. For example, scenarios could be formulated to investigate the impact of adopting ESI as a pathway for the energy transition, under exogenous constraints such as delivering net-zero carbon emissions or under varying conditions of energy supply and demand. Those conditions can be affected by changes to the RES capacity or the peak demand levels. Thus, the scenario formulation stage

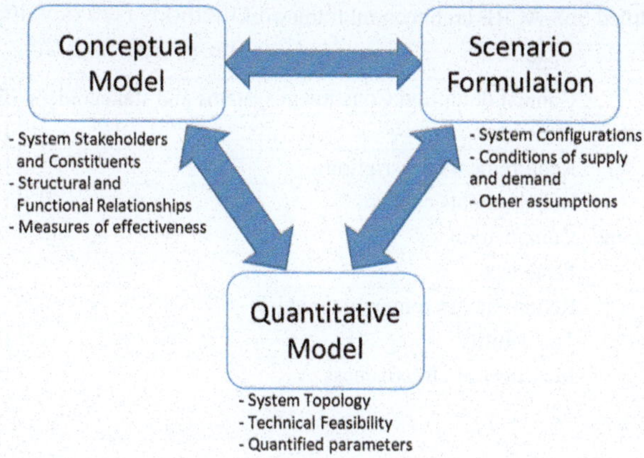

Fig. 3.3 Whole energy systems evaluation framework implementation

provides the description of the system that is then modelled conceptually and quantitatively.

The conceptual system model portrays the WES through its stakeholders and CSs, the structural and functional relationships within and across its CSs and the measures of effectiveness for evaluation. Measures of effectiveness are the indicators used eventually for the social, technical, economic, environmental and/or political evaluation. The conceptual model is developed following a new architectural framework established for the purposes of this analysis, applying a SoS modelling approach and using MBSE techniques. The architectural framework employed in this analysis is a modified version of the system-of-systems approach to context-based requirements engineering (SoS-ACRE) framework [29, 32], which is presented in Table 3.1. This architectural framework is used as it allows decomposing the system under study into different levels for the SoS approach, in addition to showing the structural and functional interactions between different components at different levels. In this case, the integrated energy system is divided into four levels: the context, SoS, CS and whole system levels. On each level, several views are developed as specified in the architectural framework to show the system structure, composition, stakeholders, requirements and measures of effectiveness using SysML diagrams. Those are further demonstrated in Sect. 3.4.

Accordingly, appropriate evaluation criteria are deduced based on the systems analysis, and corresponding indicators are assigned to measure the state of the criteria. The evaluation criteria are related to system requirements at different levels and are traced to the relevant system components or functions that contribute to its satisfaction or fulfilment. Indicators are then assigned by considering what parameters are indicative to measure this extent of satisfaction. Evaluation reflects both contextual objectives at the SoS level and the functional requirements at the CSs level. This shows the performance of the energy systems in delivering capabilities independently and as a whole. This approach enables evaluation considering

Table 3.1 Modified SoS-ACRE architectural framework

Level	View	SysML diagram
Context	Context definition:Constituent systems and stakeholders	Block definition
System-of-systems	Structure	Internal block
	Requirements description	Requirements
	Requirements context	Use case
Constituent systems	Composition	Block definition
	Structure	Internal block
	Requirements context interaction	Use case
Whole system	Traceability	Requirements
	Measures of effectiveness	Parametric

different system levels and multiple perspectives. Thus, this stage identifies the indicators and the underpinning relationships to be quantified by the quantitative model.

The quantitative model, in turn, combines inputs from the two stages with the given system topology and physical constraints in mathematical terms, to identify the technical feasibility and provide quantified output parameters. The output parameters from the quantitative model are those needed to calculate the indicators as assigned in the conceptual model. Upon quantification, indicators are analysed and combined in a dashboard to present and compare findings without masking trade-offs. The effectiveness of ESI is evaluated either absolutely against set targets or relatively as improvements to a baseline scenario, for example, one with no integration between energy systems.

This is an iterative process with feedback loops between the three stages. For instance, during the conceptual modelling, scenarios could be modified to ensure variability between scenarios while maintaining comparability. Also, scenarios could be modified if deemed infeasible in the quantitative model. Essentially, assumptions across the three activities should be checked for compatibility. Moreover, the feedback between the conceptual model and the quantitative model is mainly around the input and output parameters required to calculate the indicators. This could be affected by the data available, the exogenous variables set to the quantitative model and the nature of the model itself. Therefore, early communication between the two models is essential to ensure common understanding of what is available and possible and what changes might need to be done.

3.4 Case Study

A case study is presented in this section to demonstrate the application of the developed evaluation framework. The case study is based on the local energy system in the North of Tyne (NoT) region in the UK. The case study aims to explore the value of ESI as a future pathway for the energy transition in the region. The scenario analysis is carried out with different network configurations (electricity, heat, gas) and vector-coupling technologies (CHP, P2G, HPs) and under varying conditions of energy supply and demand. Note that the transport system and the use of EVs are excluded from this analysis. The analysis is supported by an optimisation model for integrated energy networks operation.

3.4.1 Case Study Area Description

The NoT region is a recently established combined authority that covers the local authority areas of Newcastle upon Tyne, North Tyneside, and Northumberland in the North East of England, UK. The region covers an area adding up to around 5277 km^2 and has an estimated population of 833,000 with more than 360,000 households.

The region has a variety of features in terms of energy such as a legacy of high rates of fuel poverty, high proportion of off-gas properties, above-average domestic gas consumption per meter, a number of existing district heating schemes and a significant RES capacity expansion. The region houses a number of energy research and demonstration facilities [33]. Moreover, the region represents an urban energy system with different sectors for energy demand (residential, commercial, industrial, transport).

The total energy consumption in the region for the domestic, commercial, industrial and transport sectors was estimated to be 1486.4 ktoe in 2018, which is around 17.3 TWh. This is around 1% of the UK total energy consumption in 2018. The commercial and industrial demand makes up around 37% of the total consumption, the domestic sector consumes 35%, while the transport demand stands at 28%. The domestic consumption is mainly supplied by gas (75%) followed by electricity (20%) and other fuels (5%). Similarly, the commercial and industrial sectors are mainly supplied by gas (43%) followed by electricity (35%) and other fuels (22%) [34].

The renewable energy capacity has increased in the region by around 240% from 245 MW in 2014 to 837 MW in 2019, while generation from RES has increased by around 121% from around 517 GWh in 2014 to around 1146 GWh in 2019. The increase in both RES capacity and generation has been dominated by the expansion of solar PV, onshore wind and offshore wind in Northumberland [35]. Accordingly, carbon emissions in the region have seen a decrease between the years 2014 and 2018 by around 15%. The wider North East region experienced the largest percentage reduction in carbon emissions in the UK from 2005 to 2018, in part due to industrial closures. In 2005, the industrial and commercial sectors were responsible for the majority of carbon emissions in the region with 58.2% compared to 23.9% by the domestic sector and 17.9% from transport. In 2018, the transport sector takes the lead by 34.7%, while the industrial and commercial sectors come next with 33.5% followed by the domestic sector with 31.9% of emissions [36].

This overview of energy figures highlights the need to decarbonise heating in the region, which makes up the majority of gas use in residential consumption, in addition to the commercial and industrial consumption. This could benefit from the potential for RES expansion in the region as well as for district heating. Again, this raises the prospects of ESI as a pathway to achieve net-zero carbon emissions targets through coupling the multiple vectors of electricity, gas and heat present in the region. It is worth noting that the transport sector is another priority for decarbonisation as evident by the carbon emissions figures, which can be boosted by ESI as well, but this is outside the scope of this chapter.

3.4.2 Scenario Formulation

The case study scenarios are designed to capture the value of integration under extreme conditions of change and constraint to the whole energy system, assuming no major network upgrade is made. Therefore, the baseline was chosen to be the coldest day recorded in November 2019, where the total energy demand was at its annual peak. Other scenarios consider higher RES supply and higher daily electric load peak due to the penetration of HPs.

The scenario analysis here does not aim to make projections based on the local authorities' plans or to critique them, but rather is explorative to consider the possible impact of ESI as a direction for the energy transition to decarbonisation. While we aim for a whole systems approach to the evaluation, it is still not possible to model the whole economy in this case. Scenarios are not designed to consider the temporal dynamics of the transition over the years up to 2030 or 2050, but as a steady-state snapshot of the energy system at one point of time. This study is focused on the physical system architecture, and thus the scenario analysis reflects technical changes without considering different aspects related to governance and markets that may be necessary for the transition.

The case study is designed to investigate the impacts of integration between the electricity and gas systems through CHP and P2G and the increased electrification of heat through HPs on the whole energy system. Six scenarios were formulated as a combination of different configurations and supply and demand conditions. The different system configurations are:

- Unidirectional integration via CHP.
- Unidirectional integration via CHP + HPs.
- Bidirectional integration via CHP and P2G.
- Bidirectional integration via CHP and P2G + HPs.

Note that unidirectional integration refers to the vector shifting capability from one network to another, in this case from the gas network to the electricity network through CHP. On the other hand, bidirectional integration indicates the capability of the system to shift the energy vector in both directions, i.e. from the gas network to the electricity network and vice versa, as in the case with both CHP and P2G.

For each configuration, different cases of supply and demand conditions are explored:

Table 3.2 North of Tyne scenario formulation

Scenario	Configuration	Supply and demand conditions
1	Unidirectional integration	Baseline RES, baseline load
2	Unidirectional integration + HPs	Baseline RES, high load
3	Bidirectional integration	Baseline RES, baseline load
4	Bidirectional integration + HPs	Baseline RES, high load
5	Bidirectional integration	High RES, baseline load, high P2G
6	Bidirectional integration + HPs	High RES, high load, high P2G

- Baseline RES: current supply conditions in the area.
- High RES: increase in available wind generation output, 700% increase.
- Baseline Load: current demand conditions in the area.
- High Load: increase of total electric load (15%) and peak electric load (20%) due to the penetration of HPs.
- High P2G: wide expansion of P2G capacity.

Table 3.2 summarises the six scenarios.

A number of assumptions have been considered as follows:

- The change in the high load scenarios is assumed to reflect a 20% increase in HPs use in the domestic sector [37].
- The 700% increase in the high RES scenarios is assumed given the huge potential for offshore RES in the region, the increase the region has seen in RES capacity in recent years (240%) and the region's decarbonisation ambitions.
- The changes in supply and demand conditions are assumed to apply uniformly to the region's local authorities.
- The actual NoT system has three gas-fired generators with total capacity of 110 MW. In our scenarios, those are replaced by three CHP plants with the same capacity increasing the total capacity of CHP in the region to 120 MW.
- Getting higher rates of RES generation into the system was not possible without much higher P2G capacity due to local network constraints, thus scenarios 5 and 6.
- The additional P2G assets in scenarios 5 and 6 are placed next to the wind farms which are added into the model to increase the penetration level of the renewable energy.
- Single vector storage technologies are not included in this analysis.

3.4.3 Conceptual Modelling

As mentioned in Sect. 3.3.2, the core of the evaluation framework is in developing the conceptual system model for the different scenarios formulated leading up to a set of indicators for evaluation. Eventually, the conceptual system model provides a set of indicators along with a traceability diagram that reflects the structural and functional interactions between different system components at different levels. This helps reduce the complexity of the WES for the purposes of the evaluation and allows for a socio-technical evaluation of the system against stakeholders' requirements.

In this case study, four configurations are considered with different combinations of vector-coupling technologies. Accordingly, four system models are required to carry out the analysis. All four models have the same CSs (electricity, gas, heat, coupling system) but have different coupling system components. This would mainly affect the structure, composition and traceability views of the architectural framework described in Table 3.1. Otherwise, the context definition, requirements and measures of effectiveness diagrams are similar across all configurations and scenarios. The scale of integration is considered at the distribution network level, and accordingly the system boundaries are determined to be the local energy system. The conceptual system model is developed using SysML stencils on Microsoft Visio, and colour coding was used for facilitating traceability.

3.4.3.1 Context Level

At this level, the system CSs and stakeholders are specified. The CSs in this case include the electricity, heat and gas systems, in addition to the coupling system defined as the system that includes vector-coupling technologies (Fig. 3.4). It should be noted here that the vector-coupling technologies are grouped into one CS, which is the coupling system, given that they share similar functionalities that contribute to the higher system level or the WES. The system stakeholders are defined as groups of those who directly influence the system planning and operations of the local energy system, including in this case the local government, system operators, end users, prosumers and local community (Fig. 3.5). The detailed composition of those

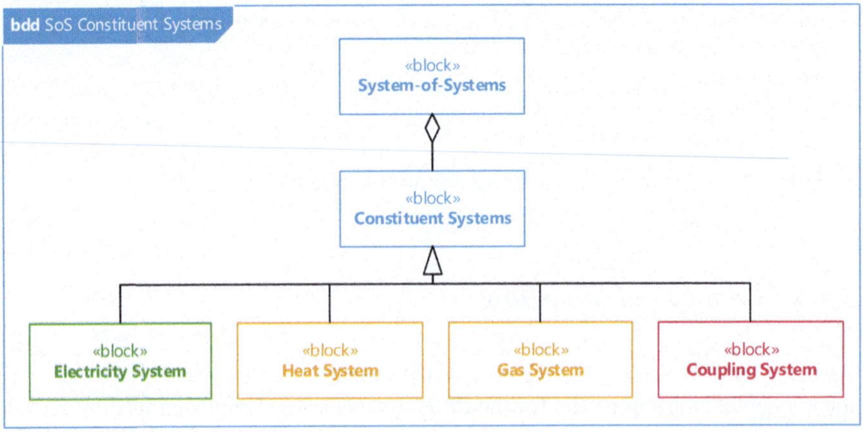

Fig. 3.4 North of Tyne context level: constituent systems

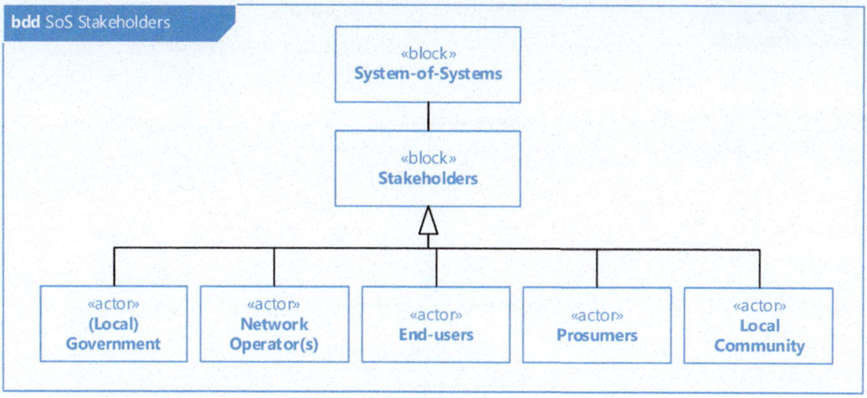

Fig. 3.5 North of Tyne context level: stakeholder groups

groups is out of the scope of this chapter. Other stakeholders such as developers, vendors, manufacturers, media and research are not considered. Moreover, since the market layer is not considered in this analysis, stakeholders such as the regulator and other market players are not included.

3.4.3.2 System-of-Systems Level

The next level in the architectural framework is the SoS level, where the structure and requirements of the system are identified. Figure 3.6 shows the structural relationships at the SoS level (highest level of abstraction) for all configurations. The diagram simply shows that the electricity, gas and heat CSs are linked through the coupling system in this case, while each of those energy CSs still also operates separately to satisfy energy demand, electrical and heat.

The system requirements for all scenarios are defined in the requirements description diagram in Fig. 3.7 and are then related to the corresponding stakeholders in Fig. 3.8. The requirements follow from the energy trilemma notion, including objectives for environmental sustainability, social and economic acceptability and technical security. These objectives are requirements typically sought by the local council representing the government, the local community pushing for environmental and social values, the network operators working to maintain a secure energy system and end users aiming for an affordable and reliable service. Additionally, prosumers require access to the grid and a requirement to retain their comfort and convenience. Requirements can be extended or constrained as shown in Fig. 3.8. The requirements at this level are the same across all scenarios given that the system stakeholders are the same.

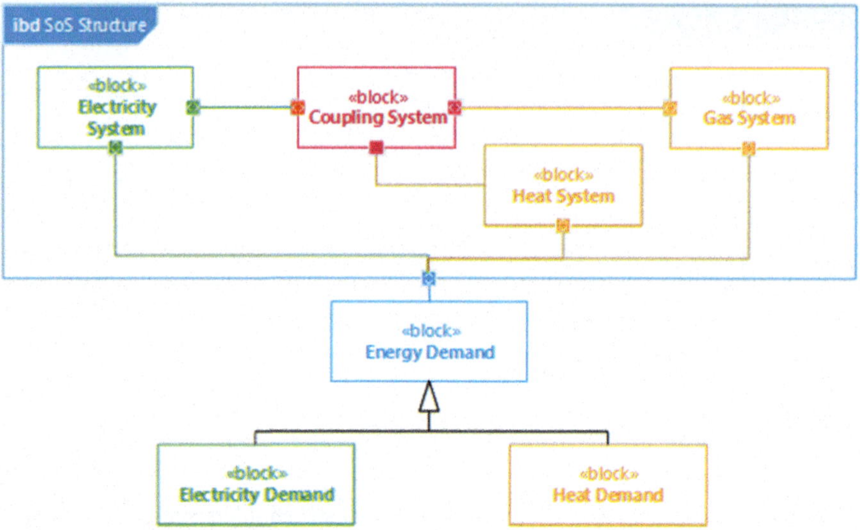

Fig. 3.6 North of Tyne SoS level: structure

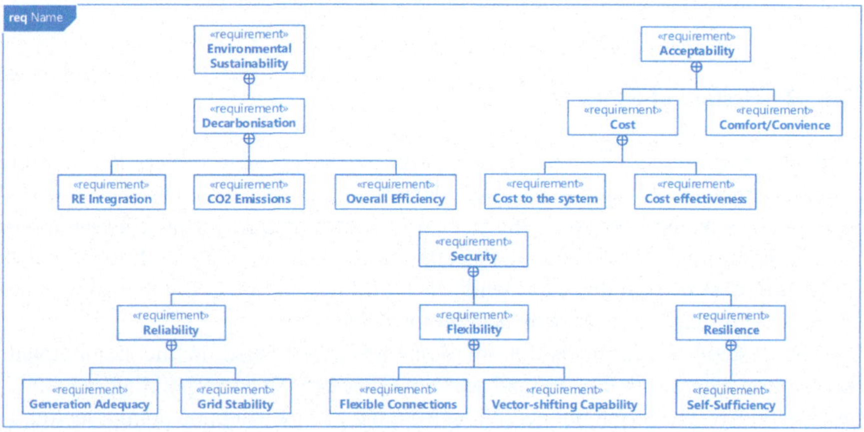

Fig. 3.7 North of Tyne SoS level: requirements description

3.4.3.3 Constituent System Level

The differences between scenarios are realised at this level, where the coupling system composition is different across the given scenarios. The composition of the electricity, gas and heat systems shown in Fig. 3.9 is the same across all scenarios. The upstream component in the electricity and gas system relates to the respective national transmission networks.

The coupling system composition for the different scenarios is presented in Fig. 3.10 as follows:

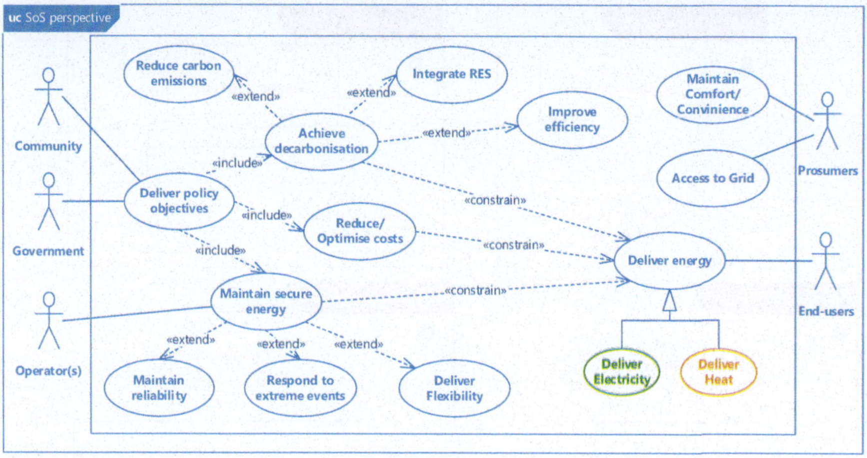

Fig. 3.8 North of Tyne SoS level: requirements context

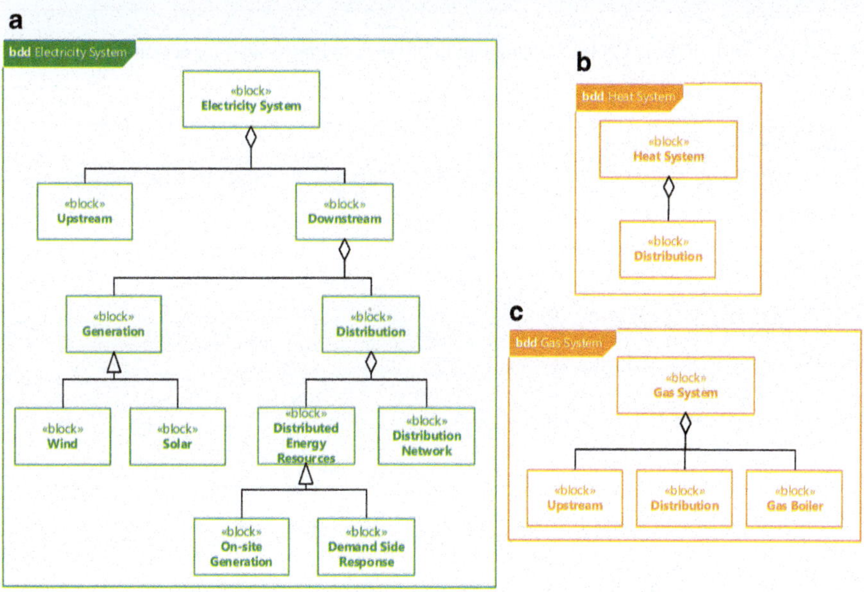

Fig. 3.9 North of Tyne CS level: composition of (a) electricity system, (b) heat system and (c) gas system

- Scenario 1: only CHP is included (Fig. 3.10-a).
- Scenario 2: CHP and HPs (Fig. 3.10-b).
- Scenarios 3 and 5: CHP and P2G (Fig. 3.10-c).
- Scenarios 4 and 6: CHP, P2G and HPs (Fig. 3.10-d).

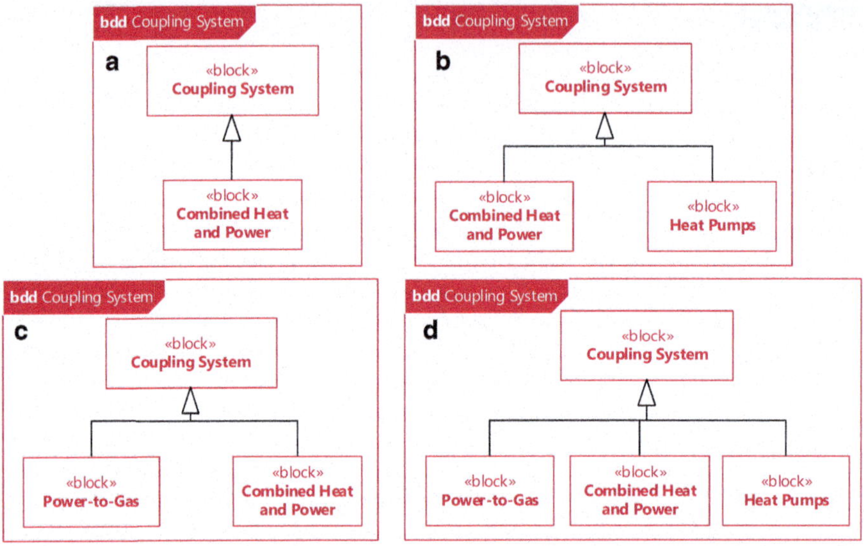

Fig. 3.10 North of Tyne CS level: composition of the coupling system for scenarios (**a**) 1, (**b**) 2, (**c**) 3 and 5 and (**d**) 4 and 6

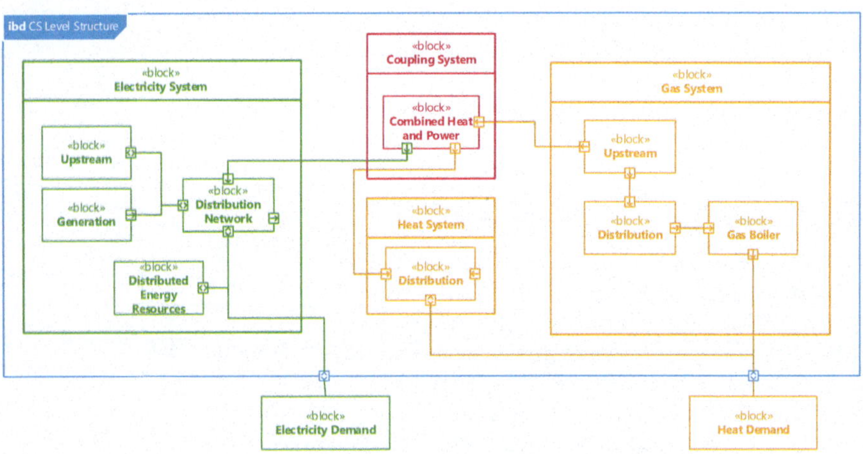

Fig. 3.11 North of Tyne CS level: structure for scenario 1

The system structure at the CS level is meant to show the structural relations between components identified in Figs. 3.9 and 3.10, within CSs and across CSs, in line with the structure at the higher SoS level. Accordingly, Figs. 3.11, 3.12, 3.13 and 3.14 show the system structure for the four different configurations. The main difference between the four diagrams is the connection between the energy CSs through the different available vector-coupling technologies.

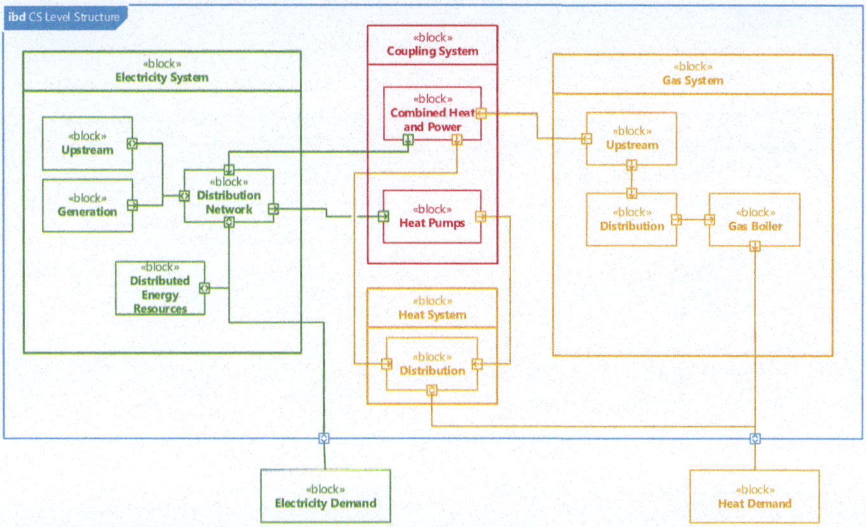

Fig. 3.12 North of Tyne CS level: structure for scenario 2

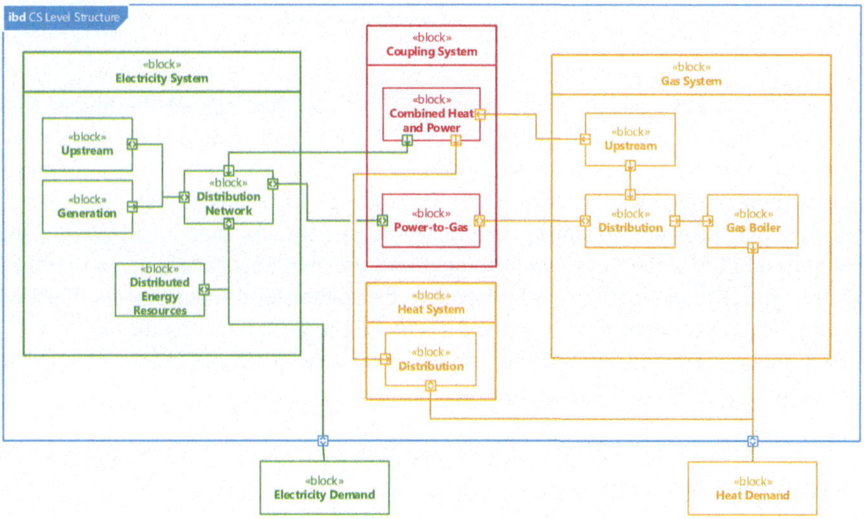

Fig. 3.13 North of Tyne CS level: structure for scenarios 3 and 5

The requirements context interaction view for the CS level for all scenarios is presented in Fig. 3.15. This diagram shows the functional interactions between the CSs and with the SoS to fulfil the system requirements. The CSs and the SoS are presented as actors in this case. The electricity, gas and heat systems mainly have the requirement to deliver their respective services to satisfy energy demand. The coupling system receives electricity and gas as inputs from the other CSs,

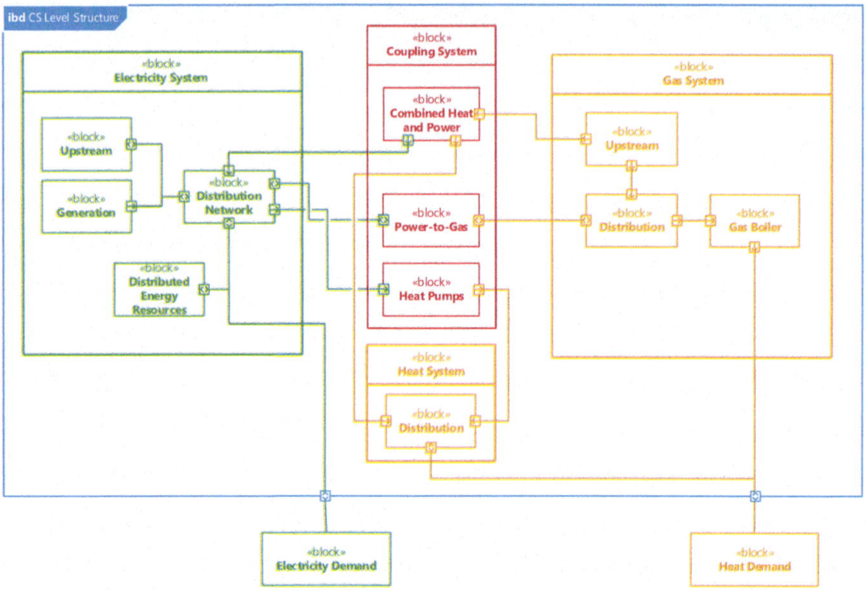

Fig. 3.14 North of Tyne CS level: structure for scenarios 4 and 6

for CHP and P2G, and in return provides functions that support the higher SoS requirements identified in Figs. 3.7 and 3.8. These include reducing the energy use that improves the overall system efficiency, providing grid services that supports the system reliability (particularly for the electricity system) and shifting energy vectors that provides flexibility to the whole system. As mentioned previously, vector-coupling technologies are grouped into the coupling system given that they share similar functionalities with respect to the higher-level requirements. Shifting energy vectors also contributes to the delivery of energy services by the energy CSs, for instance, through CHP feeding into the electricity system, P2G feeding into the gas system and HPs feeding into the heat system.

3.4.3.4 Whole System Level

Finally, on the whole system level, the traceability and measures of effectiveness views are presented. The traceability view presented in Fig. 3.16 combines different system components and functions from different levels to trace their contributions in satisfying system requirements and measuring this level of satisfaction. The traceability view can therefore support the realisation of possible trade-offs and synergies between the different system components. Accordingly, measures of effectiveness are shown in Fig. 3.17. This again applies to all scenarios, which is important for comparability in this case.

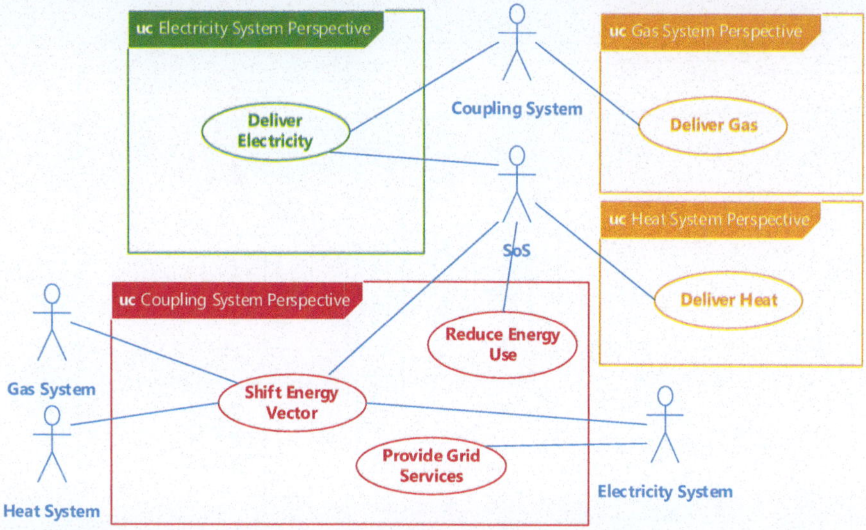

Fig. 3.15 North of Tyne CS level: requirements context interaction

Moving from the traceability diagram to assigning the measures of effectiveness is not an automatic step. It involves looking at suitable indicators that are used to measure the state of the evaluation criteria guided by the traceability view and considering the data availability required to calculate those indicators supported by the quantitative model. Accordingly, indicators presented in the measures of effectiveness view in Fig. 3.17 are believed to be suitable for the evaluation in this case, although they do not present an exhaustive or restricted list. The indicators and relevant output parameters are discussed further in Tables 3.3 and 3.5.

As mentioned previously, the evaluation criteria reflect multiple stakeholders' requirements and functional requirements at different levels, allowing for a socio-technical evaluation of the WES. This also provides a flexible, modular approach for the evaluation where if new stakeholders or different requirements are realised, this is reflected throughout the conceptual model diagrams. In this context, note that some elements included in earlier views are omitted from the analysis presented later, such as distributed energy resources, due to the lack of data around it in this case.

3.4.4 Quantitative Modelling

3.4.4.1 Network Model

The analysis is supported by an optimal power and gas flow simulation model for the operation of the integrated electricity and gas networks with different coupling

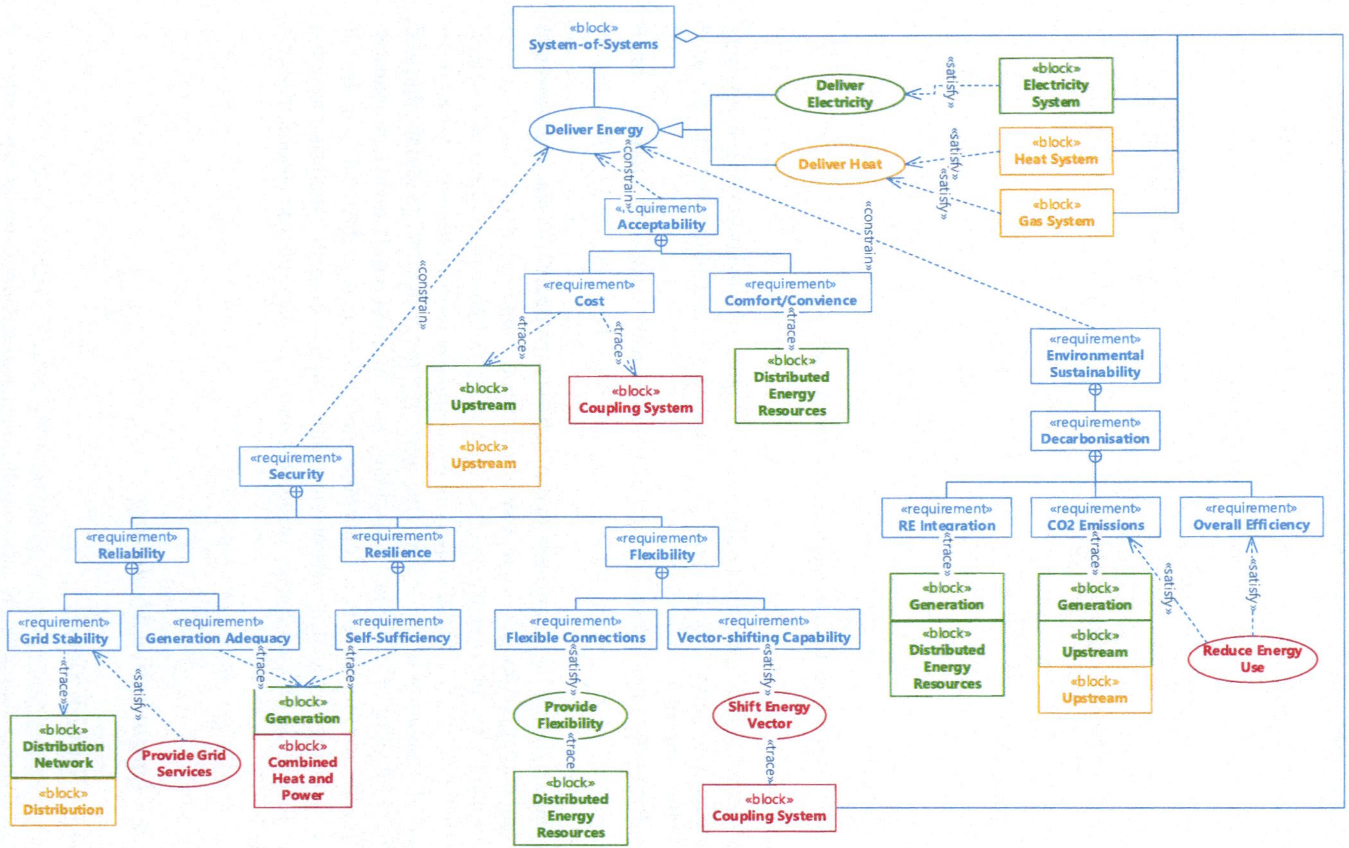

Fig. 3.16 North of Tyne whole system level: traceability

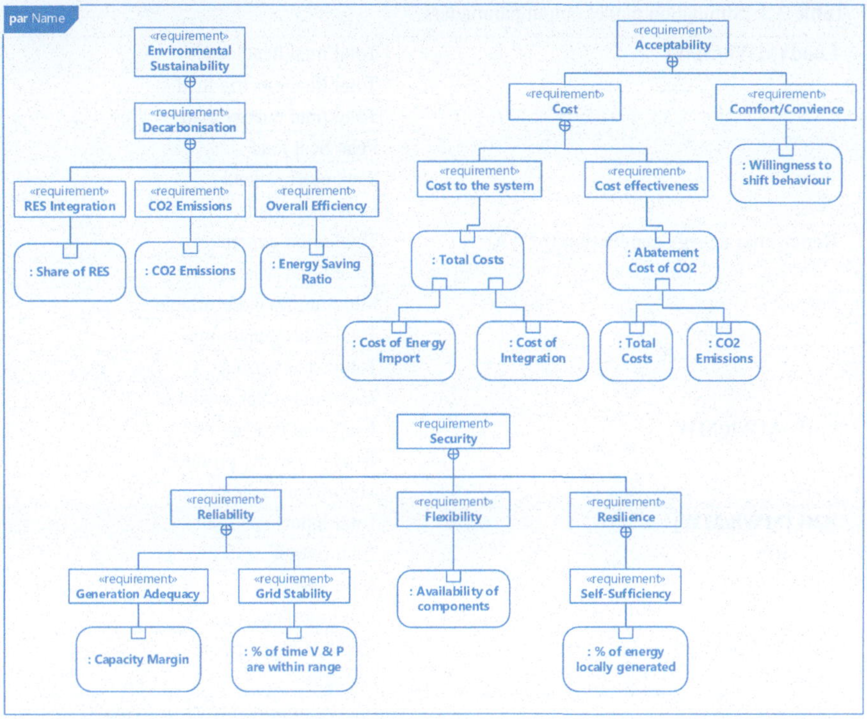

Fig. 3.17 North of Tyne whole system level: measure of effectiveness

components at the transmission and distribution levels. This type of model is chosen to evaluate the impact of the optimal operational settings for integrated energy networks under different scenarios of generation and load profiles and considering multidimensional aspects. This model is an updated version of the models developed by the authors and presented in [19, 38, 39]. The model used in this chapter considers the optimal operation of the network, while the models developed in the previous works are only operational aiming at evaluating the economic, technical, environmental parameters of the integrated energy systems. This model also outperforms other models in the literature [5] by facilitating the consideration of all the parameters affecting the optimal operation of integrated energy systems, such as different gas mixtures, gas temperature, pipeline characteristics and the electricity network topology. The model is developed in MATLAB and includes a set of nonlinear equations constrained by voltage and pressure balances for electric and gas network nodes, respectively. The optimisation is based on a cost minimisation objective function subject to physical constraints. The cost function includes the cost of electricity generation from different sources and the cost of gas supply from upstream networks. The inputs, outputs and mathematical formulations of the model are described in Fig. 3.18.

Table 3.3 Simulation model output parameters

Load (MWh/MW)	Total final heat load
	Total final electric load
	Total final energy load
	Peak heat load
	Peak electric load
	Peak total load
Renewable energy generation (MWh)	Total wind generation
	Total wind wasted
	Maximum wind available
	Total solar generation
	Total solar wasted
	Maximum solar available
CHP (MWh/MW)	Total heat production
	Total electricity produced
	Maximum capacity
P2G (MWh/MW)	Total input
	Total output
	Total waste
	Maximum capacity
HPs (MWh/MW)	Energy consumption
	Capacity
Upstream energy (gas network, electricity network, total)	Energy imported from upstream (MWh)
	CO_2 emissions from upstream (tCO_2)
	Cost of import from upstream (£)

The mathematical formulation of the gas and electricity optimal power flow can be expressed as follows:

$$\text{Min (cost of nongas electric generation} + \text{cost of gas electric generators supplied from another gas network} + \text{cost of gas supply)} \quad (3.1)$$

subject to:

$$\mathrm{h}(\mathrm{x}) = 0 \quad (3.2)$$

$$\mathrm{g}(\mathrm{x}) \leq 0 \quad (3.3)$$

where x is the state vector which includes the angle and amplitude of the voltage of all the electrical network buses, the active and reactive power of all the generators including the renewable resources and CHP, the power set points of the considered P2G assets, the pressure of the different nodes in the gas networks and the amount

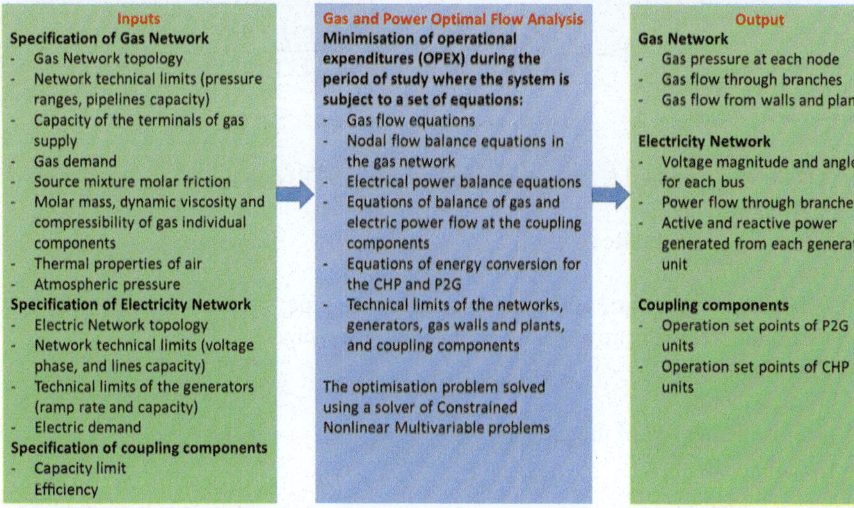

Fig. 3.18 Quantitative model inputs, outputs and algorithms

of the gas imported from upstream networks and h(x) and g(x) are the equality and inequality constraints, respectively.

The equality constraints are given by eqs. (3.4), (3.5), (3.6), (3.7), (3.8), (3.9), (3.10), (3.11) and (3.12):

$$\mathrm{Aq_p + W = 0} \tag{3.4}$$

where

q_p Vector of flow rate through pipelines
A Branch-nodal incidence matrix.
W Vector of gas supply and demand at each node.

The vector of gas injections W is obtained by

$$\mathrm{W = W_S - W_L} \tag{3.5}$$

where

W_S A vector of gas supplies at each node
W_L A vector of gas demands at each node.

The flow rate through a pipeline is calculated using the generalised gas flow equation used for calculation of flow of the branch based on the pressures of the two ends of the branch, neglecting the elevation difference, as follows. Note that more details on the gas flow model and the algorithms developed to solve the gas flow balance equations are provided by the authors in ref. [39]:

$$q = \pi\sqrt{\frac{R_{air}}{8}} \times \frac{T_n}{p_n} \times \sqrt{\frac{(p_1^2 - p_2^2) \times D^5}{f.S_{mix}.L.T.z_{mix}}} \quad (3.6)$$

where

q Gas flow in standard temperature and pressure (STP) conditions
R_{air} Air constant
T_n Standard temperature
p_n Standard pressure
p_1 Absolute gas pressure at the sending end of the pipe
p_2 Absolute gas pressure at the receiving end of the pipe
D Pipe diameter
f Friction factor
S_{mix} Specific gravity of the gas
L Length of the pipe
T Gas temperature
z_{mix} Compressibility factor.

Friction factor (f) is calculated based on the value of the Reynolds number (Re):

$$\text{Re} = \frac{D.v.\rho_{mix}}{\mu_{mix}} \quad (3.7)$$

where the value of the velocity of the gas flow (v) is calculated using the pipe cross-sectional area (A) as

$$v = \frac{q}{A} = \frac{q}{(\pi/4).D^2} \quad (3.8)$$

and the value of density of the gas mixture (ρ_{mix}) is calculated using

$$\rho_{mix} = S_{mix} \times \rho_{air} \quad (3.9)$$

where the specific gravity of the gas mixture (S_{mix}) is as follows:

$$S_{mix} = \frac{z_{air}.\sum_{i=1}^{N_c}(y_i.M_i)}{M_{air}.z_{mix}} \quad (3.10)$$

and the value of compressibility factor of the gas mixture (z_{mix}) is obtained by

$$z_{mix} = 1 - \left(\sum_{i=1}^{N_c}(y_i.c_i)\right)^2 \quad (3.11)$$

Also, the value of dynamic viscosity of the gas mixture (μ_{mix}) is computed as follows [my paper with hamid]:

$$\mu_{mix} = \frac{\sum_{i=1}^{N_c} \left(y_i.\mu_i.M_i^{0.5} \right)}{\sum_{i=1}^{N_c} \left(y_i.M_i^{0.5} \right)} \tag{3.12}$$

The explanation of the parameters of the above formulations is as follows:

Re Reynolds number of the gas flow
v Velocity of the gas flow
ρ_{mix} Density of the gas mixture
μ_{mix} Dynamic viscosity of the gas mixture
A Cross-sectional area of the pipe
ρ_{air} Density of air in STP condition
z_{air} Compressibility factor of air in STP condition
N_c Number of components of the gas mixture
y_i Molar fraction of the component *i* in the gas mixture
M_i Molar mass of the component *i* in the gas mixture
M_{air} Molar mass of air
c_i Summation factor (on page 11 of the standard ISO 6976:1995 [3])
μ_i Dynamic viscosity of the component *i* in the gas mixture.

Substituting the eqs. (3.10) and (3.11) in (9) and replacing the eqs. (3.8), (3.9) and (3.11) in (7), the equation for calculation of Reynolds number will be equal to

$$\mathrm{Re} = \frac{\rho_{air}}{(\pi/4)}.\frac{q.S_{mix}}{D}.\frac{\sum_{i=1}^{N_c} \left(y_i.M_i^{0.5} \right)}{\sum_{i=1}^{N_c} \left(y_i.\mu_i.M_i^{0.5} \right)} \tag{3.13}$$

Once the value of the Reynolds number is calculated, the value of the friction factor (f) can be calculated based on the regime of the flow as follows:

- Laminar flow ($Re < 2300$):

$$f = \frac{64}{\mathrm{Re}} \tag{3.14}$$

- Turbulent flow ($Re > 4000$):
  In this case, which frequently happens in the gas networks, the friction factor is calculated using Colebrook's equation, which has empirically been developed based on Moody chart:

$$\frac{1}{\sqrt{f}} = -2 \times \log_{10} \left(\frac{\epsilon/D}{3.7} + \frac{2.51}{\mathrm{Re}.\sqrt{f}} \right) \tag{3.15}$$

in which ϵ is the roughness of the internal surface of the pipe. The equality constraints also include

$$P_{G_i} - P_{L_i} - P_i\,(V,\theta) = 0 \tag{3.16}$$

$$Q_{G_i} - Q_{L_i} - Q_i\,(V,\theta) = 0 \tag{3.17}$$

where

P_{G_i} Real power generation at bus i
P_{L_i} Real power load at bus i
Q_{G_i} Reactive power generation at bus i
Q_{L_i} Reactive power load at bus i
P_i Real power injection at bus i
Q_i Reactive power injection at bus i
V Bus voltage magnitude vector
θ Bus voltage angle vector.

In addition to these equations, the following equality equations are used to calculate the amount of gas input to CHP estimated in m^3 and the amount of gas output from the P2G also estimated in m^3.

- Amount of gas input to CHP coupling the gas node k and the electric bus i:

$$W_L(k) = \frac{P(i) * 3600}{\zeta_{CHP} * GCV} \tag{3.18}$$

where

$W_L(k)$ Amount of the gas input to the CHP connected to the gas node k
$P(i)$ Amount of the output power from the CHP connected to the bus i
ζ_{CHP} Efficiency of the CHP
GCV Gas caloric value.
Amount of gas out from the P2G coupling the gas node k and the electric bus i:

$$W_S(k) = \frac{P(i) * 3600}{GCV} \tag{3.19}$$

where

$W_S(k)$ Amount of the gas output from the P2G connected to the gas node k
$P(i)$ Amount of the input power to the P2G connected to the bus i
GCV Gas caloric value.

The inequality constraints $g(x)$ are

$$\begin{aligned} & V_{i\,\min} \le V_i \le V_{i\,\max} \\ & I_k \max^2 \ge I_k * I_k^* \\ & P_{Gi\,\min} \le P_{Gi} \le P_{Gi\,\max} \\ & Q_{Gi\,\min} \le Q_{Gi} \le Q_{Gi\,\max} \\ & 1 \le P_{(p2g,k)} \le P_{p2g,\max,k} \\ & p_{i\,\min} \le p_i \le p_{i\,\max} \\ & Ws_{i\,\min} \le Ws_i \le Ws_{i\,\max} \end{aligned} \tag{3.20}$$

where

V_i Voltage magnitude at bus i

$V_{i\,min}$, $V_{i\,max}$ Minimum and maximum voltage magnitude at bus i, respectively

I_k, I_{kmax} Current through branch k and the maximum value of this current

P_{Gi}, Q_{Gi} Active and reactive power of the generator G_i

$P_{Gi\,min}$, $P_{Gi\,max}$ Minimum and maximum active power generated by the generator G_i

$Q_{Gi\,min}$, $Q_{Gi\,max}$ Minimum and maximum reactive power generated by the generator G_i

$P_{p2g,\,k}$, $P_{p2g,\,max\,,\,k}$ Power set point of P2G k and the maximum value of this set point, respectively

p_i Pressure at node i

$p_{i\,min}$, $p_{i\,max}$ Minimum and maximum pressure at node i

Ws_i Supplied flow from gas source i

$Ws_{i\,min}$, $Ws_{i\,max}$ Minimum and maximum flow supplied from the source i, respectively.

The model represents the same energy system topology abstracted for the conceptual system model. Note, however, that the jurisdictional boundaries of the NoT combined authority district may not be the same as the area covered by the energy networks. For instance, the network extends into parts of Gateshead and County Durham, which are not officially within the jurisdiction of the NoT combined authority.

The coupling components considered in the model are CHP and P2G, while HPs are considered as additional electric load without considering the provision of demand-side response (DSR). District heating networks are not explicitly modelled in the system, but heat output from CHP is assumed to feed into those. Furthermore, energy storage technologies are not considered in the model. In fact, the region currently lacks energy storage assets due to the relatively limited amount of RES generation.

3.4.4.2 Network Data

The model is populated by actual data from the NoT energy system for energy supply and load profiles and network flows. For the gas network, Northern Gas Networks, the gas distribution network operator in the region, has provided data of the hourly demand at each network node for a set of typical days. For this study, the demand data for a cold winter day is interpolated to get the half-hourly demand dataset. For the electricity network, Northern Powergrid, the electricity distribution network operator in the region, has provided datasets of half-hourly metered data for the primary substations on the network. The data provided covered the period from January 2016 to June 2020, of which the 15th of November 2019 was chosen for this study showing the peak demand. Along with the gas network data on the typical cold winter day, this day is expected to be a time of system stress.

Figures 3.19 and 3.20 show the load profiles for the electric and gas demands for the baseline and high load scenarios, aggregated at the electric substation level. That is considering the effect of the increased electric demand for heating through HPs on the electric substation rather than individual households. The change in the electricity load profile for the high load scenarios represents a 20% increase in the peak load and involves a shift in the morning peak hours as estimated in ref. [37].

Figure 3.21 shows the profiles of the total wind generation for the baseline RES and high RES scenarios, while Fig. 3.22 shows the profile of the total PV generation for all scenarios. Baseline data were provided by Northern Powergrid.

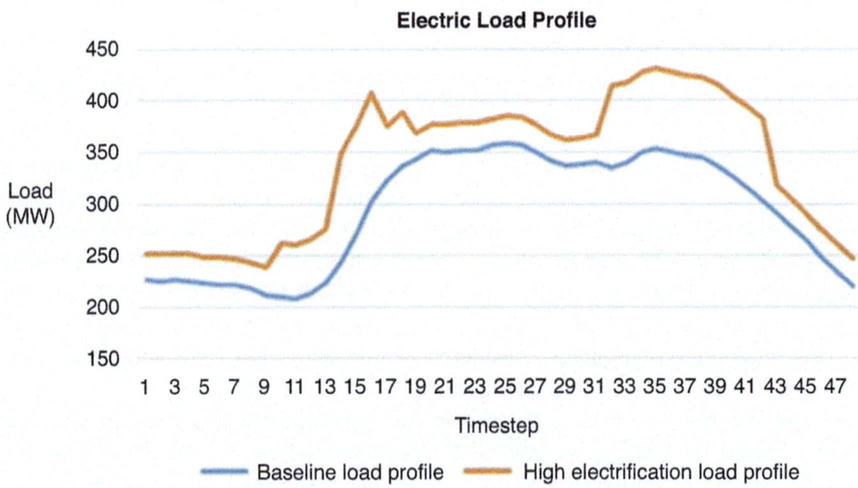

Fig. 3.19 Electricity load profiles for the baseline and high load scenarios

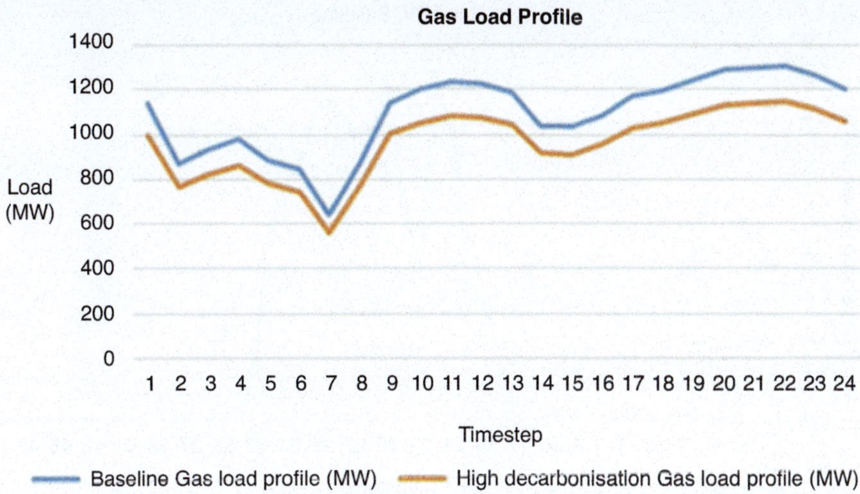

Fig. 3.20 Gas load profiles for the baseline and high load scenarios

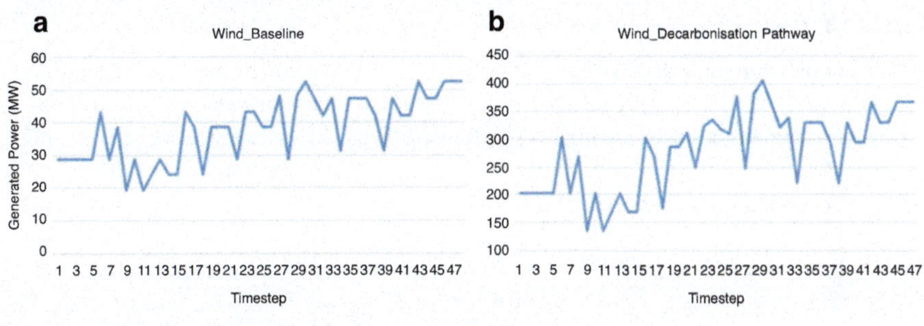

Fig. 3.21 Wind generation profile for (**a**) baseline and (**b**) high RES scenarios

3.4.4.3 Model Parameters

After running the simulations for each scenario, parameters related to the heat and electric loads, renewable energy generation, vector-coupling components outputs and capacity, electricity and gas upstream imports are obtained (Table 3.3). Those are the output parameters required to calculate the identified indicators described in detail in Table 3.5. The numerical values of the model output parameters for all scenarios are provided in ref. [40].

In addition to those parameters, the following factors for cost and carbon emissions are considered (Table 3.4).

The cost of energy import from upstream reflects an annual average of the sum of network charges on the customer for the electricity and gas transmission networks [41]. This is the cost accounted for in the optimisation as operational cost for electricity generation and gas supply, while capital and carbon costs are

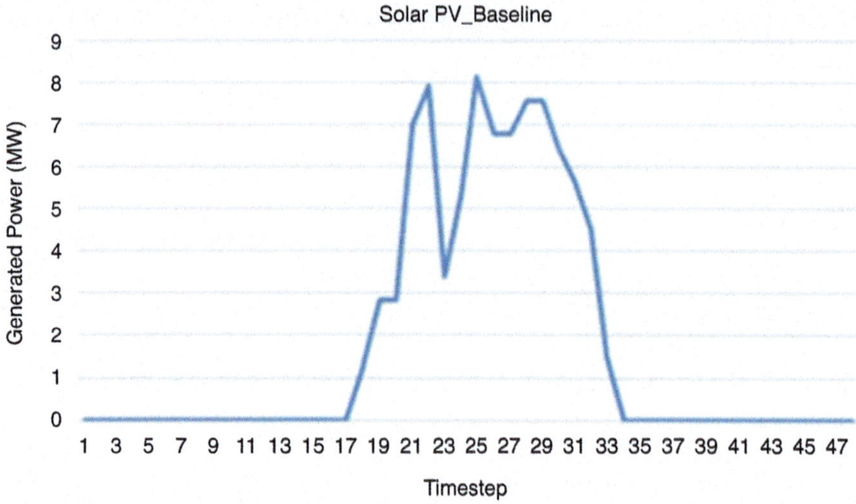

Fig. 3.22 Solar PV generation profile for all scenarios

Table 3.4 Cost and emission factors

Cost of upstream import(£/MWh)	Electricity network	28.06	[41]
	Gas network	9.42	
Carbon emissions of upstream import (kgCO_2eq/MWh)	Electricity network	253	[42]
	Gas network	203	
Cost of coupling components (£/MWh)	CHP	25	[43]
	P2G	55	[44]
	HP (air-sourced)	36	[45]

not included. The carbon emission factors are published average values for the UK supply of electricity (including T&D losses) and natural gas (including limited amount of biogas) [42]. The cost of coupling components reflects the levelised cost of energy for those technologies excluding the fuel cost (electricity or gas), which is already accounted for in upstream costs and is taken to be minimal for local RES generation. This was used here to be in line with operational costs accounted for in the model as the cost of upstream energy import.

3.4.5 Case Study Results

After running the simulations for each scenario, indicators are quantified as described in Table 3.5. Note that reliability indicators are not included in this case as this is a condition for convergence in the quantitative model used in this analysis.

Firstly, the values of all indicators for all scenarios are presented in Table 3.6.

Table 3.5 Indicators description

Dimension	Indicator	Description	Model parameters
Environmental	CO_2 emissions intensity	Amount of CO_2 emissions per energy supplied (kg-eq/MWh)	Total CO_2 emission/total energy input
	RES integration	% of RES supply from total final energy load	Total used RES/total final energy load
	Overall efficiency	% of total final energy load to total energy input (100 – Losses)	Total final energy load/total energy input and waste RES
Economic	Total cost	Cost of energy import and integration assets (CHP, P2G, HPs)	Total cost of energy import from upstream + costs of CHP, P2G, HPs
Resilience	Self-sufficiency	% of local energy generation from total energy supply	Local energy generation/total energy input
Flexibility	Flexible capacity	% of maximum available capacity of flexible assets (CHP, P2G, HPs) from total peak load	(maximum capacities of CHP, P2G, HPs)/peak total load

A radar chart is produced to compare the different scenarios with respect to the six indicators (Fig. 3.23). To do this, the indicators values were normalised to fit into the uniform scale of the radar chart. This is based on the mean normalisation method given as $x' = \frac{x - \text{average}(x)}{\max(x) - \min(x)}$. Normalised values are available on [40].

It can be noticed from Table 3.5 and Fig. 3.23 that scenarios 1 and 3 and scenarios 2 and 4 show very similar values. This means that adding P2G to the system with small capacity has a very minimal impact in all terms including RES integration. This conclusion has been reached earlier in the analysis, and as mentioned previously it is due to local network constraints that prevent getting more RES into the system. Conversely, scenarios 5 and 6 with major increase in P2G capacity and consequently in flexible capacity have shown an increase in RES integration that affected other indicators such as carbon emissions and self-sufficiency. This can be explained by the fact that electricity is more expensive than gas; thus drawing electricity to convert into gas is not economic. However, in the case where RES is abundant, using the surplus low-cost electricity becomes competitive.

To realise the impact of HPs, scenarios 1 and 2, 3 and 4 and 5 and 6 are compared. For scenarios 1 and 2 and scenarios 3 and 4, similar trends are noticed with HPs increasing the total system cost, reducing the overall system efficiency, while not showing major improvements in terms of RES integration, self-sufficiency and CO_2 emissions despite the increase in flexible capacity. On the other hand, comparing scenarios 5 and 6 shows improvements in most indicators including RES integration, but with higher costs and lower efficiency.

Table 3.6 Indicators values for all scenarios

Scenario	CO_2 emissions (kg-eq/MWh)	RES integration (%)	Efficiency (%)	Total cost (£)	Self-sufficiency (%)	Flexible capacity (%)
1	201.65	2.72	94.99	506,995.64	16.43	7.39
2	202.97	2.88	94.58	551,015.41	17.82	12.47
3	201.65	2.72	94.97	508,806.05	16.49	7.89
4	203.05	2.89	94.52	554,291.41	17.98	12.99
5	167.26	18.59	93.07	394,723.89	27.30	17.25
6	164.37	20.35	92.64	448,491.90	31.99	22.81

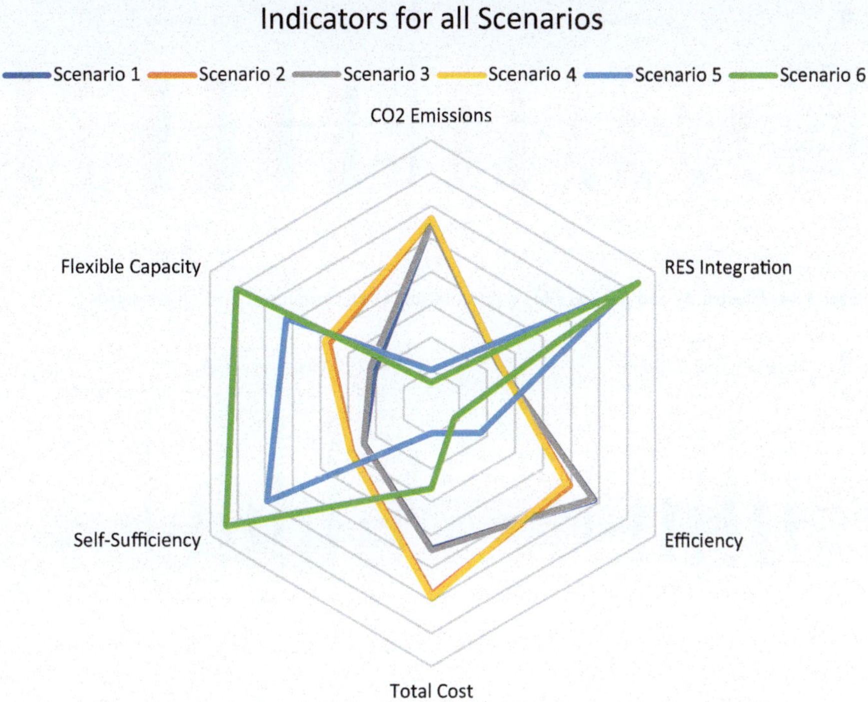

Fig. 3.23 Radar chart of indicators for all scenarios

To draw a better picture on the relations between indicators and the different system components, some indicators were further broken down and shown relative to other indicators following from the relations in the traceability view in Fig. 3.16.

3.4.5.1 Environmental Indicators

The first indicator is the intensity of carbon emissions that is calculated as the ratio of the total CO_2 emissions over the total energy input to the system. The latter in this case includes generation from RES (solar and wind), generation from CHP plants and imported energy from upstream. In addition to the carbon intensity value, we are interested in tracing the contribution of different vectors, and thus we compare the carbon emissions and intensity of electricity and that coming from gas (Fig. 3.24).

For electricity, scenarios 1–4 show slight variations for the amount of emissions and the carbon intensity, while for scenarios 5 and 6, the amount of emissions becomes negligible (Fig. 3.24-a), and the intensity sees a steep decrease (Fig. 3.24-b). This is clearly due to the major increase in RES supply. On the contrary, the amount of emissions from gas encounters a decrease for scenarios with HPs (2,4,6), but the intensity stays the same across all scenarios. The absolute decrease

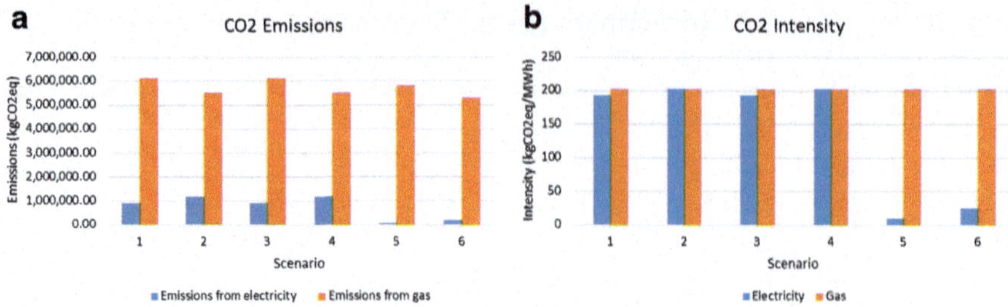

Fig. 3.24 Electricity and gas (**a**) CO_2 emissions and (**b**) CO_2 intensity for all scenarios

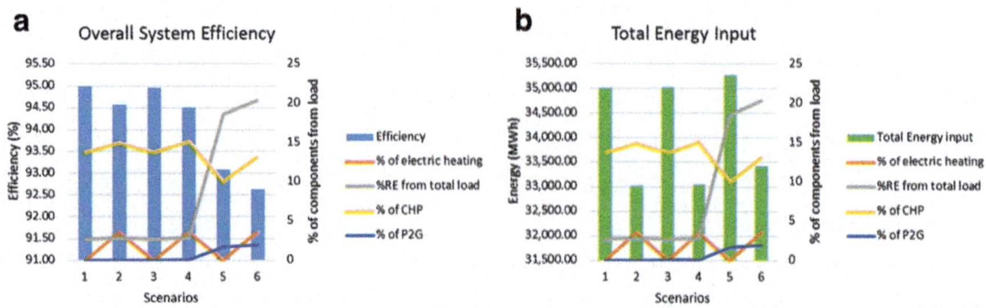

Fig. 3.25 (**a**) Overall system efficiency values and (**b**) total energy input compared with system components for all scenarios

in emissions is mainly due to a fraction of heating being satisfied by electric heating (around 12%), which means that the amount of gas imported from upstream is decreased. Note that the decrease in scenario 5 emissions is due to the lower CHP generation with more electricity demand being met by RES. However, the intensity stays the same across all scenarios because the amount of renewable gas produced via P2G is still minimal (around 2%) compared to the amount of upstream gas used for heating or for generation through CHP plants. Note that the combined CO_2 intensity for electricity sources becomes lower than that of gas when combined with local RES generation, again highlighting the prospects of heat decarbonisation through ESI.

The second key indicator is RES integration, which is calculated as the % of energy supply from RES over the total final energy load. The total final energy load is defined by the load profiles described earlier. This indicator is presented on its own, but is also shown in correlation with other indicators including efficiency and flexibility (Figs. 3.25 and 3.27).

The third environmental indicator is the overall efficiency, which represents the system losses due to RES curtailment, network losses and technology efficiencies. To understand the sources of losses to which it could be traced to, this indicator is shown in relation with the percentages of RES, CHP, P2G and HPs from the total final energy load (Fig. 3.25).

From Fig. 3.25-a, it could be said that system losses increase slightly with the expansion of HPs, RES, CHP and P2G. However, this is not to be confused with the expected requirement of HPs and CHP to reduce the overall energy use as evident in Fig. 3.25-b, while delivering the same energy service, which in this case is heat. Thus, the system losses can be attributed to the electricity network losses driven by more flow from the increased RES supply and the increased HPs demand, in addition to losses from RES curtailment and P2G efficiency losses.

3.4.5.2 Acceptability Indicators

The cost to the system is calculated as the sum of energy imported from upstream (electricity and gas) and the cost incurred by the coupling components (CHP, P2G, HPs). The cost breakdown is shown in Fig. 3.26. Note that the cost is limited to the operational costs, since the analysis is based on an operational model that doesn't account for capital and carbon costs. Thus, the total cost is dominated by the cost of energy imported from upstream.

Another indicator of cost is the abatement cost of CO_2, which relates economic and environmental aspects and represents cost-effectiveness, which is typically important for decision-making. It is calculated as the ratio of additional costs incurred to the amount of CO_2 reduced in a scenario compared to the baseline scenario. However, this indicator faces two limitations in this case. First, the cost calculated includes only the operational costs and doesn't include the capital costs, which are significant when considering the implementation of P2G, for example. Although cost unit factors used for coupling components take into account the levelised cost of those technologies, this cost although indicative would still not be sufficiently representative in this analysis for decision-making on cost-effectiveness. Second, some scenarios show an increase in emissions from the baseline scenario

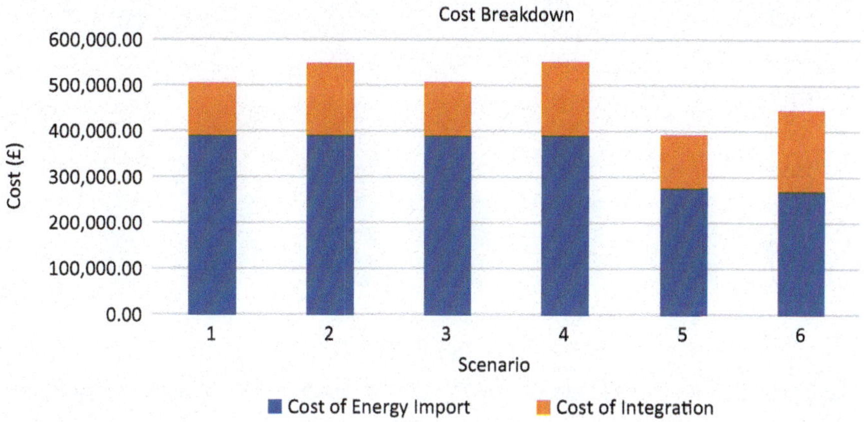

Fig. 3.26 Cost breakdown for all scenarios

or at least no change. Thus, it is not sensible to compare the cost-effectiveness of reducing emissions between scenarios with different directions.

Note also that an indicator for comfort and convenience, such as the willingness to shift behaviour, was not realised in this analysis. This is important to evaluate the quality of the energy service, such as heat, when it is being delivered in different forms. However, such indicators are typically qualitative and require direct involvement with consumers and prosumers on the ground through surveys or interviews in the region, which is outside the scope of this project.

3.4.5.3 Security Indicators

Reliability indicators, such as generation capacity margin and networks capacity, are not applicable in this case given that reliability constraints are a condition for the model convergence. The requirement of providing grid services was therefore not evaluated since DSR was not considered in the model.

Self-sufficiency is calculated as the % of local generation to the total energy input and is considered an indicator of resilience to supply disruptions.

The system flexibility is considered as the ratio of the maximum available capacity of flexibility assets to the peak energy load. In this case, flexibility assets are limited to the coupling components since there is no electric or gas storage and no DSR. To trace the contribution of the different coupling components, the capacities are broken down and shown in relation to the RES integration indicator (Fig. 3.27).

As mentioned previously, P2G is the main contributor to flexibility and RES integration as evident in Fig. 3.27. A major expansion in P2G capacity is necessary to realise the impact. On the other hand, HPs on its own seem to have a limited

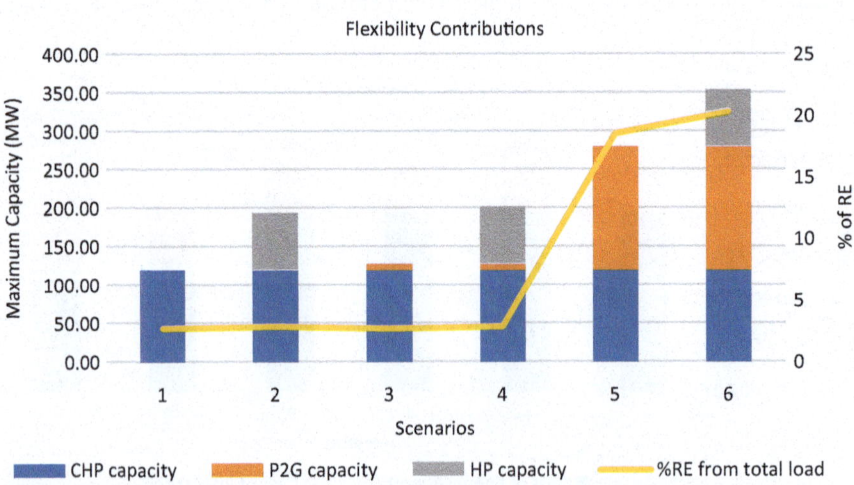

Fig. 3.27 Flexibility contributions by coupling components for all scenarios

impact on RES integration. This is due to how the technology is accounted for in the model as a load component without the provision of DSR. However, when HPs are used in combination with high P2G capacity as in scenario 6, HPs are able to absorb more of the RES into the system and to reduce curtailment.

Finally, an indicator that was not included in this analysis is the % of heat electrification (uptake of HPs) since it is an exogenous variable to the model. However, given that net-zero plans typically set objectives for electrification of heat (and transport), then this is an indicator to consider in other cases where technology diffusion and adoption is considered dynamically.

3.5 Discussion

3.5.1 Findings

The results of the case study presented in Sect. 3.4.5 show that ESI provides a direction towards achieving the energy transition objectives. The results are evaluated in line with the traceability view presented in Fig. 3.16, which includes the system requirements at both the SoS and CSs levels. Thus, the level of satisfaction of the identified system requirements at different levels indicates the level of ESI effectiveness. This is only relative in this case comparing between scenarios rather than with respect to an absolute target.

Overall, the scenario analysis has shown that integration through vector-coupling technologies, particularly P2G, is an effective measure to enable more RES into the system while providing means for a viable network management. Scenarios 5 and 6 with high P2G capacity have seen improvements in all indicators to achieve the set objectives, except for overall efficiency. This is still an acceptable reduction of around 2% and is to be expected with the increased electrification, given that losses from the electricity network are naturally greater than those associated with the gas network.

In terms of system requirements and starting with the requirement to deliver energy in different forms to end users, this requirement was physically satisfied within the quantitative model. In terms of the requirements to achieve the trilemma policy objectives, with regard to environmental sustainability, the requirement to integrate RES into the system was delivered with increased system integration, particularly with high capacity of P2G. In terms of overall efficiency, the overall energy use of the system was lower due to CHP and HPs despite increasing the system losses relative to the total energy input to the system. This was reflected numerically with the CO_2 emissions reduction in scenario 5 with bidirectional integration and even lower in scenario 6 with the addition of HPs.

Getting more RES generation to cover the electric load including for heating also reduces the cost to the system, being dominated by upstream energy imports. This is allowed by the flexible capacity to shift energy vectors provided by the vector-coupling technologies, mainly P2G. While the reliability and resilience of the system were not examined in response to sudden, extreme events, those criteria were physically satisfied in the simulation model under conditions of system stress.

As mentioned earlier, prospects of DSR are not included in the quantitative model; thus the requirements of prosumers in relation to comfort and convenience and access to the grid can't be evaluated.

3.5.2 *Complexity*

The developed framework, particularly the conceptual system modelling stage, aims to facilitate the understanding of the integrated energy system and the deduction of criteria and indicators leading up to the evaluation. In this regard, the framework delivers on this objective in a complex energy system like in the NoT region. This is done through the abstraction and decomposition of this system into its different levels and components and focusing on the interaction across the different levels and components.

During the framework implementation and while dealing with the scenarios and modelling, it was clear the high level of detail involved in the NoT case study. The framework aims to reduce complexity by abstracting the system at high levels, but in practice lower-level challenges should still be considered. For instance, with increasing RES, local network constraints were realised and had to be dealt with in the quantitative model by adding P2G close to the point of wind generation. This highlights the importance of communication and feedback between the conceptual system model and the quantitative model to manage technical complexities.

In terms of results, the dashboard approach for presenting results is supported by the traceability view to realise the contribution of different system components to different system requirements. The traceability view is a condensed representation of the interactions within the system and a reference point to the other system diagrams that show the structural and functional relationships in more detail. Having the traceability view and the complete conceptual system model behind the indicators set also reduces the system complexity that would otherwise be reflected in the results.

3.5.3 *Limitations*

The work presented in this chapter shows a number of limitations. First, the quantitative model used in this analysis is an optimal operation model. This type of modelling is valuable for ensuring the system operation on the short term, but planning features are also needed to realise the optimal sizing, costing and localisation of assets for energy generation as well as the vector-coupling technologies. Second, energy storage and DSR are not considered in this analysis as other forms of flexibility provision to the WES. While the presented framework can account for energy storage and DSR due to its flexible nature, the challenge here is around data availability for those technologies in this particular case study. Finally, this work has started before the COVID-19 pandemic, and the data involved are precedent to the times of COVID, which has caused disruption to the energy generation and demand patterns. It is yet to be seen if the energy system goes back to the normal pre-COVID patterns or if the pandemic will have a lasting impact on the energy system.

Due to the aforementioned limitations and the number of assumptions included in the analysis, the findings discussed are deemed applicable to the case study under the stated conditions and assumptions and might not be necessarily generalisable. However, in relation to the developed methodological framework, the case study demonstrates its applicability and usefulness to a complex, local energy system given its generic nature and the modular approach it provides.

3.6 Conclusion

In summary, this chapter presents a methodological framework to evaluate whole energy systems (WES), which is demonstrated through a case study for a local energy system. The framework is based on the identified WES evaluation principles, namely, being multidimensional, multivectoral, systemic, futuristic, systematic and applicable. The framework provides a socio-technical approach for the evaluation at different levels, combining the system stakeholders' requirements with the system components and functions in a system-of-systems architecture methodology.

The framework involves three methodological stages. First, the scenario formulation describes the system under study in terms of system configurations, conditions of energy demand and supply and any other assumptions. Second, the conceptual system modelling represents the system architecture at different levels of abstraction and decomposes the system into its different components. The conceptual system model shows the system stakeholders and constituents, the structural and functional relationships and measures of effectiveness. Finally, the quantitative system modelling includes the mathematical formulations representing the system topology and technical feasibility.

The three stages interact iteratively. Scenarios are first specified to be modelled conceptually and quantitatively. Those scenarios could be modified upon the conceptual modelling to ensure variability between scenarios while maintaining comparability or if deemed infeasible in the quantitative model. The final outcome of the conceptual modelling stage is the measures of effectiveness, which make the set of indicators used for the evaluation. Accordingly, parameters and relationships that need to be quantified by the quantitative model are identified. The feedback between the conceptual and quantitative models is thus around those parameters, which could be affected by the data availability, the exogenous variables and the nature of the quantitative model. The described framework for the WES evaluation is demonstrated through a case study for the local energy system in the North of Tyne region in the UK. This includes the provision of energy systems integration (ESI) through vector-coupling technologies, such as combined heat and power, power-to-x and heat pumps. Hence, scenarios include different system configurations (combinations of energy networks and vector-coupling technologies) under different conditions of energy supply and demand. The case study demonstrates the applicability of the evaluation framework on a complex energy system and shows the potential of ESI as a direction to achieve the energy transition objectives at a regional level.

Further improvements to the work presented in this chapter can be realised in future work. First is investigating the impact of uncertainty on the evaluation. Sources of uncertainty could include system parameters (performance and properties), unit factors (cost and carbon), load and generation data and environmental effects (e.g. ambient temperature). Second is incorporating investments and carbon costs to the quantitative model optimisation function. This is to account for energy systems planning features and the future interaction between the local and the national energy systems, which would partly depend on the electricity and gas networks emission factors and decarbonisation efforts at the national level. Finally, it is involving stakeholders with the evaluation in practice by directly eliciting their requirements and getting their feedback as part of the iteration between the framework stages.

References

1. Lund, P. D., Lindgren, J., Mikkola, J., & Salpakari, J. (2015). Review of energy system flexibility measures to enable high levels of variable renewable electricity. *Renewable and Sustainable Energy Reviews, 45*, 785–807.
2. Singh, H. V., Bocca, R., Gomez, P., et al. (2019). The energy transitions index: An analytic framework for understanding the evolving global energy system. *Energy Strategy Reviews, 26*, 100382. https://doi.org/10.1016/j.esr.2019.100382
3. Jamasb, T., & Llorca, M. (2019). Energy systems integration: Economics of a new paradigm. *Economics of Energy and Environmental Policy, 8*, 7–28. https://doi.org/10.5547/2160-5890.8.2.tjam
4. Guelpa, E., Bischi, A., Verda, V., et al. (2019). Towards future infrastructures for sustainable multi-energy systems: A review. *Energy, 184*, 2–21. https://doi.org/10.1016/j.energy.2019.05.057
5. Hosseini, S. H. R., Allahham, A., Walker, S. L., & Taylor, P. (2020). Optimal planning and operation of multi-vector energy networks: A systematic review. *Renewable and Sustainable Energy Reviews, 133*, 110216.
6. Hanna, R., Gazis, E., Edge, J., et al. (2018). *Unlocking the potential of Energy Systems Integration*. https://imperialcollegelondon.app.box.com/s/0sil57fndc5tn9gfy6ypzp8v61qnv3mg
7. Mancarella, P. (2014). MES (multi-energy systems): An overview of concepts and evaluation models. *Energy, 65*, 1–17.
8. Abeysekera, M., Wu, J., Jenkins, N. (2016). *Integrated energy systems: An overview of benefits, analysis methods, research gaps and opportunities*. In: HubNET position Pap. Ser. https://orca.cf.ac.uk/133441/1/HubNET Position paper-Multi Energy-Revised version.pdf.

9. Berjawi, A. E. H., Walker, S. L., Patsios, C., & Hosseini, S. H. R. (2021). An evaluation framework for future integrated energy systems: A whole energy systems approach. *Renewable and Sustainable Energy Reviews, 145*, 111163. https://doi.org/10.1016/j.rser.2021.111163
10. Kroposki, B., Garrett, B., Macmillan, S., et al. (2012). *Energy Systems Integration: A Convergence of Ideas.*
11. O'Malley, M., Kroposki, B., Hannegan, B., et al. (2016) *Energy systems integration: Defining and describing the value proposition.* Nrel/Tp-5D00–66616 9. https://doi.org/10.2172/1257674.
12. Ramsebner, J., Haas, R., Ajanovic, A., & Wietschel, M. (2021). The sector coupling concept: A critical review. *Wiley Interdisciplinary Reviews: Energy and Environment, 10.*
13. Kriechbaum, L., Scheiber, G., & Kienberger, T. (2018). Grid-based multi-energy systems-modelling, assessment, open source modelling frameworks and challenges. *Energy, Sustainability and Society, 8.*
14. Heendeniya, C. B., Sumper, A., & Eicker, U. (2020). The multi-energy system co-planning of nearly zero-energy districts – Status-quo and future research potential. *Applied Energy, 267.* https://doi.org/10.1016/j.apenergy.2020.114953
15. Wang, J., Zong, Y., You, S., & Træholt, C. (2017). A review of Danish integrated multi-energy system flexibility options for high wind power penetration. *Clean Energy, 1*, 23–35.
16. Witkowski, K., Haering, P., Seidelt, S., & Pini, N. (2020). Role of thermal technologies for enhancing flexibility in multi-energy systems through sector coupling: Technical suitability and expected developments. *IET Energy Systems Integration, 2*, 69–79. https://doi.org/10.1049/iet-esi.2019.0061
17. Cambini, C., Congiu, R., Jamasb, T., et al. (2020). Energy systems integration: Implications for public policy. *Energy Policy*. https://doi.org/10.1016/j.enpol.2020.111609
18. Leitner, B., Widl, E., Gawlik, W., & Hofmann, R. (2019). A method for technical assessment of power-to-heat use cases to couple local district heating and electrical distribution grids. *Energy, 182*, 729–738. https://doi.org/10.1016/j.energy.2019.06.016
19. Hosseini, S. H. R., Allahham, A., & Adams, C. (2021). Techno-economic-environmental analysis of a smart multi-energy grid utilising geothermal energy storage for meeting heat demand. *IET Smart Grid, 4*, 224–240. https://doi.org/10.1049/stg2.12020
20. Mancarella, P., Chicco, G., & Capuder, T. (2018). Arbitrage opportunities for distributed multi-energy systems in providing power system ancillary services. *Energy, 161*, 381–395. https://doi.org/10.1016/j.energy.2018.07.111
21. Moslehi, S., & Reddy, T. A. (2018). Sustainability of integrated energy systems: A performance-based resilience assessment methodology. *Applied Energy, 228*, 487–498. https://doi.org/10.1016/j.apenergy.2018.06.075
22. Neely, K., Bortz, M., & Bice, S. (2021). Using collaborative conceptual modelling as a tool for transdisciplinarity. *Evid Policy, 17*, 161–172. https://doi.org/10.1332/174426419X15468578119304
23. Energy Systems Catapult (2019). *Systems thinking in the energy system: A primer to a complex world.* https://es.catapult.org.uk/wp-content/uploads/2019/05/System-thinking-in-the-energy-system-dec-final.pdf. Accessed 28 Aug 2020.
24. Bale, C. S. E., Varga, L., & Foxon, T. J. (2015). Energy and complexity: New ways forward. *Applied Energy, 138*, 150–159.
25. Mangoyana, R. B., Smith, T. F., & Simpson, R. (2013). A systems approach to evaluating sustainability of biofuel systems. *Renewable and Sustainable Energy Reviews, 25*, 371–380. https://doi.org/10.1016/j.rser.2013.05.003
26. Ingram, C., Payne, R., Perry, S., et al. (2014). Modelling patterns for systems of systems architectures. In *8th annual IEEE international systems conference, SysCon 2014 - proceedings* (pp. 146–153).
27. Geyer, P., & Buchholz, M. (2012). Parametric systems modeling for sustainable energy and resource flows in buildings and their urban environment. In *Automation in construction* (pp. 70–80).

28. Nielsen, C. B., Larsen, P. G., Fitzgerald, J., et al. (2015). Systems of Systems Engineering: Basic concepts, model-based techniques, and research directions. *ACM Computing Surveys, 48*, 1–41. https://doi.org/10.1145/2794381
29. Holt, J., Perry, S., Payne, R., et al. (2015). A model-based approach for requirements engineering for systems of systems. *IEEE Systems Journal*. https://doi.org/10.1109/JSYST.2014.2312051
30. Perry, S., Holt, J. (2014). *Definition of the COMPASS Architectural Framework Framework*. http://www.compass-research.eu/Project/Deliverables/D21.5b.pdf. Accessed 29 Nov 2020.
31. Holt, J., Perry, S. (2011). *SysML for Systems Engineering*.
32. Holt, J., Perry, S., Hansen, F. O., Hallerstede, S. (2012). Report on guidelines for sos requirements. In: COMPASS Deliv. http://www.compass-research.eu/Project/Deliverables/D211.pdf
33. North East LEP (2019). *North East Energy for Growth*. https://www.northeastlep.co.uk/wp-content/uploads/2019/09/full-strategy-energy-for-growth-strategy.pdf. Accessed 8 Mar 2021.
34. BEIS (2020). *Sub-national electricity consumption data - GOV.UK*. https://www.gov.uk/government/collections/sub-national-electricity-consumption-data. Accessed 8 Mar 2021.
35. BEIS (2020). *Regional Renewable Statistics - GOV.UK*. https://www.gov.uk/government/statistics/regional-renewable-statistics. Accessed 8 Mar 2021.
36. BEIS (2020). *UK local authority and regional carbon dioxide emissions national statistics: 2005 to 2018 - GOV.UK*. https://www.gov.uk/government/statistics/uk-local-authority-and-regional-carbon-dioxide-emissions-national-statistics-2005-to-2018. Accessed 10 Mar 2021.
37. Love, J., Smith, A. Z. P., Watson, S., et al. (2017). The addition of heat pump electricity load profiles to GB electricity demand: Evidence from a heat pump field trial. *Applied Energy, 204*, 332–342. https://doi.org/10.1016/j.apenergy.2017.07.026
38. Hosseini, S. H. R., Allahham, A., & Taylor, P. (2018). Techno-economic-environmental analysis of integrated operation of gas and electricity networks. In *Proceedings - IEEE international symposium on circuits and systems*.
39. Hosseini, S. H. R., Allahham, A., Vahidinasab, V., et al. (2021). Techno-economic-environmental evaluation framework for integrated gas and electricity distribution networks considering impact of different storage configurations. *International Journal of Electrical Power & Energy Systems*, 125. https://doi.org/10.1016/j.ijepes.2020.106481
40. Berjawi, A. E. H, Allahham, A. (2021) NoT Parameters. https://doi.org/10.25405/data.ncl.14614179
41. OFGEM (2021). *Estimated network costs per domestic customer (GB average) | Ofgem*. https://www.ofgem.gov.uk/data-portal/estimated-network-costs-domestic-customer-gb-average. Accessed 12 Apr 2021.
42. BEIS (2020). *Greenhouse gas reporting: conversion factors 2020 - GOV.UK*. https://www.gov.uk/government/publications/greenhouse-gas-reporting-conversion-factors-2020. Accessed 12 Apr 2021.
43. BEIS (2020). *BEIS Electricity Generation Costs (2020) - GOV.UK*. https://www.gov.uk/government/publications/beis-electricity-generation-costs-2020. Accessed 12 Apr 2021.
44. McDonagh, S., O'Shea, R., Wall, D. M., et al. (2018). Modelling of a power-to-gas system to predict the levelised cost of energy of an advanced renewable gaseous transport fuel. *Applied Energy, 215*, 444–456. https://doi.org/10.1016/j.apenergy.2018.02.019
45. Hansen, K. (2019). Decision-making based on energy costs: Comparing levelized cost of energy and energy system costs. *Energy Strategy Reviews, 24*, 68–82. https://doi.org/10.1016/j.esr.2019.02.003

Chapter 4
Targeting and Design Multigeneration System Through Total Site Integration Approach

Matin Karbasioun, Arash Esmaeilzadeh, and Majid Amidpour

Table of Variables

Variables	Sign	Unit
Temperature	T	°C, K
Pressure	P	bar
Heat load	Q	kW
Shifted temperature	T^*	°C
Minimum temperature difference	$\Delta T_{\min}$	°C
Enthalpy	H	kJ
Enthalpy per mass	h	kJ. kg^{-1}
Specific heat capacity at constant pressure	C_P	kJ. K^{-1}. kg^{-1}
Heat capacity rate	C_P	kJ. K^{-1}
Mass flow	$\dot{m}, m$	kg. s^{-1}
Power	W	kW
Willan line slope, polytropic coefficient	n	–
Willan line Y-axis intercept point	W_{INT}	kW
Willan line coefficients	a, b, c	–
The boiler discharge flow to the output flow ratio	R_{BD}	–
Steam preheater enthalpy to steam enthalpy ratio	R_{PH}	–
The site power consumption to heat consumption ratio	R	–
Net calorific value	NCV	MJ. kg^{-1}
Gross calorific value	GCV	MJ. kg^{-1}

(continued)

M. Karbasioun · A. Esmaeilzadeh · M. Amidpour (✉)
Energy Systems Department, K. N. Toosi University of Technology, Tehran, Iran
e-mail: amidpour@kntu.ac.ir

© The Author(s), under exclusive license to Springer Nature Switzerland AG 2022
V. Vahidinasab, B. Mohammadi-Ivatloo (eds.), *Whole Energy Systems*, Power Systems, https://doi.org/10.1007/978-3-030-87653-1_4

Variables	Sign	Unit
Lower heating value	LHV	$MJ.\ kg^{-1}$
Carbon emission factor per unit	EF	$tCO_2.t_{fuel}^{-1}$
Greenhouse gases	GHG	$tCO_2.\ year^{-1}$
Rate of steam utilized by process	U_{steam}	$t.\ h^{-1}$
Compensatory water	W_{makeup}	$t.\ year^{-1}$
Number of hours the system operates in a year	time	h
Condensate return rate	CRR	–
Vent rate	VR	–
Blowdown rate	BR	–
Annualized factor	AF	–
Annualized capital costs	ACC	m$
Interest rate	IR	–
Maintenance cost	MC	m$
Capital cost	CC	m$
Ratio of the price	CI	–
Total annualized cost	TAC	m$
Concentration	C	ppm

Table of Greek Signs

Variables	Signs	Unit
Efficiency	η	–
Volume per mass	υ	$m^3.\ kg^{-1}$
Steam discharge rate	α	–
Density	ρ	$kg.\ m^{-3}$
Heat capacity ratio	γ	–

4.1 Status Quo, Challenges, and Outlook

The annual increase in population and human desire to make the most of technology to improve life quality has led to an increase in energy consumption in recent years. On the other hand, limited resources, especially fossil fuels (which play a major role in energy supply), as well as increasing environmental concerns such as global warming and climate change (which cause storms, floods, and droughts), have led us to resources crisis in the future. Hence, efforts in recent years to find a way out of the current path have increased. These efforts can be divided into three categories: first, efforts to find new energy sources such as renewable energy, nuclear energy, biomass, and so on; second, efforts to reduce consumption through culture building, such as encouraging the use of bicycles, taxing for carbon emission, and holding international conferences to raise public awareness; and, last, efforts to reduce consumption through optimization and increase productivity and efficiency. This is done in two ways: one is the use of new technologies to reduce energy consumption and increase equipment efficiency, and the other is by exploring and finding new ways for interaction between equipment such as utility systems.

In the field of optimization and improving efficiency, cogeneration systems can be named. The Total Site approach is one of the various methods that got attention in recent years as an answer to the energy crisis in the future. In general, the basis of this method is the use of waste energy of industrial units, including waste energy in oil industries or power plants.

Due to the tendency of countries to decentralization today, cogeneration systems will also change due to lack of massive energy resources with significant losses; instead, smaller amounts of energy in decentralized units will be seen more in which energy is found in different forms, such as systems in clouding solar energy, fuel cells, and water desalination systems. These changes increase system interactions and thus the complexity of the analysis while improving efficiency to whole new levels.

4.2 The Role of Efficiency

Energy and water consumption, climatic changes and global warming, hunger and poverty, pollutant emissions, and environmental changes are introduced as the most fundamental problems of the modern world in various scientific, political, and cultural assemblies. These problems have led to the establishment of many international organizations and contracts to investigate different impacts and aspects of these phenomena and uncover approaches for confronting their hazards. That is why numerous researchers in the world addressed energy production and consumption optimization, energy efficiency enhancement, water, and pollution reduction. The outcome is many papers, standards, rules, and agreements in most countries concerning water and energy consumption improvement toward sustainable development and the prevention of environmental destruction and climatic changes. Thus, the significance of optimizing and reducing energy and water consumption and the necessity of improving the environment has become more serious over time. According to specialists and policy-makers' opinions, the energy, water, and environmental problems have been pounded as serious crises in the world.

According to the reports presented by the International Energy Agency [6], the world faced a severe reduction in energy consumption during the corona pandemic in the first season of 2020. This event declined energy consumption demand by

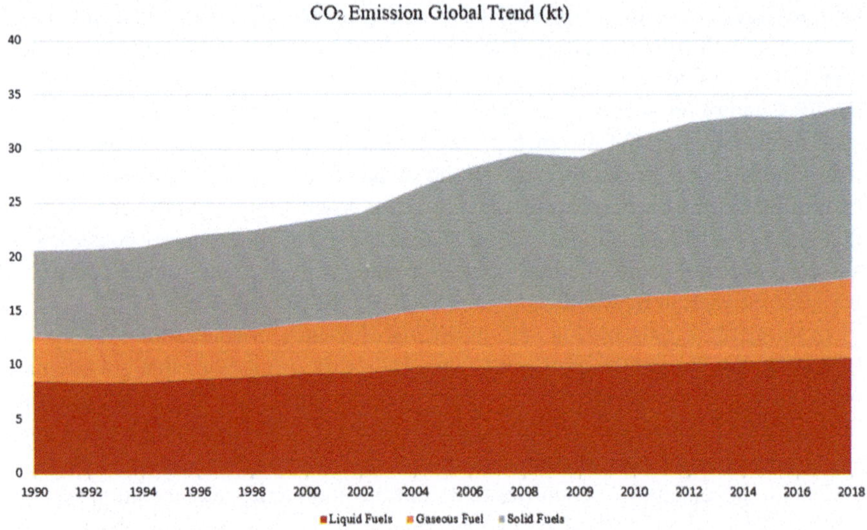

Fig. 4.1 The growing trend of global CO_2 emission based on World Bank data [22]

6%, being the most significant percentage in the past 70 years and the largest value until the present. Demand reduction has influenced the consumption of diverse fuels, among which we can refer to an 8% reduction in coal and a 5–10% electricity use reduction, leading to a decline in gas consumption in some areas. An 8% reduction, approximating 209 billion tonnes (Gt), is expected for the global emission of CO_2 based on inspections of the International Energy Agency. These numbers depict a direct relationship between energy consumption and greenhouse gas emissions, giving rise to global warming.

On the other hand, based on the published data of the International Energy Agency in 2019 [4, 5] and the World Bank data indicators [22], as illustrated in Fig. 4.1, fossil resources emit the maximum CO_2, two-thirds of which relate to the power generation sector. This emission rise is associated with the upsurge of energy consumption due to global economic growth and climatic changes in some parts of the world. The degressive slope of CO_2 emission since 2014 is linked to the enhancement of low-carbon technologies and improvement in energy efficiency, resulting in a reduction in carbon consumption.

Concerning the growing trend of global CO_2 emission in Figure 4.2, the global production and emission rate of CO_2 should have increased by 3.8% from 2017 to 2018, owing to economic growth. However, due to the fulfillment of diverse approaches, such as changing coal power plants to gas burner ones, increasing energy efficiency, developing nuclear power plants, increasing the renewable energies contribution to electricity generation, and so forth, this rate has decreased to 2.9%. Among the implemented approaches, increasing the efficiency of energy consumption has maximally contributed to reducing greenhouse gases. This indicates

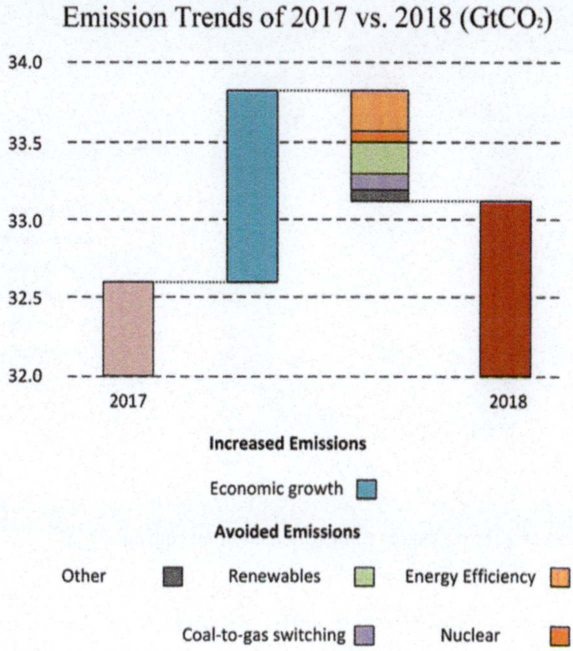

Fig. 4.2 Compare emission trends of 2017 vs. 2018 based on the IEA report [4]

the importance of efficiency enhancement in the industry and power generation sectors.

Now that the significance of optimization and its role in improving the environmental quality and greenhouse gas emission reduction has been illuminated, we need to determine the energy-consuming sectors wherein optimization has been more influential and has further decreased fossil fuels and resources consumption leading to reduce CO_2 emission.

As shown in Fig. 4.3, the reports of the International Energy Agency in 2020 [6] indicated that the industry sector had globally allocated the maximum share of energy consumption to itself in 2018. Hence, we can perceive that the biggest energy consumer in the world is the industry sector, including the industries of iron, steel, chemicals, petrochemistry, minerals, and so on, which constitute about 73% of the entire world total energy consumption. Among these industries, the largest share belongs to the process industries wherein physical and chemical processes are applied to convert raw materials to worthy final products or exploitable in other industrial sectors. These industries encompass a significant part of the industry sector. Hence, it can be conceived that the process industry optimization crucially affects the energy consumption and the global emission rate of CO_2 gas.

Today, in addition to considering the society's needs for different process industries and the production of various products, it is necessary to pay attention to the initial costs associated with an industry establishment and current costs during

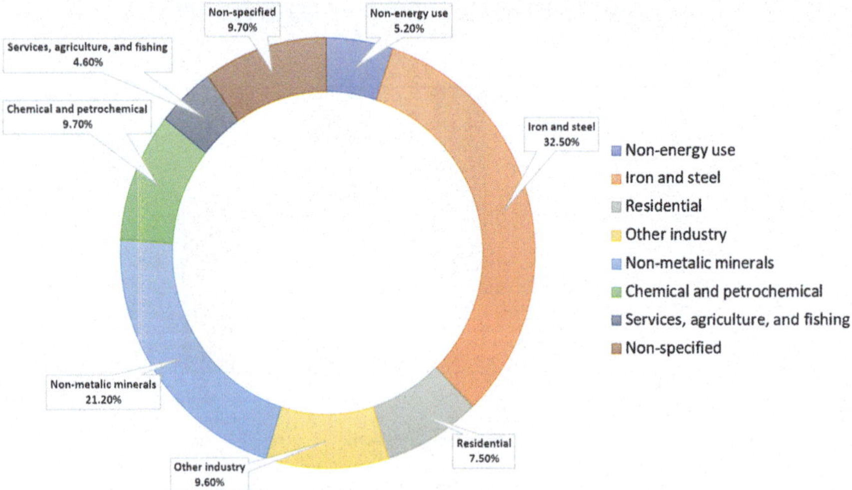

Fig. 4.3 Emission of 2018 by sectors (million tonnes of oil equivalent) [5]

its life. Many process industries have suffered a loss owing to their extreme energy requirements and energy cost fluctuations. These losses, especially during special periods, the same as economic stagnation in 2009 and the pandemic and quarantines in 2020, were more evident than ever due to the shutdown of many small and big industries due to the unprofitability of their production costs. Besides paying attention to the process industry economy, environmental factors and the generation rate of pollutant emissions are other determinant parameters in the industry life. As mentioned, the maximum portion of energy consumption and, as a result, the significant share in the global emission of CO_2 belong to the industry sector in the world. Based on the rules of some countries, environmental parameters also influence the industry's economy through taxes and penalties determined by central governments. From the perspectives of experts and researchers of this field, global warming due to greenhouse gas emissions has endangered the life of organisms on the one hand and reduced drinking water besides extending droughts in many parts of the world on the other hand. For this reason, it is evidently necessary to decrease the consumption of fossil fuels and modify industries to minimize the production of greenhouse gases.

Various approaches have been posed for developing industries sustainably in all aspects so far. An example is employing renewable clean fuels instead of the current fossil fuels. However, these fuels do not enjoy the requisite and sufficient potential of energy generation in initiating and implementing the present processes in the process industries; moreover, the current equipment of renewable power generation cannot supply the required energy of a process industry sustainably and at a minimum time. Hence, the employment of fossil fuels to supply the needed energy of the process industries is still prevalent. The consumption upsurge of these fuels is proportionally related to increases in costs, pollutants, and greenhouse gases.

For this purpose, different methods are introduced for energy consumption reduction in diverse process industries.

Among the presented methods for decreasing energy consumption in the process industries, we can refer to consumption reduction and optimization resultant from implementing energy management standards and online energy management in the process industries, employing energy and thermal insulation in different industrial sectors, updating the technologies and equipment used in the existing industry, modifying the present thermal network and implementing heat integration in the considered process industry, and establishing and initiating a cogeneration network of heat and power for sustainable supply and reduction in the generation costs of the process industries. These methods are different in terms of their capital cost and difficulty of implementation. However, the more are the expenses of these methods, the higher is the optimization rate in terms of fuel consumption. Based on the fuel price, current cost, capital cost, and the present risks and contributions in every process industry, we can choose the most suitable method for optimizing fuel consumption.

4.3 What Is the Utility System?

The most fundamental requirements of a process industry are its units' needs for power, heating, and cooling. For example, a processing unit burns fuel to supply the needed heating consumptions in different processes by transferring heat from fuel combustion. However, this heat transfer is not easily fulfilled because heat is consumed in different sectors of a process unit, and it is impossible to situate a heat-generator device, like a boiler, in every sector. Likewise, the materials needed for the direct heat transfer of the combustion-stemming gases to varying heat flows are very costly, and it is impossible to implement them in the whole unit. In small process units, heat is transferred within these units by average- or low-pressure steam or hot water. However, heat is supplied differently in large process sites, consisting of several small and big units. These are the required services that are jointly supplied by a unit, called the utility unit, which is in charge of supplying power, heating, and cooling utilities, and some other requirements for other sites.

Looking at process industries, we encounter various production processes associated with one or more utility systems. When it comes to the design of utility systems, we are faced with a lack of fundamental views. Often, there is a single approach to each production process in designing and modifying utility systems. Each production process has its business plan, independently of other processes. Using a strategic view for all production processes, the importance of facing the whole system in one picture becomes more transparent. Therefore, it can be said that the importance of having a more comprehensive view and designing all production processes based on meeting these strategic needs is a necessity. The following reasons can be named as the importance of studying a utility system when designing processes:

- Need for retrofit because of changes in one or more processes or adding/removing some processes.
- Usually, studying the utility system is needed to determining the value of saving before implantation of energy conservation projects.
- Revamp an old utility system with new equipment.
- Changing utility system might lead to a reduction in operation cost.
- Studying the utility system is important to calculate the energy costs associated with a production expansion.
- Need for modification of an existing system (retrofit) or designing a new system (grass roots).

For designing a utility system, some questions should be asked, including:

- The amount of power required by different processes
- Existing potential for power generation, import, or export
- The potential of steam production
- Available steam at different pressure levels
- Refrigerator and water-cooling systems
- The requirements of each process

For answering these questions, various cases should be considered among the processes available on the site, including the following:

- The amount of steam and power required for each process
- How processes interact with the utility system
- How each process interacts with the generated steam levels
- Simultaneous production of energy and heat
- The efficiency of each process in interaction with each other
- The amount of heat production and consumption in each process

All of these factors make studying and designing a utility system a complex problem. Therefore, more complex and flexible methods are needed to meet this need.

Various methods have been proposed for studying and designing a utility system, each with its strengths and weaknesses.

In some studies, they are analyzing the energy and exergy of utility systems based on the defined needs by modeling ready-made cycles using various software packages and combining them. In these methods, the main purpose of optimizing the developed models is based on the efficiency of the first law and the second law of thermodynamics, the developed cycle does not accept much flexibility in terms of structure, and only the operating conditions are optimized in most cases. In completing these methods, the use of economic and environmental parameters in the optimization of exergoeconomic and exergoenvironmental of these models helps to develop more accurate models of the operating conditions of the operating cycles of a utility system [1].

The advantages of these methods are the accurate design of process performance, optimization of all components participating in the processes, the inclusion of user and designer opinions, and high design accuracy.

The limitation of these methods is in the analysis of complex utility systems and their time compared to the method presented in this chapter. Also, it is not easy to get an overview of all processes and systems in previous methods. Also, structural optimization in these methods is difficult to apply. On the other hand, optimization methods in this type of modeling are often nonlinear, which increases the cost. These methods can be easily used to design utility systems with smaller scale and less complexity.

Another method for designing utility systems is to create superstructures and consider all possible scenarios for different relations between power and heat generation and production processes. A superstructure of all possible states is created in these methods, and after simplification, the resulting superstructure is optimized using nonlinear optimization methods. As a result, an optimal network is developed [14].

One of the advantages of this method is that it is automated and the fact that the final network will be optimal in terms of modeling. In contrast to these advantages, the complexities of designing a mathematical model and its optimization, the lack of convergence problems in optimization due to the nature of its nonlinear optimization, and the lack of participation of designers and decision-makers in designing networks can be considered as its disadvantages.

The designing method, which employs the Total Site approach, uses conceptual design to target the potentials in the utility system based on the potential of heat production and consumption and needs assessments performed in each process. In this method, first, all units are optimized based on pinch technology or another method like that, and then considering the general interactions between all separate processes, an optimal utility system is designed using linear optimization [7].

This approach uses another form of the second law of thermodynamic named R-ratio, and, as Kenney [8] said: "although it does not use the available energy concept directly, is a direct practical application of the second law thinking. As fuel utilization increases, less fuel (with its attendant losses of available energy) is fired."

The advantages of the proposed method are simplicity of modeling and using linear optimization instead of nonlinear, considering all interactions independent of the type of processes in each unit, flexibility in adding complexity, and changing the design structure of the system. Also, in this method, the designer's decision is effective in designing the system. It is also practical to use this method to target and design cogeneration systems in utility systems. Another feature of this method is considering root changes and observing the second law of thermodynamics in the design of structures. Furthermore, in this approach, the designer can use exergoeconomical and exergoenvironmental indicators to optimize the whole system [9].

Some of the disadvantages of this method, such as accuracy and attention to conditions of each process separately and user feedback in the design process, have been improved by using designers' decisions and combination of this method with other methods, which leads the results closer to reality.

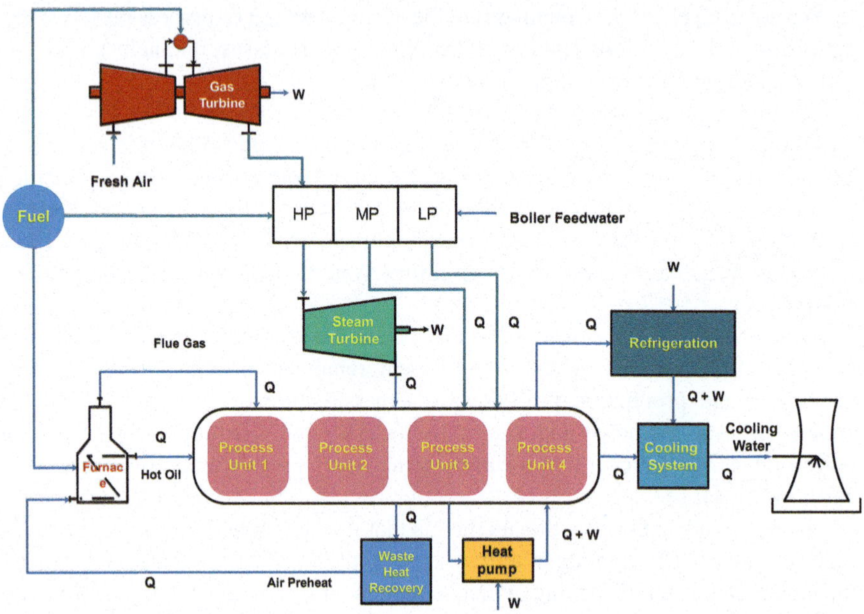

Fig. 4.4 Schematic view of power, heating, and cooling utilities in a process site

Power utility employs power generation components, such as steam turbines, gas turbines, small and medium internal combustion engines, and other power-generating equipment, to generate power, and direct drivers transfer the power generated in the form of electricity. The important point is that the heating and cooling utilities, themselves, fall into several categories.

We can observe the schematic views of different utilities present in a processing unit. As Fig. 4.4 illustrates, the heating utilities in a unit involve high-pressure (HP), medium-pressure (MP), and low-pressure (LP) steam, condensates, hot water, flue gas, gas turbine exhaust, hot oil, and heat pump reject. Furthermore, the cooling utilities of the site include cooling water, cooling air, air preheat, boiler feedwater (BFW) preheat, heat pump heat sink, molten salt, and refrigeration.

As stated in the utility explanations, among the present hot utilities, the exploitation of steam at various pressure levels is one of the most typical methods for heat transfer. Of the most important reasons for the prevalence of using steam in the process industries are the following unique features of steam:

- High latent heat (high thermal capacity)
- High operating temperature range
- Generating heat at one point and transferring it to another point
- Generating power in the steam turbine and steam ejector
- Cheap equipment for production and distribution
- Being nonpoisonous and the controllability of the steam temperature

One of the remarkable points concerning the steam used as a utility at a site level is its superheated temperature. The use of superheated steam prevents droplets and liquid creation due to pressure drops inside the steam tubes. If this steam is employed for power generation, the steam in the tube should be superheated by 10–20 degrees before it enters the turbine; otherwise, the moisture can damage the turbine.

4.4 What Is the Energy and Process Integration?

One of the methods for optimizing thermal cycles in the process industry is the integration of its existing processes and equipment with other processes. Based on Smith [18], integration in the process industries means the aggregation of producers and consumers within a system. For example, many flows and equipment in process industries generate heat or transfer it from an environment to a water cooler. Opposite to these heat generators, there are heat consumers as heat sinks in the same industries and need to receive heat. If the temperature of the heat source is higher than that of the heat sink and if other parameters like spatial proximity, heat transfer, heat exchanger for heat recovery designing, and cross the pinch point are appropriately possible, the system can transfer heat from the heat source to the heat sink. This makes the supply of the requisite heat for the heat sink by the system itself rather than the excess fuel, and the total rate of the unit's fuel consumption reduces. Thus, the heat integration of different processes can decrease utility consumption within the system and consumed fuel.

Likewise, besides the integration of the processes, another operational approach to reducing fuel consumption and pollutants and increasing efficiency in the process industries is to employ cogeneration systems.

4.5 What Is the Cogeneration of Heat and Power?

An approach to increasing the site's efficiency and reducing fuel consumption and costs is establishing a cogeneration system based on the site context. The cogeneration system means the simultaneous generation of power and heat from a joint heat resource. In a cogeneration system, the required consumed power of a site can be supplied in two ways. The power generated in a process site can be supplied indirectly either by a generator connected to gas and steam turbines or by connecting a direct drive to rotary equipment like compressors. Meanwhile, another benefit of the cogeneration systems is a reduction in the emission of pollutant gases. This subject should be examined against power generation in power plants outside the site.

To raise efficiency and reduce fuel consumption, we can simultaneously implement all these methods more favorably and optimally. The Total Site view has also been developed based on these principles.

4.6 What Is the Total Site Integration?

A Total Site, whose framework is illustrated in Fig. 4.5, consists of several isolated sites and a conventional utility unit. By relying on this approach and employing energy integration among the unit of a Total Site, we can enhance energy efficiency and storage. By developing a Total Site, we can supply the required heat of the consumer units by exchanging the heat between them and heat-generator units via a thermal network and decrease fuel consumption accordingly. Moreover, another crucial advantage of a Total Site is implementing cogeneration systems and supply of power utility by the site itself.

When introducing a Total Site, we should first assume that several process units come together. Based on the nature of the processes of every unit, some units are endothermic, and some are exothermic. In designing an independent process unit by methods like pinch technology, engineers embark on optimizing process flows and creating heat exchanges between their flows. Thus, the units that come together in a Total Site are optimized in terms of energy consumption. Now, heating and cooling utilities are required to supply heat and its removal from some of these units.

At first sight, we can assume that we consider an independent heat generator, like a boiler, for every one of the units and supply the heat of each internally. Hence, we should have considerable expenditure for establishing boilers and heat removal systems per site. Some of these units need heat with high quality and temperature, and some require heat with medium quality and temperature. Therefore, we have to

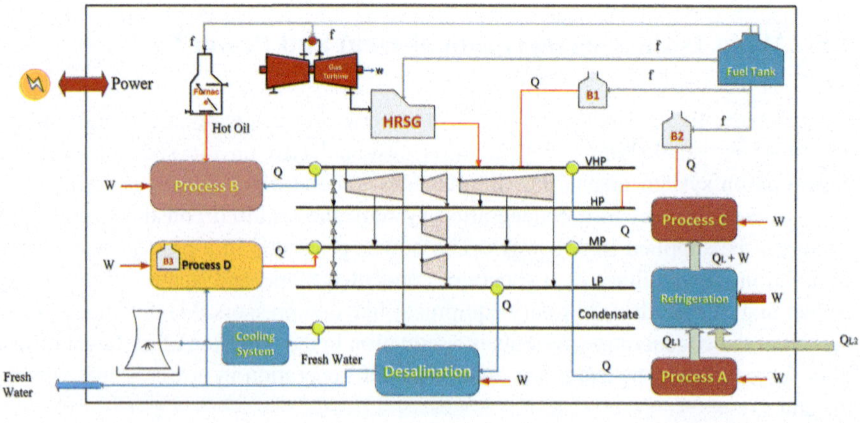

Fig. 4.5 General scheme of a Total Site

prepare several kinds of heat generation systems with varying dimensions. Except for the high expenses of this method, its complex maintenance and overhaul and inflexibility of changes in the capacity of every unit cause this approach not to seem optimal.

In another approach mentioned above, to reduce costs and enhance the efficiency and stability of the heat supply for process units, a heat network, which is usually a steam grid, is developed among the process units. Accordingly, heat is transferred to other units by the heat network in an integrated way in a generation utility unit instead of being produced in different sectors. Considering different requirements of every unit in terms of the required heat quality, temperature, and steam pressure and precluding the multiplicity of boilers with various pressures, we must generate steam at the highest pressure required for heat transfer and thermal need of each unit. After that, we should employ pressure-reducing equipment, such as letdown stations to transfer heat in varying pressures and temperatures. Therefore, an integrated heat network flows among different units in a Total Site, wherein heat is produced and consumed by different sectors with varying needs in an integrated way. However, is heat generation at high pressure and its consumption at low pressure cost-effective? Is it possible to use this network more optimally? Can we reduce fuel consumption by this network?

To answer these questions, we should state that one of the main potentials of developing a joint heat network is to make the heat exchange between heat-producing and heat-consuming units possible. Thus, all heat-producing units give their heat to the heat network. This decreases the heat load rate of heat-generating systems of the utility unit and, in turn, increases the efficiency of the Total Site and reduces fuel consumption. However, we can explain that generating heat at high pressure and temperature and diminishing its pressure to medium and low pressures destroys the steam exergy and ruins its work generation potential. Therefore, this approach is still far from the ideal condition regarding the optimality of energy generation and consumption. On the other hand, one of the most paramount needs of the process units is power. Now, we should ask from where is the power of a Total Site supplied?

To answer this question, we should also claim that one of the approaches to supplying power is buying electricity from the national and local networks or generating power by the non-concentrated or concentrated generation systems inside the site itself. However, we can express that the power generation cost surpasses the heat generation cost with the same energy capacity. Hence, the provision of power from national and local networks or producing it inside the site, in addition to imposing a high capital cost, increases fuel consumption and the rate of pollutants having roots in the Total Site processes.

Here, we can refer to one of the most important potentials of a Total Site. The development of a joint heat network, generation of heat at high temperature and pressure, and heat consumption at low temperature pave the way for employing pressure-reducing equipment such as steam turbines rather than the ones that destroy steam exergy. Thus, by decreasing steam pressure and setting it to the desired rate of every consumer, we can generate power through the existing energy in the steam of higher quality. This kind of power and heat generation is called power and heat cogeneration due to the commonality of the consumed fuel, reducing the costs of a Total Site and increasing its efficiency.

Among important features of the cogeneration systems are having consumption-based flexibility in their capacity rate, needing repairs in the unit, and changing the capacities of other units. Furthermore, another advantage of cogeneration systems is the entry and exit of power to the system. This enables us to sell the excess power of a Total Site to the distribution or local networks or supply the requisite power from the national or local networks when it lacks.

Two approaches are examined to designing a heat network and generating the power of a Total Site. The first approach, known as the grass roots, is to optimally design a heat network and cogeneration system for a Total Site assuming that this system does not exist in the Total Site. The second, known as the retrofit approach, deals with modifying the heat network and cogeneration system present in the Total Site. This chapter suffices to explain the grassroots approach to present a designing method for a cogeneration system in a Total Site.

Concerning optimizing the process units, the following section deals with technology introduction and pinch analysis, as well as fuel consumption and power generation targeting in a Total Site. Afterward, it discusses the instruments needed for evaluating the efficiency and designing a Total Site and explains the designing of cooling and desalter systems and their integration in a Total Site.

4.7 What Is Pinch Analysis?

One of the conventional methods for conducting integration in heat networks is to employ pinch technology. This method was introduced by Linhoff in 1993 [11]. The introduced tools in this method help recognize thermal flows, increase recovery, and reduce fuel consumption. Pinch analysis is currently the most widely used process integration approach in the industry. As it has certain limitations as also all other approaches (pure heuristic-based, insights-based, optimization-based, and hybrid approaches combining heuristics, thermodynamics, and/or optimization), it has been beneficial to combine various approaches in a way to take the best from each of them [3, 7].

For the design and integration of a heat network, the processes posed in a site are first identified, as shown in Fig. 4.6. Then, every process flow that needs heat exchange is specified, and the rate of exchangeable heat and its transfer direction is determined. Next, after the exchangeable heat among the process flows is specified, the minimum allowable temperature is determined for the transfer of heat among flows by considering the intra-unit exchangers' structures and the economy of a process unit.

Next, the minimum heating and cooling consumptions of a system are determined by the problem table. Later, via analytical-graphical tools of grand composite curve (GCC), the hot and cold flows are delineated compositely by considering the

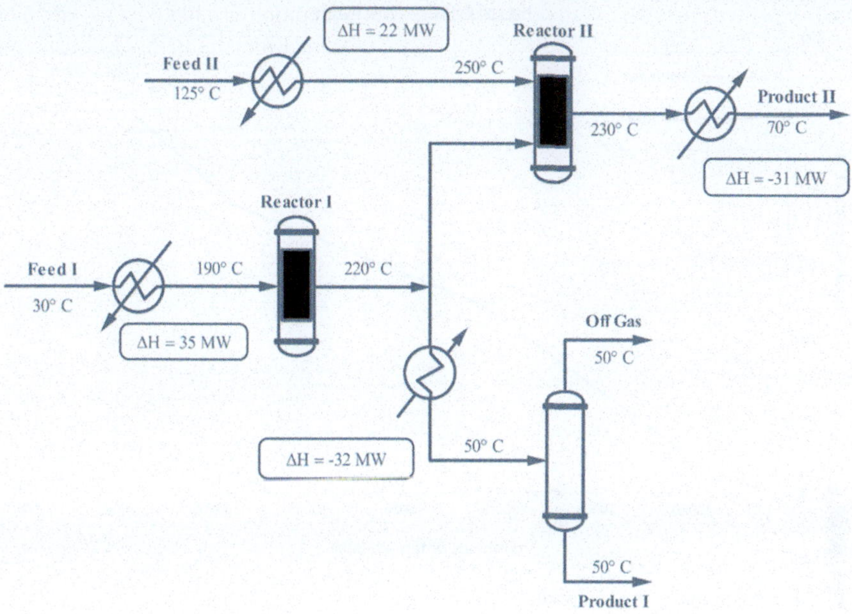

Fig. 4.6 Process flows in the flow diagram map (PFD) of a process unit

constraints related to the allowable heat transfer or pinch temperatures. Another analytical-graphical instrument, called composite curve (CC), wherein hot and cold flows are drawn separately by considering the minimum allowable heat transfer temperature, is used. Accordingly, the heat recovery potentials in the system are identified, and the interaction manner of the components of the process industries with existent flows is determined in the considered industry. Figure 4.7 shows an example of a composite curve.

The purpose of using the pinch technology in designing a process unit is to reduce heat exchanger costs by optimally designing or modifying a heat exchange network and increasing heat recovery among the unit flows. Various instruments are used for analyzing a site by pinch analysis; however, since designing a service-providing unit and a Total Site necessitates dominance over the pinch analysis rules, the next part of the chapter reviews these rules.

4.7.1 Targeting Steam Consumption and Generation

As mentioned in the previous section about cogeneration systems, after extracting the flow data related to every process inside a process unit, we can study the targeting of the utility generation and consumption in a Total Site in two ways. The first method is to draw the Total Site composite curve (TSCC) for all units together and

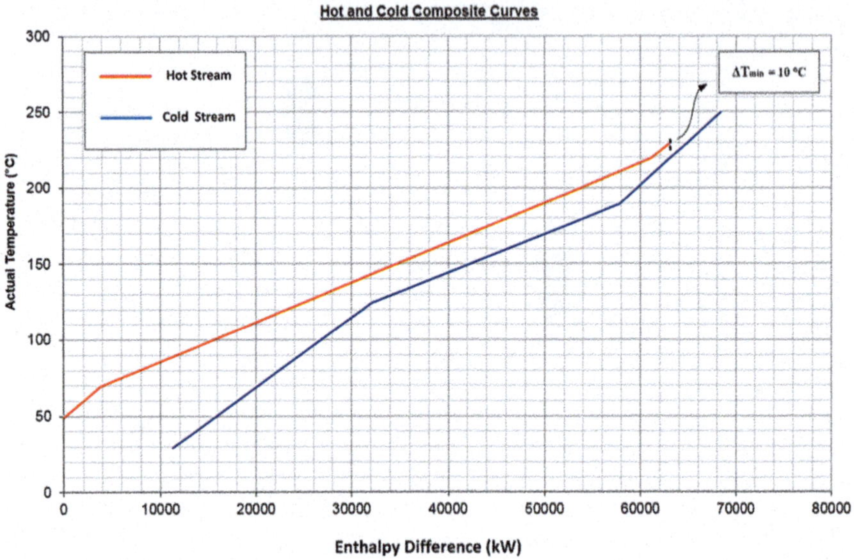

Fig. 4.7 An example of a composite curve

target the heating and cooling services in this site. The second method is to draw the site composite curve (SCC) separately for every site and aggregate the targeting accomplished for the heating services [7].

As Khoshgoftar Manesh et al. [12] clarify in their study, separate targeting of every site often displays the consumption of fuel and heating and cooling services higher or equal to the targeting in the TSCC drawing condition. However, the important point in this relation is disregarding the existing constraints in the targeting based on the TSCC graph. The reason is the accumulation of the present flows in every thermal range, leading to the aggregation of even two spatially far flows and their simultaneous generation and consumption targeting. For example, if generation targeting is based on the aggregation of two flows, the generation of a heating service may practically be fulfilled by the flows of two distant units, which may be impossible.

The methods for drawing the TSCC and SCC graphs are similar. The mere difference is that the method is used for all units of a Total Site in TSCC designing, and a single graph is developed for a Total Site, while, in the SCC method, the drawing process is implemented separately for every site, and the SCC graph of that site is delineated. Hence, for every process unit in a site, an SCC graph is drawn, and all SCC graphs are aggregated for the TSCC graph production.

4.7.1.1 Drawing and Analyzing TSCC and SCC Graphs

To draw TSCC and SCC graphs, we have to exploit the GCC or CC tools in pinch technology. Since the present study aims to present a designing method, it needs to use the GCC graph for designing based on the explanations posed in the pinch analysis section when optimizing a heating system.

After extracting the data of the flows in a unit, solving the problem table associated with that unit, and drawing the GCC graph, we first specify all possible recoveries in a site according to the GCC graph of that site. Next to considering all possible recoveries, we omit the recovered flows from the GCC graph. The noticeable point respecting recovery is the presence of large energy pockets in the GCC graph. Based on the explained principles in the pinch analysis section, similar to Fig. 4.8, we have to heed the generation and consumption targeting of heating services according to the temperatures of the regarded services and remove other remaining parts from the graph for recovery.

All cold and hot flows are separated after thermal recoveries. All temperatures of the hot flows in GCC are transferred in the negative direction as much as half of the minimum allowable temperature difference. It means that hot flows are transferred in a negative direction at their real temperatures as much as half of the allowable temperature difference. Likewise, all cold flows are transferred in a positive direction in the same size. The reason for this topic is the use of the real temperature of the heating and cooling services and observance of the minimum allowable temperature difference between exchangers at the time of generation and consumption targeting of the heating and cooling services considered in the Total Site of a unit.

After transferring the temperatures of the flows in positive and negative directions, we should draw cold flows at the right side of the vertical axis of the SCC graph and hot flows at the left side of the vertical axis of this graph. The horizontal axis of the SCC graph shows enthalpy, and the vertical axis depicts temperature. Thus, hot flows are drawn in negative enthalpies, and cold flows are drawn in positive enthalpies. The reason is that the thermal capacity of the hot flows is considered negative in the SCC graph. Figure 4.9 illustrates an example of an SCC graph.

Now, to draw the TSCC graph, we have to implement this process for all existing units. Thus, we first draw the GCC graph of all units. Then, after carrying out the available recoveries of every unit, similar to Fig. 4.9, separating hot and cold flows from each other, and transferring the real temperatures of the flows as much as half of the minimum value of the allowable temperature difference in every unit, we add them up in every thermal range in teams of their thermal capacity. The noticeable point in the formation of the TSCC graph is the possibility of varying the value of minimum allowable temperature difference in every site compared to others.

Hence, a site's hot and cold flows are transferred in relation to the minimum value of the allowable temperature difference of that site, and then their thermal capacities are added in every thermal range.

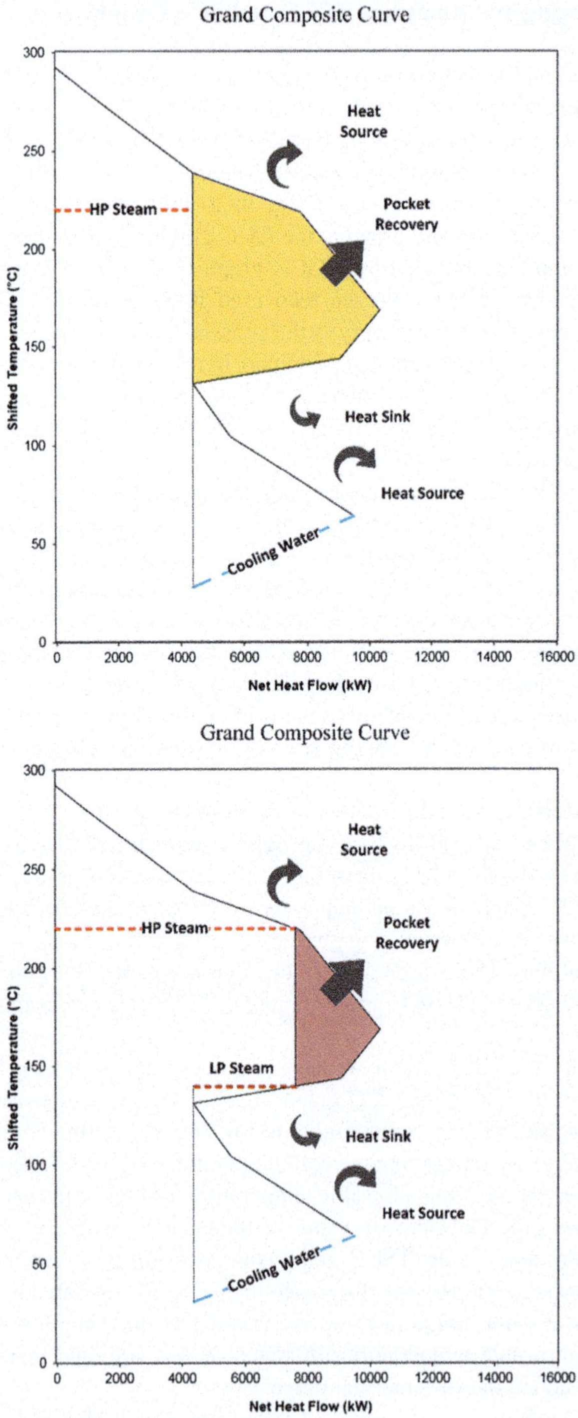

Fig. 4.8 Heat recovery in large pockets

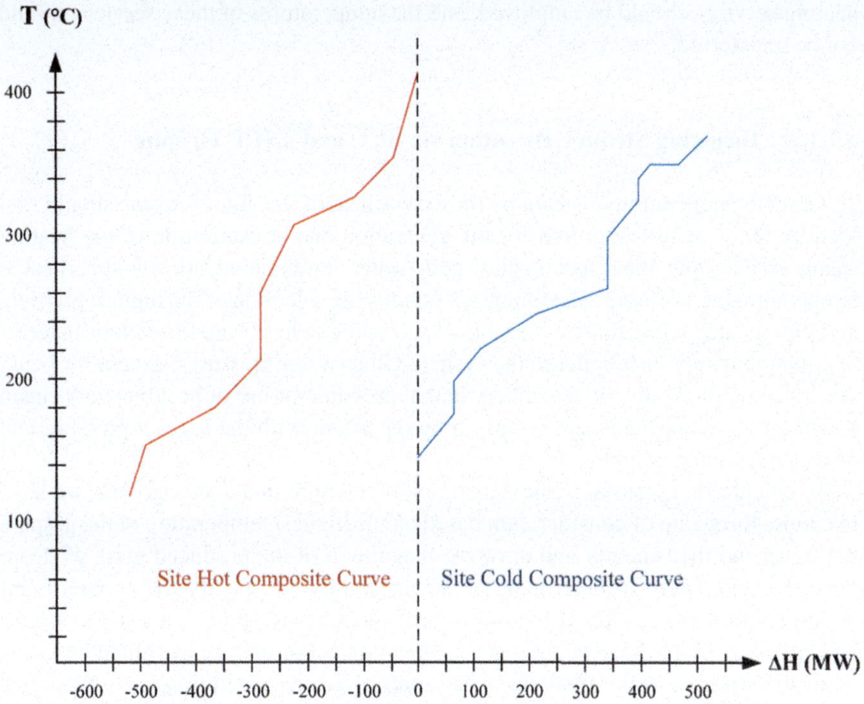

Fig. 4.9 Site composite curve for a sample site

4.7.1.2 Targeting the Generation and Consumption of Heating and Cooling Services Using SCC and TSCC Graphs

The most important application of the SCC and TSCC graphs is to target the generation and consumption of the utilities required for a unit and Total Site. As mentioned at the first of chapter, the most significant heating service valuable for heat transference between the flows of a Total Site or power generation from this site is aqueous water at various temperatures. In the present study, the sole heating service used for designing a Total Site is also aqueous water at various temperatures. The generation and consumption of aqueous water were targeted at saturation temperature in the past. This method lacked the necessary accuracy for determining the rate of the produced aqueous water and the final consumed fuel. In 2015, Sun [20] introduced a modified method for targeting the rate of generated and consumed steam. In this new method, the boiler feedwater preheats, the phasic change from saturated liquid to saturated steam, and steam preheat are considered at targeting time. However, for computational simplicity and acceleration in targeting, the thermal capacity of steam is also considered constant.

Furthermore, during the targeting of cooling and heating services in the SCC and TSCC graphs, it should be remembered that the real temperatures of the cooling and heating services should be employed, and the temperatures of these services should not be transferred.

4.7.1.3 Targeting Steam Generation via SCC and TSCC Graphs

To target the generation of steam by the excess heat of the Total Site, we should first identify the pressures required for the generation and consumption of the heating steam service and then specify the input water temperature and its superheated temperature per pressure. The important point is the selection of the highest pressure and superheated temperature, determined according to hardware limitations in terms of equipment such as a boiler in the system. Other water pressures, except for being generated by the boiler, heat recovery in that pressure, or use of heat recovery steam generator in gas turbines, are produced by the steam with the highest pressure and pressure-reducing equipment.

To embark on targeting steam generation in the SCC and TSCC graphs, we have to choose the steam of consideration based on the highest temperature of the graph's hot flows and then specify and draw the magnitude of the produced mass of steam using try and error. To determine the steam points, we specify the contact point upon the hot flows' graph at the boiler feedwater temperature. Later, by using try and error regarding the generated mass, other contact points are identified based on their temperatures and enthalpies. The target of steam generation is to maximize generation mass. After targeting steam generation at a single pressure, we can also examine the steam generation in other pressures using the reminiscent of the hot flows' graph. To discharge the thermal load of the flows at temperatures lower than the boiler feedwater at its minimum pressure, we should exploit cooling water, air, or refrigeration systems at temperatures below the environment. Figure 4.10 shows examples of steam generation targeting in a saturated condition irrespective of the preheated boiler water and superheated steam, and Fig. 4.11 displays steam generation targeting regarding the preheated boiler water and superheated steam.

As observed, steam generation targeting in a saturated condition reveals a larger rate of generated mass flow compared to the other condition.

4.7.1.4 TSCC and SCC Targeting of the Consumption of Heating Services Using Graphs

Cold flows need heat; thus, steam consumption targeting, as a heat transfer factor, is another purpose of using the TSCC and SCC graphs in a unit or Total Site. Targeting steam consumption is not as simple as targeting steam generation. It is because of the presence of various entries for the losses of steam transfer per pressure. In a cogeneration network, different components such as the passing flow of the top-down loss, steam turbine inputs, steam inputs produced by the processes, boiler input, and heat recovery steam generator in that pressure and other inputs enter the head, and every one of them differs in its enthalpy and entrance temperature

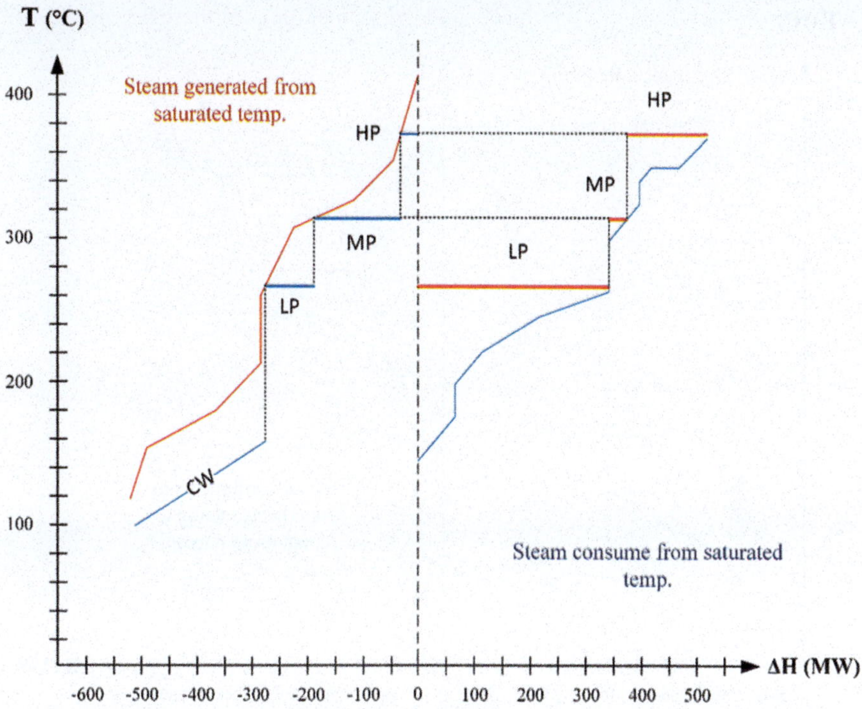

Fig. 4.10 Targeting steam generation and consumption by not preheating the boiler water and superheating the steam

at a similar pressure. Hence, every loss' enthalpy magnitude and temperature are determined with respect to the mixed enthalpy of the input flows per loss. Therefore, depending on the mass flow imported by every input, the temperature of the losses differs, and the consumption is targeted based on the existing temperature. For network analysis in terms of consumption, it is necessary to perform several try and error or numerical computations in every system condition.

Another intricacy of the consumption targeting relates to the aqueous water conditions regarding heat transfer with the system's cold flows. As shown in Fig. 4.10, steam consumption was targeted in a saturated condition in the past. However, Sun et al. also modified the type of steam consumption targeting and examined it in several conditions. On the consumption side, each steam can exchange heat with cold flows in four conditions:

- By considering the superheated, saturated, and subcooling parts of the aqueous water
- By considering the superheated and saturated parts of the aqueous water
- By considering the saturated and subcooling parts of the aqueous water
- By considering the mere saturated part of the aqueous water

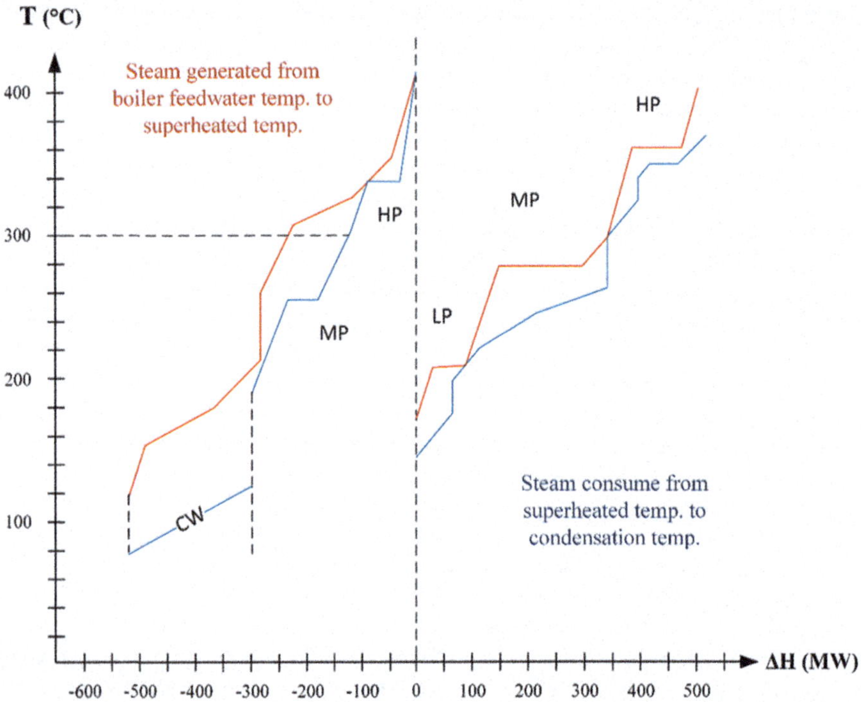

Fig. 4.11 Targeting steam generation and consumption by preheating water and superheating steam

After steam consumption targeting, the remaining cold water is supplied by the highest steam pressure directly obtained via heat transfer to the boiler-entering water. Likewise, the other targeted consumption steam is integrated; the rate of the generation steam, targeted from the graph, is subtracted from the consumption rate; and the minimum rate of consumption fuel is estimated in a full recovery condition.

Sun targeted the steam consumption, power generation by a cogeneration system, fuel consumption, and mass flow rate passing from the passage port between two losses in different scenarios and in two decomposed and integrated conditions for a model site by considering heat transfer between the steam and cold flows of a site. Table 4.1 displays the results of their investigation. As the table shows, the best kind of heat transfer for the decomposed and integrated conditions is considering the superheated and saturation conditions and disregarding the heat transfer condition for condensed steam. The reason is the addition of the compensating water to the deaerator for the purpose of making up the reduced condensed steam return to the deaerator, needing to be heated by direct heat transfer from the inlet temperature to the saturation point. On the other hand, due to the upsurge of the mass flow rate of the condensed steam in the network, the loss rate of the system increases after the steam fulfills its heating responsibilities within the network.

Table 4.1 Targeting heat and power in Sun et al. research [20]

	Boiler flow rate	Letdown flow rate	Power generation	Power generation per unit boiler
Unit	[kg. s^{-1}]	[kg. s^{-1}]	[MW]	[MW. kg^{-1}s]
Integrated options	Superheated steam use + condensate heat recovery			
	50.29	3.6	20.83	0.414
	Superheated steam use + condensate heat recovery			
	49.95	3.9	22.68	0.454
	Superheated steam use + condensate heat recovery			
	52.66	3.3	21.35	0.405
	Superheated steam use + condensate heat recovery			
	52.06	3.7	23.10	0.444
Decomposed options	Superheated steam use + condensate heat recovery			
	53.35	3.1	21.48	0.403
	Superheated steam use + condensate heat recovery			
	53.25	2.8	23.18	0.435
	Superheated steam use + condensate heat recovery			
	55.95	2.8	21.84	0.390
	Superheated steam use + condensate heat recovery			
	55.48	2.6	23.68	0.427

4.8 Cogeneration System

Targeting fuel consumption in cogeneration systems based on the generation and consumption of heating and cooling services cannot render a true picture of fuel consumption in a cogeneration system. It is because the power generation components and their effects are neglected in system targeting.

Therefore, for the targeting of a cogeneration system, suitable models are needed for modeling the components of a cogeneration system.

4.8.1 Modeling Heat and Power Components

Since all power and heat components do not have ideal efficiency in every condition, we need to accurately model a system's components and consider its efficiency variations owing to the changes in the heat load and mass flow of the system to obtain a proper picture of a cogeneration system. This section deals with modeling power components, including steam turbines, gas turbines, as well as heat components, such as boilers and heat recovery steam generators.

4.8.1.1 Modeling Steam Turbine

The steam turbine is the most paramount component of power generation in a cogeneration system. Steam turbines fall into different categories, the most prevalent of which are associated with the kind of turbine and the number of its outlets. Steam turbines have two condensing and back-pressure kinds in terms of the type of output pressure. In a condensing turbine, the output carries a trivial value of the steam mass fraction while it changes into a saturated liquid in the ideal condition. These turbines can have more than one input and output and are called extraction and induction turbines, respectively. Figure 4.12 shows the different kinds of extraction and induction turbines.

Equation (4.1) can be used to calculate the passing flow through each turbine where $\dot{m}$ is in [kg. s^{-1}]. The efficiency of the steam turbine changes according to the designing and passing mass flow rate. The maximum efficiency of a steam turbine happens when its mass flow is maximal.

$$\dot{m}_{\text{level } i} = \dot{m}_{\text{steam turbine,input}} - \sum_{j<i} \dot{m}_{j,\text{output}} + \sum_{j<i} \dot{m}_{j,\text{input}} \tag{4.1}$$

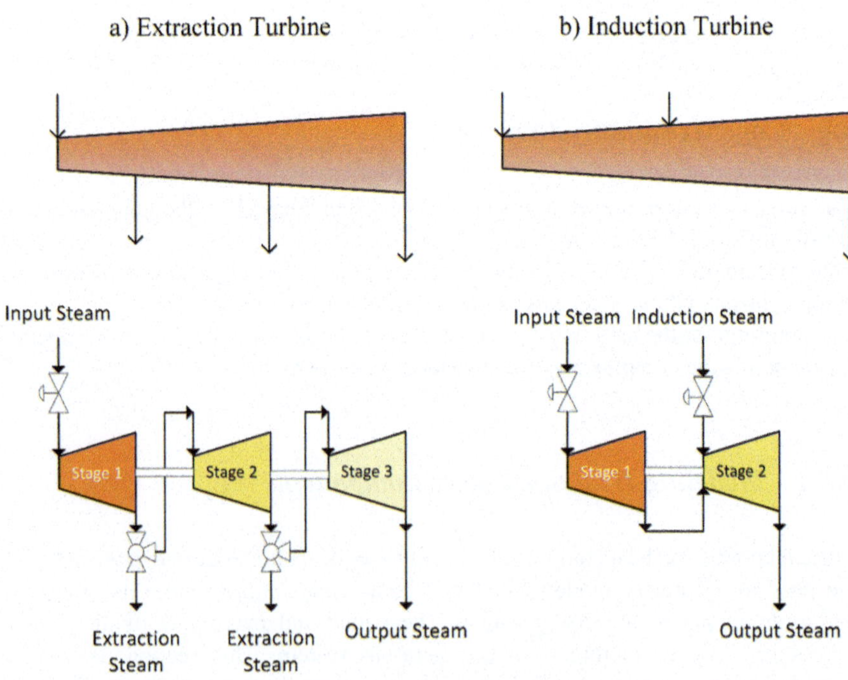

Fig. 4.12 (**a**) Extraction steam turbine, (**b**) induction steam turbine

In a design of a steam network, the efficiency variations of the steam turbine are extremely influential and lead to a change in the efficiency of the whole network. Thus, the design of a cogeneration network needs more accurate modeling of the steam turbine based on the mass flow variations.

In 2004, Varbanov et al. [21] used an Eq. (4.2), known as the Willan line equation, to present an equation for modeling the variable efficiency of the steam turbine.

The Willan line equation is used as a linear model for modeling steam turbines and can show the behavior of efficiency changes and generation power of the steam turbine according to the mass flow variations.

$$W = nm - W_{\mathrm{INT}} \tag{4.2}$$

In this equation, W is in [kW], and W_{INT} [kW] is the intercept of the fitted line on the efficiency points of the steam turbine, and its value is calculated according to Eq. (4.3). Likewise, n shows the graph slope of the Willan line and is calculated by Eq. (4.4).

$$W_{\mathrm{INT}} = \frac{c}{a}\left(m_{\max}\Delta H_{\mathrm{IS}} - b\right) \tag{4.3}$$

$$n = \frac{(1+c)}{a}\left(\Delta H_{\mathrm{IS}} - \frac{b}{m_{\max}}\right) \tag{4.4}$$

In 2015, Sun and Smith [19] updated the a, b, and c coefficients by investigating the data of 70 back-pressure turbines with 1–35 MW sizes in 214 operational conditions, between 40% and 100% loading, and 104 condensing turbines with 8–60 MW sizes in 335 operational conditions between 40% and 100% loading.

According to the obtained coefficients, the developed model has a maximum of 2.71% of the mean error in the equation to its real magnitude. The constants of the Willan line equation for a steam turbine can be computed by Eqs. (4.5), (4.6), and (4.7). The constant coefficients are listed in Table 4.2.

$$a = a_1 + a_2 P_{\mathrm{in}} + a_3 P_{\mathrm{out}} \tag{4.5}$$

$$b = b_1 + b_2 P_{\mathrm{in}} + b_3 P_{\mathrm{out}} \tag{4.6}$$

$$c = c_1 + c_2 P_{\mathrm{in}} + c_3 P_{\mathrm{out}} \tag{4.7}$$

The isentropic efficiency of the steam turbine is also computed based on Eq. (4.8), where H is enthalpy in [kJ]. Thus, having the steam turbine's mechanical efficiency, whose value is often considered at 97% to 99% at hand, we can compute the isentropic efficiency of the steam turbine based on Eq. (4.9).

Table 4.2 Coefficients of Willan line equation for simulating a steam turbine [19]

Constants	Back-pressure turbines	Condensing turbine
a_1	1.1880	1.3150
a_2	-2.9564×10^{-4}	-1.6347×10^{-3}
a_3	4.6473×10^{-3}	−0.36798
b_1	449.98	−437.77
b_2	5.6702	29.007
b_3	−11.505	10.359
c_1	0.20515	7.8863×10^{-2}
c_2	-6.9517×10^{-4}	5.2833×10^{-4}
c_3	2.8446×10^{-3}	−0.70315

$$\eta_{IS} = \frac{H_{in} - H_{out}}{H_{in} - H_{IS}} = \frac{\Delta H}{\Delta H_{IS}} \tag{4.8}$$

$$\eta_{IS} = \frac{\Delta H}{\Delta H_{IS}} = \frac{W}{\eta_{MECH} m \Delta H_{IS}} = \frac{nm - W_{INT}}{\eta_{MECH} m \Delta H_{IS}} \tag{4.9}$$

Furthermore, the efficiency of the steam turbine can be shown by the use of Eqs. (4.10) and (4.11), where m unit is [kg. s^{-1}].

$$\eta_{ST} = \frac{\Delta H}{\Delta H_{IS}} = \frac{W}{W_{IS}} = \frac{W}{m \Delta H_{IS}} \tag{4.10}$$

$$\eta_{ST} = \eta_{MECH}\ \eta_{IS} \tag{4.11}$$

The output enthalpy of the steam turbine can also be calculated by Eq. (4.12).

$$H_{out} = H_{in} - \frac{W}{\eta_{MECH} m} \tag{4.12}$$

In 1985, according to data of the available turbines, Peterson and Man revealed that the efficiency of the turbines operating in their maximum possible loading enhances as the input pressure turbine power rise provided that the output pressure is constant. Varbanov et al. also showed that a linear relationship was established between the shaft and isentropic powers of turbines with different sizes when they operate in their maximum loading. Equation (4.13) is established between the maximum isentropic power and shaft power of a turbine in its maximum loading in [kW].

$$W_{IS,max} = a W_{max} + b \tag{4.13}$$

Accordingly, we can explain the efficiency of the steam turbine in its maximum loading as Eq. (4.14).

$$\eta_{\mathrm{ST,max}} = \frac{1}{a + \frac{b}{W_{\mathrm{max}}}} \tag{4.14}$$

This equation is used for designing a turbine. Moreover, we can also explain the efficiency of a steam turbine as Eq. (4.15) in case the steam turbine-passing mass flow is known instead of the required power. This equation discloses the efficiency of a steam turbine based on the maximum mass flow passing the steam turbine.

$$\eta_{\mathrm{ST,max}} = \frac{1}{a}\left(1 - \frac{b}{m_{\mathrm{max}}\Delta H_{\mathrm{IS}}}\right) \tag{4.15}$$

A very important point in modeling a steam turbine by the Willan line method is that the changes in the output enthalpy of the turbine in every stage accord with the output loss enthalpy connected to it. This variation is most significant in simulating extraction turbines since the input enthalpy to other parts of the turbine varies due to the changes in the output enthalpy of the preceding parts. To model an extraction steam turbine, we should first guess an enthalpy for the output of a steam turbine and then modify the guessed enthalpy by try and error and compute the generation power by the remaining parts of the turbine.

4.8.1.2 Modeling a Gas Turbine

Similar to a steam turbine, a gas turbine has variable efficiency proportionate to loading on it. Varbanov et al. modeled a gas turbine linearly using Willan line. Moreover, several models have been introduced by other scholars for modeling a gas turbine. However, this study employs the model introduced by Maninand and Zhou [13] due to the lack of documentable information from the manufacturers of gas turbines. It is worth mentioning that the data of the mentioned model enjoys acceptable conformity with the results of the explained model in the nominal power condition.

For industrial and aerodynamic gas turbines applied in the aerospace industry, Maninand and Zhou presented some equations to explain fuel consumption according to the turbine power, the ratio of power to the mass passing and exiting from the turbine, and the output temperature of the gas turbine. To investigate and target the rate of generation power of a gas turbine, by having the magnitude of the required steam at hand, we first guess the primary mass is exiting the gas turbine. The nominal power for the flow passing from a gas turbine is computed by Eq. (4.16). Then, the magnitude of the temperature of output gases in [K] is also estimated by Eq. (4.17).

$$W_{\mathrm{GT}} = \frac{M_{\mathrm{ex}}}{2.9} \tag{4.16}$$

$$T_{\mathrm{ex}} = 0.4 \times W_{\mathrm{GT}} + 493.42 \tag{4.17}$$

Afterward, by possessing the temperature of the acidic point of output gases (T_{adp}), the temperature of the output gases of the turbine, the specific heat of the output gases, the temperature of the input and output gases of the heat recovery steam generator, and steam mass targeted for the steam header, we can examine the rate of the required generation and consumption heat using Eqs. (4.18) and (4.19) and modify the gas turbine in case of mass inequality. This process continues until the generation and consumption of heat become equivalent where Q and h are in [kW] and [kJ. kg^{-1}], respectively.

$$Q_{\mathrm{HRSG}} = M_{\mathrm{ex}} C_{\mathrm{P,ex}} \left(T_{\mathrm{ex}} - T_{\mathrm{adp}}\right) \tag{4.18}$$

$$Q_{\mathrm{Steam}} = M_{\mathrm{Steam,out}} \left(h_{\mathrm{Steam,out}} - h_{\mathrm{BFW}}\right) \tag{4.19}$$

According to Eq. (4.20), the magnitude of the specific heat can be estimated for a large number of fuels and in an acceptable thermal range [18].

$$C_{\mathrm{P,ex}} = 1.1 \left[\mathrm{kJ} \cdot \mathrm{kg} \cdot {}^{\circ}\mathrm{C}^{-1}\right] \tag{4.20}$$

4.8.1.3 Modeling a Boiler

The most applied equipment used in the process industries for the supply of the heat needed by the processes is boilers and furnaces. Similar to steam turbines, boilers have variable efficiency in heat loadings. Thus, the selection of a suitable boiler according to the heat load variations in a cogeneration network is one of the most important decision variables under investigation. Figure 4.13 illustrates the schematic view of a sample boiler.

In a boiler, there is a tributary flow called the blowdown flow, which is used for venting precipitations and reminiscent of the previous flows of the boiler. This flow, which exits the boiler by its controller, is saturated. The ratio of the output tributary flow to the output steam flow of a boiler is called the blowdown rate. This rate is 5–10% for low-pressure boilers, 2–5% for high-pressure boilers, and <2% for boilers with extremely high pressure and net input.

The calorific value of a fuel is calculated based on the net calorific value (NCV) or gross calorific value (GCV). GCV is the rate of heat released from a certain amount of fuel in full combustion, at a standard temperature (15–25 °C), and at standard pressure (1.01325 bar), such that water is in a primary condition and fuel products take liquid forms. NCV resembles GCV, and it is assumed that the latent heat of the aqueous water is not recoverable, and water takes a steam form initially and in combustion products. The NCV and GCV equations can be explained as Eq. (4.21).

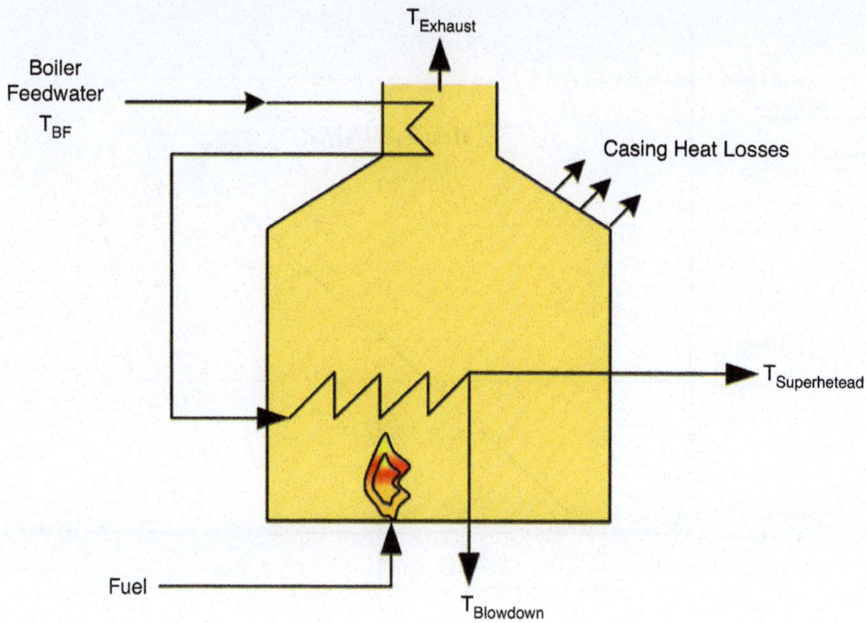

Fig. 4.13 Schematic view of a boiler

$$NCV = GCV - m_{COND}\Delta H_{VAP} \tag{4.21}$$

The rate of input heat to a boiler can be estimated by Eq. (4.22), wherein GCV [MJ. kg^{-1}] can be used instead of NCV [MJ. kg^{-1}] for the heat transfer rate.

$$Q_{INPUT} = m_{FUEL}NCV \tag{4.22}$$

Some part of the boiler-entering heat is used for water evaporation, and another part is introduced as losses. In Fig. 4.14, we can observe the consumption of the input heat to the boiler on the enthalpy-temperature graph.

The efficiency of a boiler indicates the rate of useful energy consumed from the fuel in a boiler and is explained as Eq. (4.23).

$$\eta_{Boiler} = \frac{Q_{OUTPUT}}{Q_{INPUT}} = \frac{m_{STEAM}(H_{STEAM} - H_{BFW})}{m_{FUEL}NCV} \tag{4.23}$$

In a boiler, the useful energy of the fuel is used for evaporating the boiler feedwater until the considered steam temperature. Meanwhile, we can depict the efficiency of a boiler as Eq. (4.24) according to the rate of its losses.

$$\eta_{Boiler} = 1 - \frac{Q_{LOSS}}{Q_{INPUT}} \tag{4.24}$$

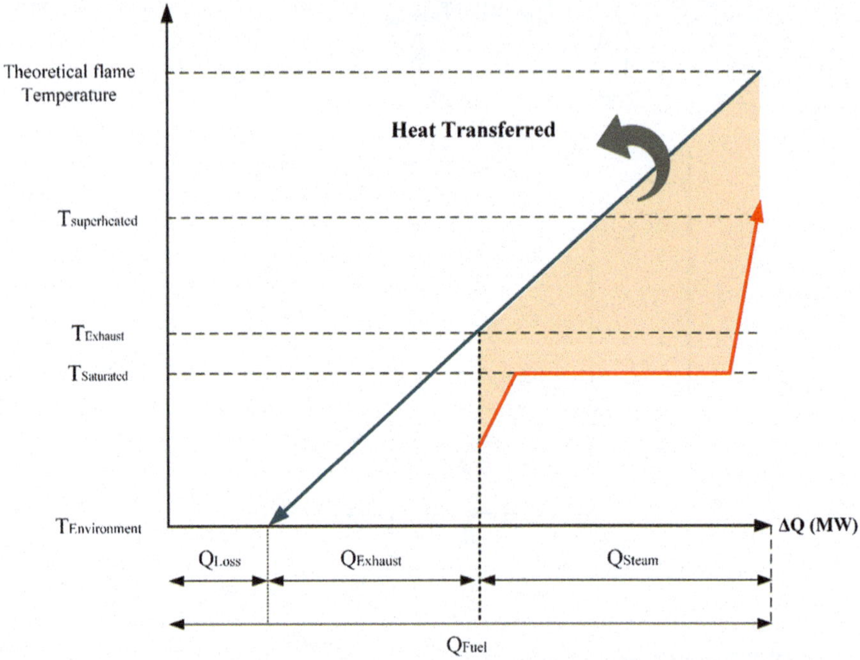

Fig. 4.14 The enthalpy-temperature graph of a boiler

The considerable point about the efficiency of boilers is their variability vis-à-vis the rate of steam generation. Thus, the efficiency is also considered variable according to the mass flow of the generated steam for boilers. Two kinds of maximum loading are heeded for a boiler [18]:

- Maximum continuous rating (MCR) when the boiler operates continuously; a boiler can operate above the MCR point in a short temporal period.
- Economic continuous rating (ECR) is considered as the maximum point of the boiler. ECR can be heeded as 0.6 to 0.8 of the MCR value. The best possible value for ECR equals 0.8 of the MCR value.

The fuel consumption of a boiler with constant efficiency can be calculated by Eq. (4.25).

$$Q_{\text{Fuel}} = \frac{1}{\eta_{\text{Boiler}}} \cdot m_{\text{Steam}} \cdot \Delta h_{\text{Steam}} \tag{4.25}$$

Concerning model, a boiler based on variable loading, Shang [17] presented the modified equation of fuel consumption as Eq. (4.26) where R_{BD} is the boiler discharge flow to the output flow ratio.

$$Q_{\text{Fuel}} = \frac{Q_{\text{Steam}}}{\eta_{\text{Boiler}}} \\ = \frac{\Delta H_{\text{Steam}} m_{\text{Steam}}}{\eta_{\text{Boiler}}} \Delta H_{\text{Steam}} \left[(b_{\text{Boiler}} + 1)\, m_{\text{Steam}} + a_{\text{Boiler}} m_{\text{Steam,MAX}} \right] \\ + R_{\text{BD}} \cdot m_{\text{Steam}} \cdot \Delta h_{\text{pre}} \tag{4.26}$$

Moreover, based on the presented equation, we can use Eq. (4.27) to calculate the boiler efficiency in a variable loading condition where R_{PH} is the steam preheater enthalpy to steam enthalpy ratio [21].

$$\eta_{\text{Boiler}} = \frac{m_{\text{Steam,0}}}{a_{\text{Boiler}} + m_{\text{Steam,0}} \cdot (b_{\text{boiler}} + 1) + R_{\text{PH}} \cdot R_{\text{BD}} \cdot m_{\text{Steam,0}}} \tag{4.27}$$

And

$$m_{\text{Steam,0}} = \frac{m_{\text{Steam}}}{m_{\text{Steam,max}}} \tag{4.28}$$

$$R_{\text{PH}} = \frac{\Delta h_{\text{pre}}}{\Delta h_{\text{Steam}}} \tag{4.29}$$

Coefficients a and b are determined for every boiler according to the type of its design, efficiency, and maintenance. However, Shang considered the proposed value of the coefficients for modeling the variable efficiency in a boiler as Eq. (4.30) [10].

$$a_{\text{Boiler}} = 0.0126 \\ b_{\text{Boiler}} = 0.2156 \tag{4.30}$$

4.8.1.4 Modeling Letdown Station of a Header

The presence of a letdown station between two steam headers is among the exigencies of designing a steam network. Although we can import all the mass that passes from a header to another to steam turbines to generate power, we can never zero the rate of mass passing from every letdown station. It is because of the importance of superheated steam inside every header. Due to the presence of pressure drops in tubes within headers, a small amount of steam can condense into liquid and thus destroy the turbine after its entrance. Hence, the temperature of headers should always be 10–20 degrees superheated for the preclusion of this event. Thus, to conserve this constraint for all headers, the steam passing the letdown station between two headers cannot be zero. Due to the existence of a pressure difference between the two headers, there is a pressure-reducing valve. Moreover, before the entrance of the flow to the pressure-reducing valve, we can import a desuperheater, which is actually the boiler feedwater, with a control valve to reduce the steam temperature. In Fig. 4.15, we can observe the scheme of a sample letdown station without a desuperheater.

Fig. 4.15 Scheme of a sample letdown station

Equation (4.31) can be used for modeling a letdown station.

$$\begin{aligned} \dot{m}_{\text{Input}} &= \dot{m}_{\text{Output}} \\ h_{\text{Input}} &= h_{\text{Output}} \end{aligned} \tag{4.31}$$

4.8.2 *Targeting a Cogeneration System*

As described in Sun et al. [20] and Ren et al. [16], it is important to target the cogeneration potential of utility systems in the process of designing or retrofitting Total Site utility systems. To target the minimum fuel consumption, maximum power generation potential, and thermodynamic properties of every steam header in a cogeneration system, we need to access suitable models for designing a steam network and cogeneration system. Afterward, the steam headers' pressure, superheated temperature, boiler feedwater temperature in that pressure, and ideal or primary efficiencies are considered. Since targeting using ideal efficiency is too far from reality, we estimate steam turbine efficiency more accurately using the equations presented in the steam turbine modeling section instead of considering ideal efficiency to increase targeting precision. After entertaining the input assumptions, we should develop a network of turbines, similar to Fig. 4.16 or Fig. 4.17, in combination with letdown stations between headers. Thus, we consider steam generators and consumers to obtain the maximum power generation potential of the network and fuel consumption.

After developing the network, we maximize the mass passing rate through every turbine and consider the rate of mass passing the letdowns between two headers at zero. For fuel consumption targeting, the power generation potential should be examined in a condition wherein the power generation network operates appropriately. For this purpose, the mass variations in every header should be

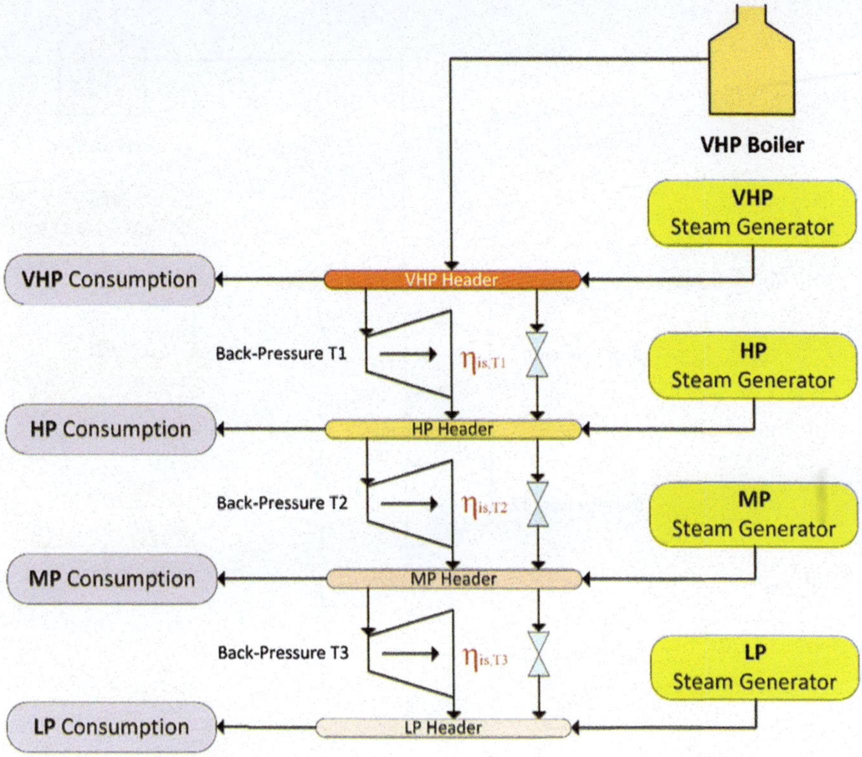

Fig. 4.16 Targeting network of a cogeneration system regardless of a condenser turbine

adjusted to its superheating rate, so the moisture formation owing to pressure drop in tubes is prevented. Moisture destroys the blades of turbines. Thus, if the last header is superheated, the superheating of other headers is also guaranteed. Hence, as the targeting constraints of a power generation network, we consider that the last header is superheated (10–20 degrees over the saturation temperature) or the specific mass fraction is about 10% of the steam mass fraction in the header of the saturated water. Then, the mass rate calculation passing the turbine starts from the last header. This mass can be estimated by the header-surrounding mass balance, similar to Eq. (4.32) in every header targeting.

$$\sum \dot{m}_{\text{Turbine}} + \sum \dot{m}_{\text{Letdown Station}} + \dot{m}_{\text{Steam Generation}} - \dot{m}_{\text{Steam Use}} = 0 \qquad (4.32)$$

When the generation and consumption rates are known in a header, we can obtain the rate of turbine-passing mass using the mass balance around the header with lower pressure. Using the energy balance around the header with lower pressure, according to Eq. (4.33), we determine the rate of enthalpy exiting the turbine and entering the header with lower pressure. The enthalpy of the header with lower

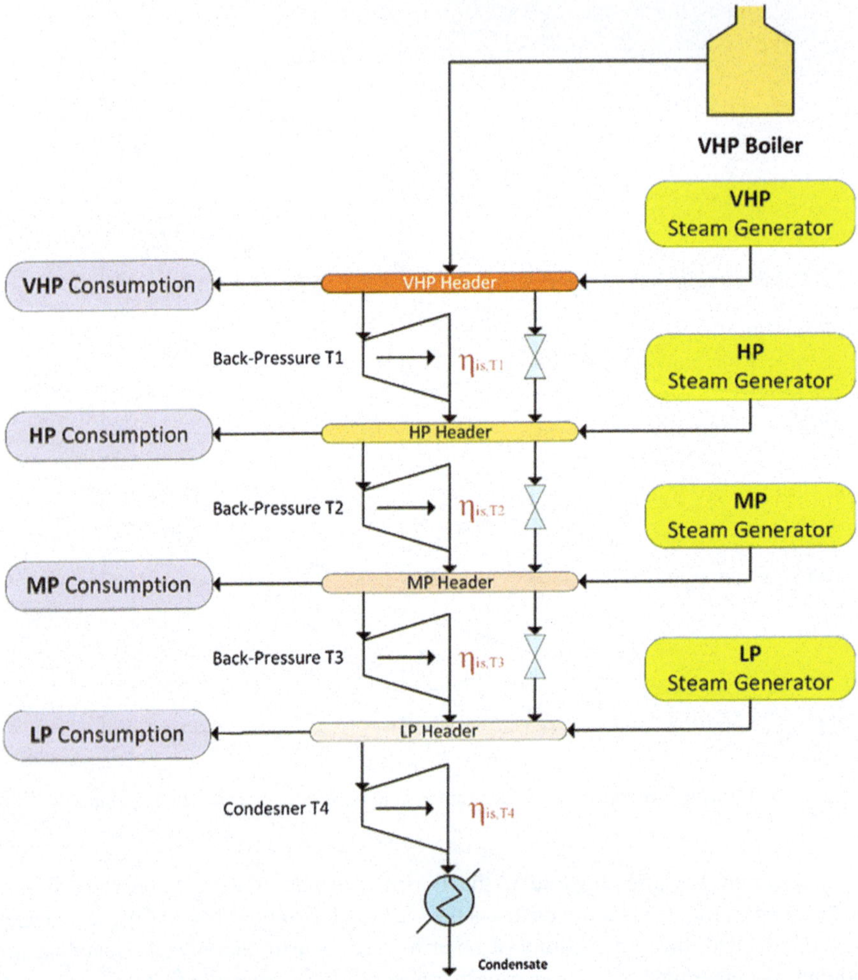

Fig. 4.17 Targeting network of a cogeneration system by regarding a condenser turbine

pressure at given pressure and temperature is obtained from previous stages or its targeting assumption.

$$\begin{aligned} & h_{\text{Turbine,input}}\dot{m}_{\text{Turbine,input}} + h_{\text{Letdown Station,input}}\dot{m}_{\text{Letdown Station,input}} \\ & \quad + h_{\text{Steam Generation}}\dot{m}_{\text{Steam Generation}} \\ & \quad = h_{\text{header}}\left(\dot{m}_{\text{Steam Generation}} + \dot{m}_{\text{Turbine,input}} + \dot{m}_{\text{Letdown Station,input}}\right) \end{aligned} \tag{4.33}$$

We can compute the maximum efficiency of the turbine by accessing the header enthalpy entering the steam turbine, according to Eq. (4.15), and the mass flow of the steam passing through the steam turbine. Next, we use Eq. (4.12) to calculate the rates of the steam turbine-exiting enthalpy and generation power of the steam turbine.

To estimate the enthalpy of the unknown header, we can use numerical methods and conjectures for the initial rate of enthalpy and modify it until it equals the output header enthalpy. We can also estimate its temperature by having the enthalpy and pressure of the unknown header. Then, the steam consumption targeting should be repeated, so the new steam generation rate is acquired according to the new header temperature. This process continues until all conditions of headers are observed. After calculating steam temperature in the header with the highest pressure, we should modify the primary steam temperature, allow the passage of less mass flow from the turbines, and maintain some part of it for the header output letdowns if the temperature exceeds the bearable temperature, which should be about <600 °C, for a steam distribution system. Accordingly, we observe the term of superheating headers by injecting steam directly from high-pressure headers. This process should continue by try and error until all system constraints are observed. We can see Fig. 4.18 depicts the targeting process briefly.

4.9 Criteria for Design and Optimization of the Total Site Utility System

According to the subjects presented in the last sections, selecting the most optimal cogeneration system in terms of its structure and efficiency needs some criteria for system evaluation. This section introduces some criteria for the design and optimization of the Total Site utility system.

4.9.1 Evaluating Cogeneration Systems

The essential criterion in optimizing a cogeneration system is that the system should be economical besides supplying the anticipated needs of the system. The efficiency of a system can be opted for as a criterion for its selection. However, system efficiency, by itself, cannot suffice the selection of the most optimal system or the most optimal condition of the existing system. For example, several fuels may be present in some systems, leading to lower efficiency due to their lower calorific value while optimizing the system economy more. Likewise, optimal fuel consumption is related to the better efficiency of a system. However, in some systems, newer technologies, which raise system costs, are employed for increasing efficiency.

The efficiency of a cogeneration system can be explained by Eq. (4.34) [8, 10].

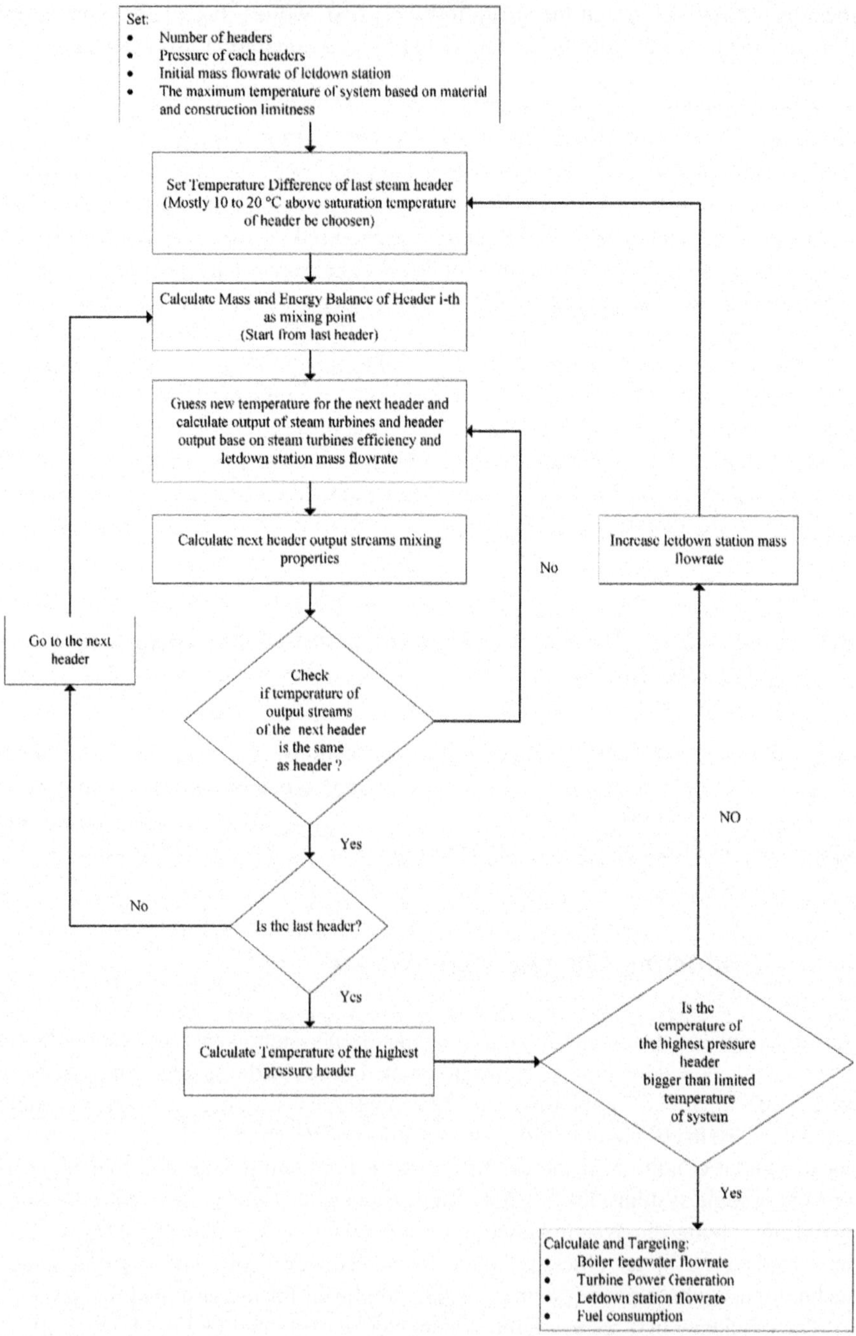

Fig. 4.18 Algorithm for targeting steam network

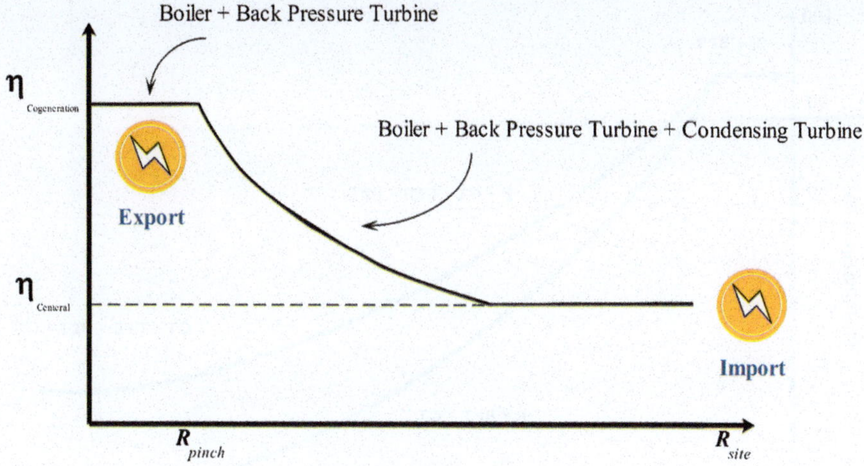

Fig. 4.19 R-curve graph for a power generation network

$$\eta_{\text{Cogen}} = \frac{W_{\text{Site}} + Q_{\text{Site}}}{Q_{\text{Fuel}}} \tag{4.34}$$

In this equation, Q_{fuel} refers to the consumed fuel of the service-provider unit. It includes both the fuel consumed for heat generation and the one used for power generation. Furthermore, all heat recoveries within the system are thermally integrated as a fuel with other system-entering fuels.

To determine the most optimal structure in terms of the system components' arrangement, we can utilize another criterion, namely, the R-curve graph. In this graph, shown in Fig. 4.19, the R-value is obtained by Eq. (4.35).

$$R_{\text{Site}} = \frac{W_{\text{Site}}}{Q_{\text{Site}}} \tag{4.35}$$

To draw this graph, we assume that the input and generated power needed by the site equals zero. Then, the *R*-value rises with an increase in the workload needed by the site. The system efficiency equals the thermal efficiency of the system at first when power is zero. Thereafter, we can raise the system efficiency to its maximum by increasing *R* and employing the full potential of steam for power generation. This efficiency is achieved due to the full recovery and use of steam potential in the system. In this condition, the *R*-value is called R_{Pinch} since heat recovery is at its maximum level. Next, an increase in the site power demand leads to further fuel injection for power generation.

The considerable point regarding the use of R-curve is the designation of the point of getting electricity from the local or national distribution network. If measured economically, the importation of electricity from the national network is

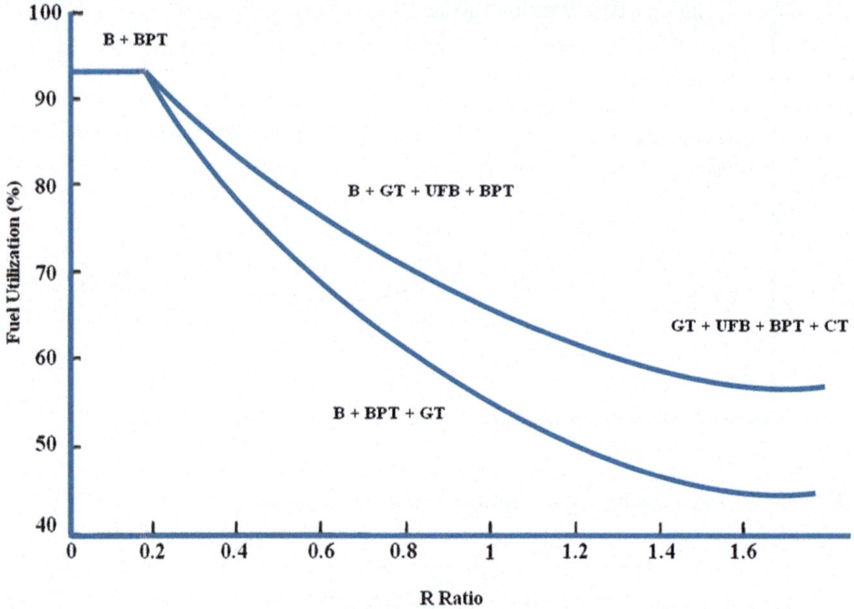

Fig. 4.20 R-curve graph for comparing different structures to supply the required power of the site

cost-effective when the utilization of the national network electricity is cheaper than electricity generation at the Total Site location in terms of expenses. Other reasons, strategical approach to power generation, risk appetite, and other relevant noneconomic reasons can economize electricity generation at the site location. However, the importation of electricity from the national network is cost-effective when the cogeneration efficiency in the site equals the efficiency of power generation in the national electricity distribution network.

Figure 4.20 compares different structures for supplying the site's excess power in relation to the R_{Pinch} condition. Accordingly, we can do as Fig. 4.20 for selecting the optimal structure of a considered cogeneration system to supply the requisite excess power [8].

4.9.2 Evaluating and Modeling the System Economically

The most important target respecting designing a Total Site and supplying the expected need is its analysis and optimization with an economic approach. Some factors, such as being strategic politically and martially besides other noneconomic factors, can blur the economic factors. However, designing a system, it is ultimately its economy that forms an important factor after need supply. In designing a

Total Site, the economy can be reckoned as the most significant factor in system optimization due to the efficiency variability of heat and power generation and other system conditions. The present study also adopts an economical approach to investigate the posed theory in targeting heat and power generation and refrigeration systems.

The economic objective function in this research is the total annualized cost (TAC). Based on this objective function, the whole equipment prices spent for the network and its efficiency and maintenance in a year are equalized. Then, a minimization of the value of the economic objective function is needed for network optimization.

The costs of the network equipment constitute the following cases in the present study:

- Back-pressure steam turbines
- Condensing steam turbines
- Gas turbine package
- Heat recovery steam generator
- Generators applied in the refrigeration network, including the condenser, evaporator, and intermediate coolers
- Boilers utilized in the network
- Deaerator cost
- Generator cost for electricity generation

For the total annualized price of the equipment used in network designing, we can employ the total annualized factor (AF), shown in Eq. (4.36), and annualized capital costs (ACC [m$]), in Eq. (4.37), to equalize the capital needed for all equipment according to their efficiency longevity and interest rate where C is capital cost [m$] [15].

$$\mathrm{AF} = \frac{\mathrm{IR} \times (1 + \mathrm{IR})^n}{(1 + \mathrm{IR})^n - 1} \tag{4.36}$$

$$\mathrm{ACC} = \mathrm{AF} \times \sum \mathrm{CC}_{\mathrm{Components}} \tag{4.37}$$

Likewise, to calculate and apply the maintenance cost (MC) to the system, according to references, we can consider 6% of the equipment investment as the annual maintenance costs. On the other hand, the cost of installing power equipment equals 14% of the equipment costs according to reference.

Since the capital cost of equipment is not up to date in different references, we can use the method presented in Baakeem et al. paper and apply Eq. (4.38) to estimate the present prices of the equipment resorting to its base price in the past and multiplying it by the price change index presented by varying references, like Marshall and Swift [2], as a reference for measuring price changes. In this equation, the α value differs for varying equipment; however, we can find it out using various references. According to a general law, namely, the law of one-sixth, the value of α is usually considered at 0.6.

$$CC_{new} = CC_{base}\left(\frac{Capacity_{new}}{Capacity_{base}}\right)^{\alpha} CI \tag{4.38}$$

Where the CI value indicates the ratio of the price stated by some references, such as Marshall and Swift, this value can be calculated by Eq. (4.39).

$$CI = \frac{\text{Cost index for required year}}{\text{Cost index for redrence year}} \tag{4.39}$$

The current expenses in the discussed network of a Total Site comprise the below cases:

- The cost of consumed fuel
- The cost of supplying and refining consumed water
- The cost of the cooling water in the system
- The cost of the electricity purchased from the national network
- The cost of electricity sold to the national network
- The cost of the tax on the carbon produced by the burned fuel in the system

We can employ the equations presented by different references to estimate the equipment prices of a cogeneration system.

We can also use Eq. (4.40) to calculate the rate of carbon produced by the fuel in a cogeneration system and the rate of tax over the emitted carbon where GHG, EF, and M are greenhouse gases [tCO_2. $year^{-1}$], carbon emission factor per unit [$tCO_2.t_{fuel}^{-1}$], and mass flow [t. h^{-1}], respectively, and time is the number of hours the system operates in a year. Moreover, the compensatory water of the system W_{makeup} [t. $year^{-1}$] can be computed by Eq. (4.41), wherein U_{steam} [t. h^{-1}] is the rate of steam utilized by the existing processes of a Total Site and CRR and BR are condensate return rate and vent rate, respectively.

$$GHG = M_{fuel} \cdot EF_{fuel} \cdot time \tag{4.40}$$

$$W_{makeup} = U_{steam} \cdot (1 - CRR + BR) \cdot time \tag{4.41}$$

After calculating the equipment prices, operating costs, and other expenses of the system, we can use Eq. (4.42) to estimate the total annualized cost [m$] of a system to exploit it as an optimization objective function.

$$TAC = AF \times \text{Capital Cost} + \text{Operating Cost} + \text{Maintenance Cost} + \text{Environmental Cost} \tag{4.42}$$

4.10 Refrigeration and Steam Network Coupling Through Total Site Integration

The energies required by every process site include mechanical, electrical, heating, and cooling energies. The cooling energy is, in fact, a kind of thermal energy with a negative potential. Due to being emphasized in the present book, it is called refrigeration.

Today, refrigeration networks and systems are extensively employed in many process industries. Some or entire parts of many process industries operate below the environment temperature and need refrigeration networks and systems to transfer heat. Among the refrigeration network applications, we can refer to the processes of liquidation and separation of gases (for instance, liquidation of natural gas, separation of ethylene, and so forth.), gas condensations, crystallization of solids, control of reactions, preservation of foods, and air conditioning.

Except for thermal, mechanical, and electrical energies that give rise to the low to high costs, respectively, one of the most expensive consumed energies is refrigeration. It is because refrigeration supply within the thermal range required by the process industries needs complicated conditions and equipment. For the supply of refrigeration, the temperature of the refrigerant is reduced by pressure-reducing equipment. With a decrease in temperature, the refrigerant can absorb the thermal energy from the heat source at temperatures significantly lower than the environment temperature and give this heat to the environment or another flow, like cooling water, using a compressor or absorption pressure-increasing cycles. The supply of energy needed by compressors for pressure enhancement is very costly in terms of investment and consumption. That is why the expenses of a process industry with refrigeration networks and systems are very high. Thus, numerous papers and studies have attempted to optimize energy consumption by refrigeration networks and systems and design fewer consuming cycles; therefore, this topic has been changed to a challenge in designing these industries. This chapter explains some methods for targeting, designing, and evaluating power generation systems in Total Sites.

4.10.1 Refrigeration Networks

The refrigeration cycle and system aim to create cooling below the environment temperature. It is because it can supply cooling at temperatures equal to and above the environment. Thermodynamically, a cooling cycle is in charge of taking heat from the low-temperature heat source and transferring it to a high-temperature heat sink, which can be a flow with a higher temperature, air, or cooling water. Heat pumps also have structures similar to refrigeration systems. The mere difference is that their target is to receive heat from a low-temperature source and transfer it to a high-temperature one.

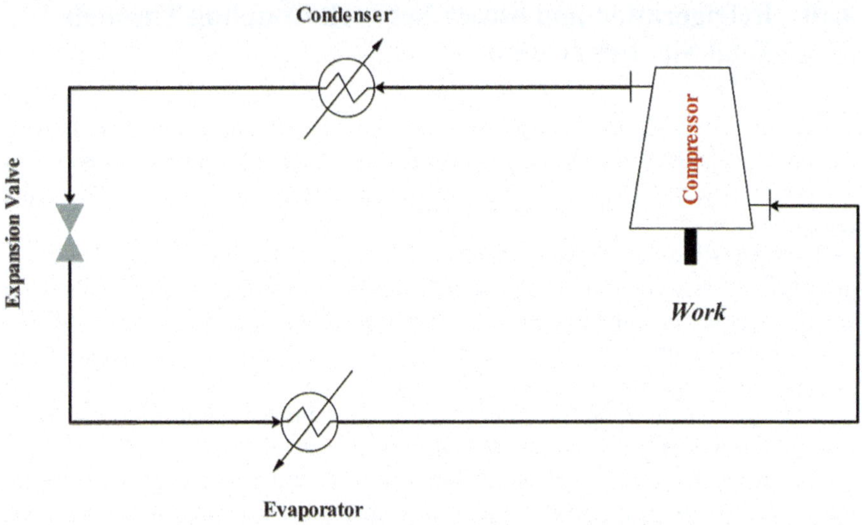

Fig. 4.21 Schematic view of a mechanical compression refrigeration cycle

Cooling cycles can be categorized into mechanical compression refrigeration, absorption refrigeration, adsorption refrigeration, desiccant refrigeration, and steam jet systems. Due to their ability to supply a wider range of temperatures lower than the environment temperature and temperatures significantly below the environment, mechanical compression refrigeration is used in low-temperature industries more.

Figure 4.21 shows a simple example of a compression cycle. As it is clear from the figure, mechanical compression refrigeration consists of four main parts:

- An operator, which receives heat from a low-temperature heat source and saturates the refrigerant
- A compressor, which raises the pressure of the refrigerant from the pressures of the operator and condenser
- A condenser, which decreases the temperature of the superheated refrigerant to the saturated liquid condition
- An expansion valve, which reduces the pressure of the refrigerant to the pressure of the operator

The refrigeration temperature and cooling rate are among the crucial factors in designing a mechanical compression refrigeration cycle. The lower is the temperature of refrigeration, the higher is the ratio of pressure required by the compressor. This increase raises the power consumption of the compressor. Likewise, as the required cooling rate enhances, the amount of power consumed by the compressor increases due to the upsurge of the mass flow of the refrigerant in the refrigeration cycle. In low-temperature process industries, multistage or cascade cooling cycles are exploited due to a reduction in refrigeration compressors' consumed power or a need for multistage cooling.

4.10.2 Types of Mechanical Compression Refrigeration

Mechanical compression refrigeration cycles are themselves categorized into several groups in terms of their cycle structures. The simplest type is a single-stage refrigeration cycle, whose scheme is shown in Fig. 4.21. These cycles are suitable for supplying cooling up to −40 °C; however, more complex structures are used for developing lower thermal or different ranges in a cooling cycle. Different structures of the refrigeration cycle are as follows:

- Multistage mechanical compression refrigeration in combination with a separator or economizer
- Multistage mechanical compression refrigeration in combination with a steam cooler or pre-saturator
- Cascade mechanical compression refrigeration
- Integrated refrigeration

The schemes of multistage mechanical compression refrigeration cycles resemble the scheme displayed in Fig. 4.22. As observed, these cycles consist of more than a compressor and generate cooling in more than one pressure. In these cycles, the refrigerant enters a separator after passing through an expansion valve. In this stage, the separator separates the steam in the refrigerant and lets it move toward the next pressure-reducing valve or operator while it is a saturated liquid.

An important point in designing these cycles is to reduce the operational work of the compressor by venting the steam in the refrigerant passing through the pressure-

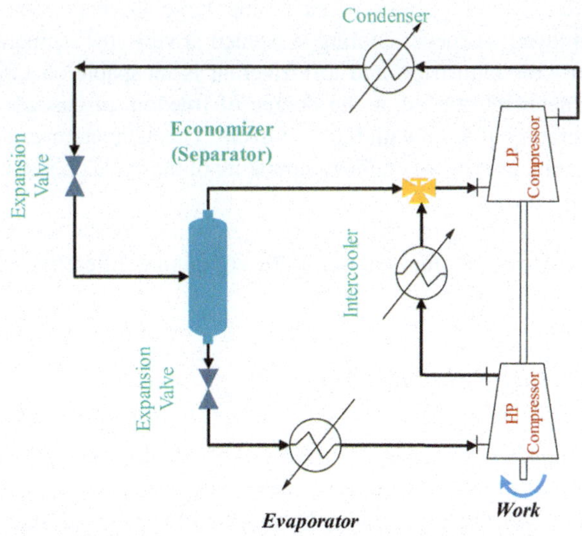

Fig. 4.22 Multistage mechanical compression refrigeration cycle, the pressure-enthalpy graph of a multistage mechanical compression refrigeration cycle

reducing valve. Since the refrigerant's steam is ineffective in absorbing heat from low-temperature flow, it's venting before it passes through the pressure-reducing valve decreases the operational work of the lower-pressure compressor. That is why the operation coefficient of these cycles increases. An approach to reducing the operation consumption rate in multistage cycles with separators is to use an intercooler, which reduces the temperature of the compressor-leaving refrigerant. Thus, the operation rate required for the next-stage compressor declines due to a decrease in temperature and an increase in the liquid phase in the refrigerant. Furthermore, another benefit of these cycles is the exploitation of many generators compared to the simple state. For example, the evaporation and boiling of the refrigerant can accompany some problems in finned plate generators.

Another type of multistage cycle is the one in combination with a pre-saturator. This cycle also operates with a separator. The difference is that the pre-saturator acts differently in this cycle. Furthermore, the compressor output in every stage enters the pre-saturator of the precedent stage. The superheated steam leaving a low-pressure compressor enters the pre-saturator, contacts the refrigerant entering from the pressure-reducing valve, and separates its pressure. Likewise, some part of the compressor-exiting steam, taking a form of a saturated liquid, moves toward the next pressure-reducing valve accompanied by the refrigerant. On the other hand, unlike multistage cycles with separators, only saturated steam moves toward the high-pressure compressor. This issue reduces the operation work of high-pressure and low-pressure compressors.

Cascade cycles are among other complex types of refrigeration cycles. These cycles are formed by the integration of several generation cycles of refrigeration. The scheme of this cycle is shown in Fig. 4.23. The purpose of using these cycles is the limitations of refrigerants in supplying large thermal ranges. Multistage cycles are employed whenever cooling is needed at very low temperatures in low-temperature process industries, and a refrigerant is incapable of supplying it. An effective parameter introduced as the degree of freedom of cascade cycles is the juncture temperature of them with the next cycle. This temperature, called partition temperature, is obtained based on the temperature of the cycle operator with a higher temperature.

Furthermore, to develop and fabricate refrigeration networks, we can obtain more complex cycles and higher thermal ranges by combining different cycles [18].

4.10.3 Selecting a Refrigerant

In refrigeration cycles, selecting a suitable refrigerant is a key parameter after its design in terms of the network structure. In respect to the selection of a suitable refrigerant, it is essential to notice three properties of a refrigerant. The first is its freezing point. A refrigerant cannot supply cooling at temperatures below its freezing point. Thus, the operator's temperature should be higher than the refrigerant's freezing point temperature, which should be examined in the operative

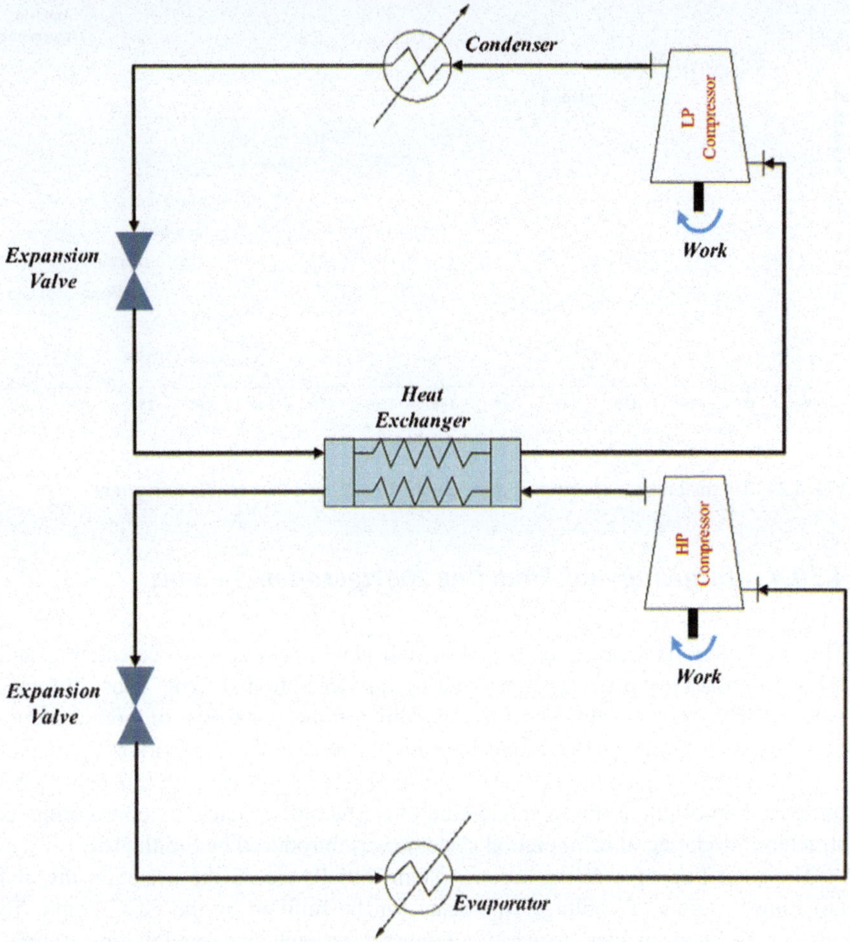

Fig. 4.23 (**a**) Scheme of a cascade mechanical compression refrigeration cycle, (**b**) pressure-enthalpy graph of a cascade mechanical compression refrigeration cycle

pressure of the operator. The second very crucial property in selecting a refrigerant is its rate of latent heat. Increases in the rate of the latent heat of a refrigerant minimize its need for mass flow to absorb heat from low-temperature flows owing to the upsurge of its heat absorption capacity. Another influential parameter in selecting a refrigerant is preventing vacuum pressure in the operator. For this purpose, the refrigerant's pressure should be above the atmospheric pressure at the operating temperature to preclude effusion hazards. The scheme in Fig. 4.24 depicts thermal ranges for the selection of a suitable refrigerant.

In selecting a suitable refrigerant, other general points should also be considered. For example, it is being nonpoisonous, inflammable, and noncorrosive, as well as enjoying low potentiality in global warming and destroying the ozone layer [18].

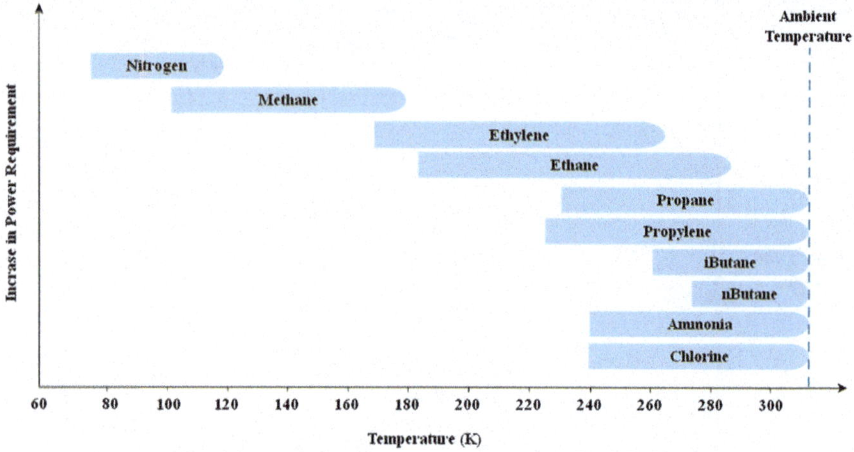

Fig. 4.24 Thermal ranges of common refrigerants for selecting the suitable refrigerant

4.10.4 Targeting and Modeling Refrigeration Systems

The most important target in modeling refrigeration systems is calculating and estimating the rate of power consumed by a refrigeration system. Thus, different methods have been developed for the modeling of these systems. In more accurate methods, the presence of the initial structure of a refrigeration network is significant to estimate or compute the network's power. Hence, to investigate this theory, the present study examines just a single kind of refrigeration system by considering its structure. Modeling of refrigeration systems was introduced by Smith [18].

The modeling of a refrigeration system initially needs the system's thermal targeting in terms of cooling. This issue can be fulfilled by the GCC graph. To this end, we can consider evaporation temperatures and condensation temperatures as cold and hot utility in the GCC graph, respectively, similar to Fig. 4.25, and then calculate the refrigeration load based on the heat surplus and deficit of streams of the site.

4.11 Total Site Integration Approach for Coupling Desalination and Steam Network

Concerning the conducted predictions, the need for water will considerably increase in the future years. This will impose further pressure on water resources. One of the introduced solutions is to use water desalination systems, in which energy is

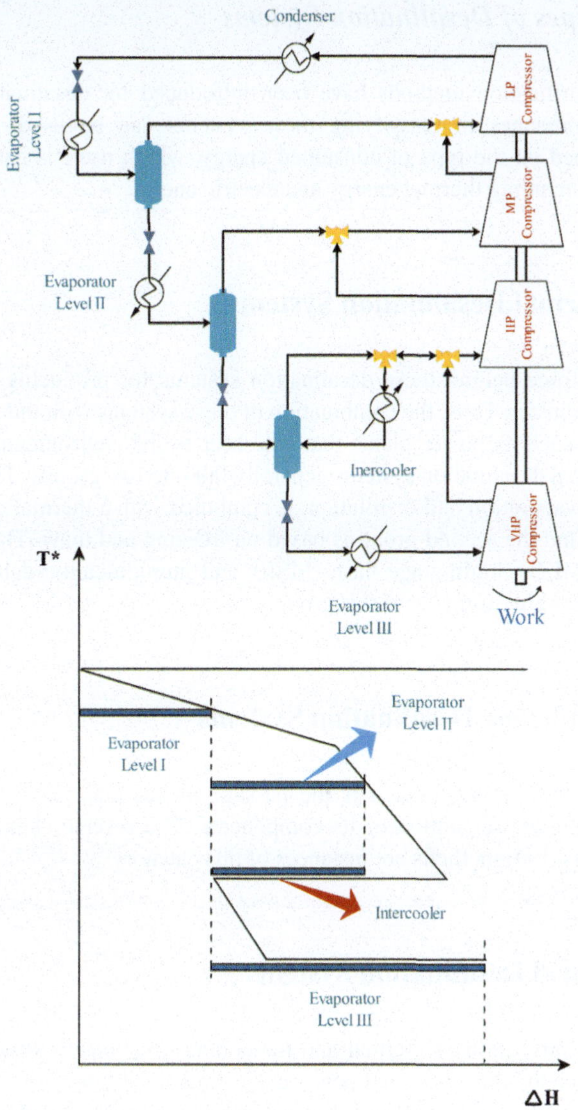

Fig. 4.25 Targeting the required cooling by GCC graph

reckoned as a key parameter in their design. It is because a large-scale consumption of energy can lead to problems in the energy sector. Total Site integration is one of the approaches to meeting two energy and water constraints while increasing efficiency and decreasing utility consumption.

4.11.1 Types of Desalination Systems

Numerous purification methods have been introduced for desalinating seawater. Different approaches to classifying these systems can be used, for example, classified based on the type of consumed energy, which itself is categorized into two groups, including thermal energy and electric energy.

4.12 Thermal Desalination Systems

The driving force behind these desalination systems for producing freshwater is heat or heat transfer. Thus, the combination of these systems with industrial devices and designs, such as power plants that lose heat to the environment, seems very useful. Thermal desalination systems generally fall into two groups. The first group, founded on evaporation and distillation, is combined with a thermal or mechanical compressor, and the second group is based on freezing and thaw. The multi-effect distillation (MED), multi-stage flash (MSF), and humidification-dehumidification (HDH) systems are examples of this group.

4.13 Membrane Desalination Systems

The fundamental of this system is the passage of the input flow from a set of membranes for the separation of its components. The reverse osmosis (RO) and electrodialysis (ED) methods are instances of this category.

4.14 Hybrid Desalination Systems

Today, the combination of thermal and membrane desalination systems has been noticed, especially in the heat and power cogeneration systems.

Considering the energy and water crises in the future and due to the high energy consumption of desalination installations, utilizing the lost energies of other industrial centers and coupling energy-centered establishments, like power plants, can be introduced as a solution for water shortage. In this condition, desalination processes can use the energy resultant from heat recoveries or be integrated with the system and lead to the economization of energy consumption.

Of course, it should be noted that the establishment of desalination systems necessitates meticulous analysis and investigation due to the presence of some factors, such as fuel price, the price of energy carriers, price of generated water, environmental and water security costs, and so on. The ignorance of these factors makes it impossible to justify the final system establishment economically.

Along with some factors such as the reception rate of input salts and concentration of the output water, energy supply is among the paramount factors in selecting a desalination system to be integrated with a Total Site. Reliance on a single type of energy not only limits balance development in a power-heat cogeneration system but also reduces the system's accessibility. In contrast, the utilization of a coupled system dependent on power and heat, in addition to increasing flexibility in designing and receiving energy, facilitates system integration. Hence, in new systems, it is attempted to use a desalination-coupled system.

In the meantime, exploiting water pinch and Total Site as instruments used for integrating a desalination unit with an energy supply system reduces the water consumption of the process units and also has noticeable effects on energy consumption.

4.14.1 Targeting Process Consumption

Water pinch can be used as an approach to reducing water consumption. In this method, the minimum need of the system for a high-quality water supply is determined with respect to the system's rate of flexibility capacity in using regeneration, reuse, and recovery waters. The feedwater of the system should observe some constraints, including the maximum input and output concentrations and minimum flow rate. For this purpose, some methods are determined so that the establishment of these terms is assured. To determine the minimum feedwater using the water pinch method, we first consider the minimum water flow rate and maximum input and output concentrations of every process. Then, all processes are drawn in the concentration-mass graph of contamination, and the concentration ranges are determined. In the next stage, the water flow rate required in every concentration range is estimated from the sum of the flows within that range. The important point is the uniformity of the contamination material of the processes, and it should be noted that processes with varying contaminations should not be put together, and a separate pinch examination should be considered for every contamination. In fact, in a process site, there are water pinch analysis graphs whose numbers equal the numbers of contamination groups, and the rate of minimum feedwater is calculated by the addition of each one's minimum. After drawing the water limiting composite curve, we draw a line from the point whose length equals the contamination mass (which is zero for the feedwater) and width equals the concentration of input water. Then, the line slope gradually increases until it contacts the water limiting composite curve. The slope of the resultant line equals the minimum consumption water of the site for a contaminator with a specific material (Fig. 4.26).

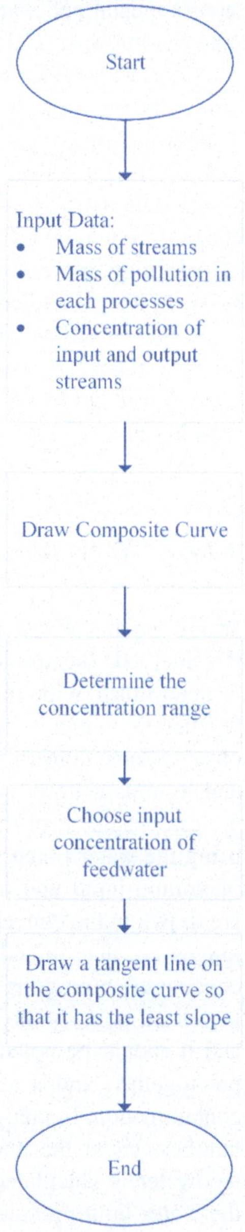

Fig. 4.26 Water pinch analysis algorithm

4.14.2 The Effect of Desalination Systems on Analyzing Cogeneration Potential

So far, we have dealt with the significance of desalination systems, their roles in cogeneration, and targeting water consumption. It should be noted that water pinch enables the determination of the minimum water required for process units. Now, if water generation and its sale to the network are economically justified, the generation of freshwater will increase to meet the other existing constraints, including the total economy.

The inclusion of desalination systems at the heart of cogeneration systems makes changes in the energy graphs and some network measurement tools. Among them, we can refer to the CC, GCC, TSCC, R-curve, and cogeneration efficiency (η_{cogen}) graphs.

The use of steam utility makes changes in the system equilibrium and takes the system to a new balance point regarding the level it is employed.

As a result of these alterations, the generation and consumption rates of the utility undergo some changes at various levels.

However, the exploitation of membrane desalination systems exerts their impact on the network not through utility consumption, but power consumption. Consequently, in both cases, the utilization of thermal or membrane desalination systems leads to changes in the rates of power and heat generation and consumption, which in turn alters the cogeneration efficiency. On the other hand, variations in the system equilibrium make the R-curve, which indicates the optimal states of the network for power and heat cogeneration, change. The magnitude of these changes depends on the energy consumption share of desalination systems out of the total energy in the utility network flow.

4.14.3 Environmental Effects

Besides the changes desalination systems bring about in heat and power cogeneration networks, their roles in the surrounding environment should also be investigated. Today, the consideration of every system's effect on the environment is noticed more than ever due to excess attention given to environmental problems compared to the past.

As Fig. 4.27 depicts, one of the most important effects of desalination systems is their brine discharge. Every desalination system encompasses one input and two output flows, one of which is freshwater flow, and the other is brine discharge with a high rate of salts. The entrance of this flow, regardless of any constraint, causes negative changes in the surrounding environment. Similarly, the entrance of uncontrolled wastewater flow increases the salinity of water resources and soil, directly affecting the lives of waterfowls, as well as agriculture. One of the suggested solutions is to treat the wastewater flow and generate usable products, such as

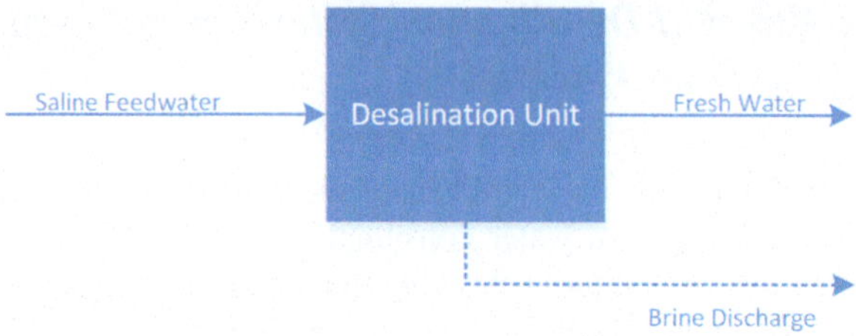

Fig. 4.27 General scheme of desalination systems

consumption salt. However, this approach itself increases the cost and complexity of a total system.

4.15 Conclusion

Concerning the pounded subjects proved to the world that the industry sectors, especially the process industries, play a significant role in producing environmental contaminators and altering the climate. Also, by having the biggest share in energy consumption, even the slightest changes in this sector cause a huge impact on world economics. On the other hand, the most important factor, which reduces energy consumption and thus leads to decreases in generation costs and environmental contaminators, is to improve and modify the efficiency of energy systems. In order to know the importance of this need in the process industries, this chapter presented the Total Site integration approach and delineated the targeting and modeling of different components in designing the heat and power network of a Total Site. Next, how a Total Site interacted with refrigeration networks and desalination systems was explored.

References

1. Ahmadi, P., & Dincer, I. (2011). Thermodynamic analysis and thermoeconomic optimization of a dual pressure combined cycle power plant with a supplementary firing unit. *Energy Conversion and Management, 52*(5), 2296–2308. https://doi.org/10.1016/J.ENCONMAN.2010.12.023
2. CoreLogic, I. (2021). Inventory Index Factors 2021.
3. Čuček, L., Boldyryev, S., Klemeš, J. J., Kravanja, Z., Krajačić, G., Varbanov, P. S., & Duić, N. (2019). Approaches for retrofitting heat exchanger networks within processes and Total Sites. *Journal of Cleaner Production, 211*, 884–894. https://doi.org/10.1016/j.jclepro.2018.11.129
4. IEA. (2020). *Energy efficiency 2019*. https://doi.org/10.1787/ef14df7a-en
5. IEA. (2020). *Global energy & CO2 status report 2019*. https://www.iea.org/reports/global-energy-co2-status-report-2019

6. IEA. (2021). *World energy outlook 2020*. https://www.iea.org/reports/world-energy-outlook-2020
7. Kemp, I. C. (2007). *Pinch analysis and process integration: A user guide on process integration for the efficient use of energy* (2nd ed.). Butterworth-Heinemann. https://www.elsevier.com/books/pinch-analysis-and-process-integration/kemp/978-0-7506-8260-2
8. Kenney, W. F. (1984). *Energy conservation in the process industries*. Academic Press.
9. Khoshgoftar Manesh, M. H., Navid, P., Blanco Marigorta, A. M., Amidpour, M., & Hamedi, M. H. (2013). New procedure for optimal design and evaluation of cogeneration system based on advanced exergoeconomic and exergoenvironmental analyses. *Energy, 59*, 314–333. https://doi.org/10.1016/j.energy.2013.06.017
10. Kimura, H., & Zhu, X. X. (2000). R-curve concept and its application for industrial energy management. *Industrial and Engineering Chemistry, 39*(7), 2315–2335.
11. Linhoff, B. (1993). Pinch analysis: A state-of-the-art overview: Techno-economic analysis. *Chemical Engineering Research and Design, 71*(A5), 503–522.
12. Manesh, M. H. K., Navid, P., Marigorta, A. M. B., Amidpour, M., & Hamedi, M. H. (2013). New procedure for optimal design and evaluation of cogeneration system based on advanced exergoeconomic and exergoenvironmental analyses. *Energy, 59*, 314–333.
13. Manninen, J. (1999). *Flowsheet synthesis and optimisation of power plants*. UMIST.
14. Mencarelli, L., Chen, Q., Pagot, A., & Grossmann, I. E. (2020). A review on superstructure optimization approaches in process system engineering. *Computers & Chemical Engineering, 136*, 106808. https://doi.org/10.1016/j.compchemeng.2020.106808
15. Oluleye, G., Jobson, M., & Smith, R. (2016). Process integration of waste heat upgrading technologies. *Process Safety and Environmental Protection, 103*, 315–333.
16. Ren, X.-Y., Jia, X.-X., Varbanov, P. S., & Klemes, J. J. (2018). Targeting the cogeneration potential for Total Site utility systems. *Journal of Cleaner Production, 170*, 625–635.
17. Shang, Z. (2000). *Analysis and optimisation of total site utility systems*. UMIST.
18. Smith, R. (2016). *Chemical process: Design and integration*. Wiley.
19. Sun, L., & Smith, R. (2015). Performance modeling of new and existing steam turbines. *Industrial & Engineering Chemistry Research, 54*(6), 1908–1915.
20. Sun, L., Doyle, S., & Smith, R. (2015). Heat recovery and power targeting in utility systems. *Energy, 84*, 196–206. https://doi.org/10.1016/J.ENERGY.2015.02.087
21. Varbanov, P. S., Doyle, S., & Smith, R. (2004). Modelling and optimization of utility systems. *Chemical Engineering Research and Design, 82*(5), 561–578.
22. World Bank. (2021). *World development indicators*. https://data.worldbank.org/indicator/EN.ATM.CO2E.KT?end=2018&start=1990&view=chart

Chapter 5
Investigating the Effective Methods in Improving the Resilience of Electricity and Gas Systems

Mohammad Mehdi Amiri, Hossein Ameli, Mohammad Taghi Ameli, and Goran Strbac

5.1 Introduction

One way to use existing infrastructure more efficiently is to consider an integrated energy system (IES) from planning and operation perspectives. Integrating the system and the cooperation between different energy vectors can lead to better management of system challenges. Researchers believe that collaboration between different energy carriers plays a positive role in improving the grid performance.

IES can be defined as a whole system for the transmission and storage of various energy vectors, an intermediary between the supplier, the consumer, and the energy transmission infrastructure. This concept can realize optimization for the operation of multi-vector energy systems. IES is an infrastructure system combined with subsystems such as electricity, heat/cooling, and gas/hydrogen. An IES is considered a pivotal option to create more flexibility in energy system operation and improving the overall efficiency for suppliers and consumers. Typically, the energy vectors are operated independently. In Fig. 5.1, an IES consisting of electricity, gas, cooling, and heating systems is presented [27].

The impressive increase in electricity production from gas has had extraordinary consequences for both the electricity and gas industries. This interdependence between the two industries requires a simultaneous operation of these two systems, which increases the energy system's efficiency.

M. M. Amiri (✉) · M. T. Ameli
Shahid Beheshti University, Tehran, Iran
e-mail: Moh_amiri@sbu.ac.ir

H. Ameli
Control and Power Group, Imperial College London, London, UK

G. Strbac
Department of Electrical and Electronic Engineering, Imperial College London, London, UK

© The Author(s), under exclusive license to Springer Nature Switzerland AG 2022
V. Vahidinasab, B. Mohammadi-Ivatloo (eds.), *Whole Energy Systems*, Power Systems, https://doi.org/10.1007/978-3-030-87653-1_5

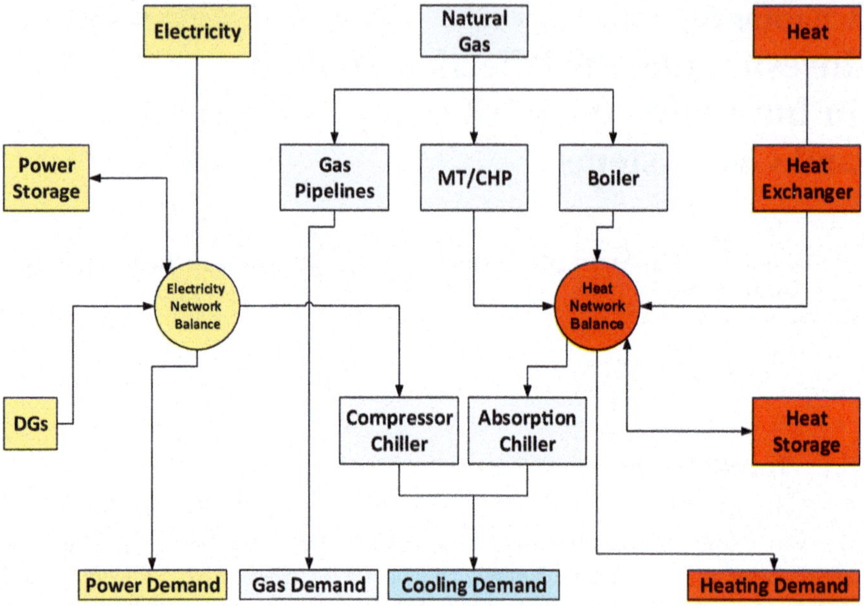

Fig. 5.1 Integrated energy system (IES)

In addition to benefits such as relatively low carbon emissions, low investment costs, and high efficiency, gas-fired power plants respond fast. It is expected that by increased use of renewable energy, the rapid response of these power plants will be able to maintain the balance between demand and supply. With renewable energy and gas-fired power plants development, the electricity and gas network (EGN) operation becomes significantly interdependent. Therefore, the simultaneous operation of EGN is of particular importance [46].

In recent years, the interdependency of EGNs has significantly increased from the transmission level to the distribution level from techno-economic perspectives due to the role of distributed gas production units [4, 18, 19, 48].

Environmental considerations to reduce greenhouse gas emissions, reduce construction time and investment costs of combined cycle production units, and increase the installed capacity of renewable generation have led to advances in gas-fired generators (GFG) technology in power grids. In some countries, gas/hydrogen is used as the primary source of electricity. Increasing the installed capacity of this technology highlights the importance of interdependence between EGNs. Due to the connection between EGNs, the lack of gas supply can pose risks to the electricity supply [3]. Disconnection of gas pipelines and extreme weather conditions can hinder the gas network's regular operation. Reducing the existing GFG capacity further endangers the security and reliability of electricity systems [34].

In this study, after examining the need for communication between system operators of EGNs, the concept of resilience and its difference with reliability in EGNs will be introduced. This study aims to investigate the methods of improving resilience and comparing these methods in EGNs.

5.2 The Resilience of Electricity and Gas Networks (EGNs)

5.2.1 Resilience Concept

Although the purpose of defining this chapter is to examine the concept and methods of increasing the resilience of EGNs, due to the simplicity of the resilience concept and the stability of its definition, first definitions and general concepts of resilience and then electricity network resilience have been proposed.

In large systems with critical equipment, which can be disconnected from the network at any time, contingency analysis studies are performed continuously to ensure system security. Maintaining the security of an infrastructure system and making its performance resistant to low-impact and high-probability (LIHP) events has been familiar as a necessity of operation.

Power grids have been vulnerable to natural disasters, and also, the rate of these events has been lower in the past (due to global warming). Human life's dependence on electricity services' reliability has not been as significant as it is today. Currently, urban and even rural life is so tied to access to electricity that the occurrence of widespread blackouts, albeit short-term, has serious economic, social, cultural, and human consequences. Also, due to the dependence of the correct operation of infrastructure systems on each other, with the blackout occurrence for several hours in a specific area, the performance of other infrastructure systems such as gas/hydrogen networks, water supply networks, and telecommunication systems is disrupted, and the consequences of electricity blackouts become broader and more intense.

Obviously, in such a situation, maintaining the security of a power system in the event of low impact and high probability is no longer sufficient, and another feature of this system that reflects its behavior to high-impact and low-probability events should be considered. This feature of the system is called resilience, which indicates the degree of robustness, vulnerability, and recoverability of an infrastructure system in the event of a severe accident [16, 56].

In order to better define the resilience concept and its difference with the two concepts of system security and reliability, possible accidents and hazards can be divided into three categories based on the events probability of occurrence and impact [1]:

1. White Swan Events

A white swan is an extraordinary festive event with three principal characteristics: it is inevitable, its effect is predictable, and after the event, we provide an explanation that confirms the certainty. However, the white swan again shifts its focus to misjudgment or other forms of human cause.

2. Gray Swan Events

The gray swan is a potentially significant event that, while unlikely to occur, is not impossible. Therefore, although the probability of occurrence is very low, it should not be ignored. It can cause a crisis in the world, the economy, and the stock market if it happens.

3. Black Swan Events

A black swan is said to be an event that is unpredictable and goes beyond what is expected from a situation. The consequences can be potentially severe. Black swan events are characterized by extreme rarity, intense impact, and persistence; however, they can only be understood only after they occur.

The power system reliability indicates the possibility of its long-term satisfactory performance. It indicates the ability to provide adequate electrical services almost continuously and with short interruptions over a long period.

Security is the ability of electrical systems to withstand sudden disruptions such as unforeseen outage of system components or short circuits. A power system's security refers to the danger level in its capability to survive impending disruptions (unforeseen events) without interrupting demand and supply.

The difference between the reliability and security of power systems is summarized below:

- The overall goal in designing and planning a power system is system reliability. For a system to be reliable, it must be secure most of the time.
- Security is a time-varying attribute that could be investigated by studying the power system's operation in different specific set of conditions. On the other side, reliability is a function of the power system's time-average operation; it can only be analyzed by considering the system's performance over a considerable time.

It should be noted that network security analysis considers only single events (and perhaps some possible or dangerous dual events). Also, the reliability assessment examines only events that are likely to occur. Therefore, it can be concluded that the events studied in security analysis and reliability assessment are limited to the first category of events, namely, white swan events, which have both the probability of occurrence and the consequences of their occurrence that are essential.

Resilience has been introduced in the social, economic, and managerial fields before being introduced in engineering. In social systems, resilience refers to a society's ability to withstand the pressures and events caused by political, economic, and social changes in that society ([33]). Different companies can protect themselves from some economic losses in the market. In the field of management, the ability of an entity to identify various risks and manage events is called resilience [14]. As can be seen, the main emphasis of these definitions is on the lack of vulnerability to these events. These events in the power systems are the same as the second and third category events, namely, gray swan events and black swan events. These events, which can be of natural or human origin, occur infrequently or have never been experienced.

In a power system whose equipment is prone to failure and various inherent, natural, and human factors constantly threaten the equipment's health, one cannot hope for the system's correct operation only with preventive measures. In these systems, maintenance and repair of equipment are among the main pillars of the operation process. After an accident and system failure, service recovery and network structure repair are inevitable and, of course, practically possible. Therefore, the definition of resilience in infrastructure systems should include both accident endurance and rapid postaccident recovery.

5.2.2 *Resilience Definitions*

Since 2016, the definition of power system resilience has converged around a combination of concepts such as absorption, adaption, anticipation, and recovery from HILF events. Some definitions of resilience are discussed below. According to the Electric Power Research Institute (EPRI), resilience means prevention, recovery, and survival [31]. The National Infrastructure Advisory Council (NIAC) considers resilience to mean preparedness, planning, recruitment, recovery, and adaptation to disasters [13]. The UK Energy Research Center (UKERC) [10] calls "the ability of an energy system to withstand error and continue to provide low-cost energy services to consumers."

The United Nations International Strategy for Disaster Reduction (UNISDR) defines resilience as the ability of a system to maintain its performance in the event of an accident through learning from previous events [54]. The recovery and endurance characteristics of a system during and after the fault from the US National Association of Regulatory Utility Commissioners (NARUC) perspective is called resilience [23, 38]. A system's resilience is defined as the prediction and rapid recovery from low-frequency high-impact events in Panteli and Mancarella [39, 40].

Resourcefulness, robustness, adaptability, and rapid recovery characteristics of a system have been defined as resilience [45]. In Hollnagel et al. [14], resilience is defined through three characteristics: predicting, understanding, and responding. From Espinoza et al. [15] and Haimes [57] perspective, resilience means resistance, rapid recovery, and system adaptability in a disaster. In National Research Council [35], resilience has been defined as the planning, complete preparedness, recovery, and adaptation to adverse events. Resilience has also been identified as:

- Anticipate, perceive, respond, and adapt in Stevens [51], Rieger [47], and McJunkin and Rieger [32].
- The system's ability to reduce the magnitude and duration of the disruption [5].
- Being "the maintenance of the nominated state of security" [52].
- The ability to sustain and recover from extreme and severe disturbances [7].
- Avoidance, survival, recovery to a high-impact rare event [17].
- The ability of a system to degrade gracefully under extreme perturbations and recover rapidly [21].
- The ability of a system to anticipate and withstand external shocks [41, 42].

Other definitions of resilience are listed in reference [8].

5.2.3 The Importance of Resilience Assessment in EGNs

As we know, the issue of resilience is fundamental in various fields. "There have also been many research" on the issue of resilience in power grids. Now, let us consider an IES with different infrastructures instead of electricity networks, for example, EGNs. We are facing a new problem: the concept of the resilience of EGNs.

Assessing the overall resilience and increasing the resilience of IES has complexities due to different infrastructures' connections. The response of energy infrastructure to unforeseen events is mainly examined by each energy system separately. The importance of simultaneously examining the resilience of EGNs can be indicated by an example, in which after an accident in the gas network and the dependence between the gas network and power plants, the electricity system may suffer severe damage and finally blackout. Increasing the interdependence between different infrastructures can increase the system failure risks [9]. These crashes can be divided into two categories [11]:

1. Chain failures: Occur when a failure in one infrastructure will impact the other infrastructures (i.e., failure of one or more components in other infrastructures).
2. Common cause failures: Occur when two or more infrastructures are simultaneously affected by a common accident.

Considering the above and the importance of resilience in IES, resilience and increasing it in IES (e.g., EGNs) are investigated in this study.

5.3 The Resilience Enhancement Methods of EGNs

5.3.1 The Concept of Resilience Triangle and Trapezoid

The two fundamental concepts in resilience analysis of IES are the concepts of triangle and trapezoid resilience, which are used for modeling and resilience studies. It can be said that the resilience triangle is only able to evaluate the single-phase resilience and, in particular, to evaluate the recovery performance of a network after an accident, which is shown in Fig. 5.2. The resistance triangle cannot be used for all events, such as a tornado that may last from a few minutes to several days. Also, in this approach, if the accident lasts like an earthquake for a few seconds, it is indicated as a sharp decrease in the system's resilience level [11, 37].

In contrast to the resilience triangle is the resilience trapezoid shown in Fig. 5.3. The phases of the network and the time sequence of these phases and related events and the type of actions available are shown. Dividing events into three phases makes it possible to evaluate multistage resilience. Another advantage of a trapezoid is that it can be used for any danger or accident, regardless of its nature. Other differences between the resilience triangle and trapezoid are also shown in Table 5.1 [41, 42].

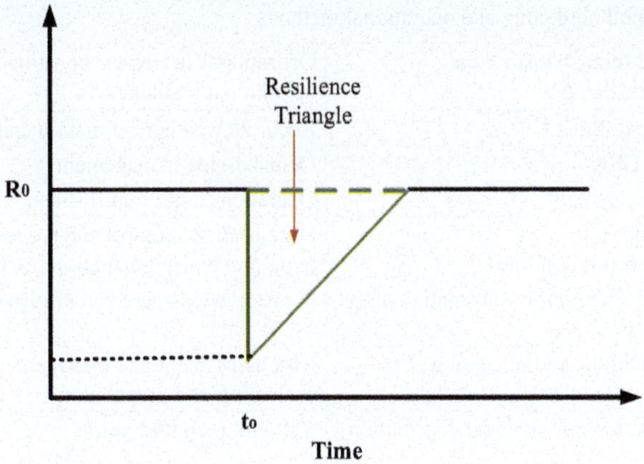

Fig. 5.2 Resilience triangle

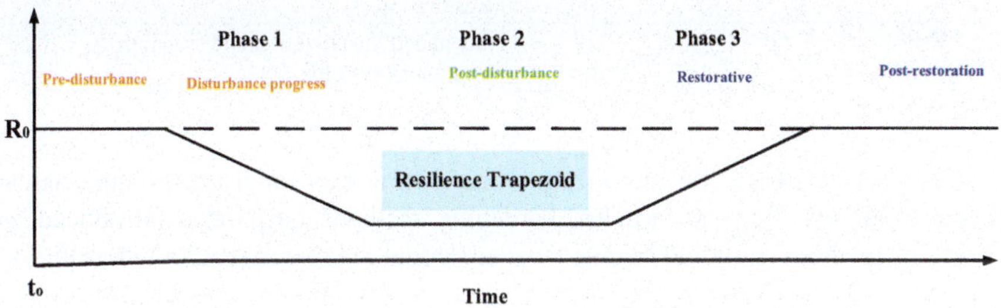

Fig. 5.3 Resilience trapezoid

Table 5.1 Differences between the resilience triangle and trapezoid

Resilience triangle	Resilience trapezoid
Assessment of single-phase resilience	Dynamic, multiphase resilience assessment
Lack of corrective actions throughout the event progress	Corrective actions
Absence of damaged condition after an accident	Post-disturbance degraded state and its duration consideration
Threat-specific	Applicable for any type of threats

5.3.2 Increasing the Resilience of EGNs

The vulnerability of EGNs is eliminated by identifying vulnerable components in electricity and gas/hydrogen infrastructure. The methods of increasing resilience used in articles can be divided into two categories including hardening and operational.

Hardening methods focus on creating network development plans to tighten transmission and distribution systems in future severe disasters. In contrast, oper-

Table 5.2 Several hardening and operational methods

Hardening measures: boosting the infrastructure resilience	Operational strategies: boosting the operational resilience
Black-start capabilities	Accurate weather estimation and forecast
Installation of DER	Demand-side management
Vegetation management	Operation of microgrid island
Undergrounding the distribution/transmission lines	Decentralized control and preventive control (e.g., preventive generation rescheduling)
Upgrading poles with more substantial, more robust materials	Disaster assessment and priority setting
Elevating substations and relocating facilities	Risk assessment and management for evaluating and preparation
Redundancy in transmission and distribution system	Network reconfiguration
	Distributed energy systems and energy storage (including gas storage)
	Restoration
	Advanced visualization and situation awareness systems

ational methods focus on optimization-based strategies for using existing assets against failures. In some articles, hardening methods have been introduced as planning methods. It should be noted that according to Preston et al. [45], retrofitting methods may make the system more resilient to operational measures but are not economically viable. Table 5.2 shows some hardening and operational methods [39, 40, 45].

It can be mentioned that most of the studies are related to increasing the resilience in the electricity networks, and fewer studies have considered EGNs simultaneously. References Jufri et al. [22], Li et al. [24, 26], Barnes et al. [6], and Nazemi et al. [36] are among the studies that have used hardening or planning methods to increase power system resilience. In contrast, several references such Poudel and Dubey [44], Li et al. [24, 26], Panwar et al. [43], Shi et al. [50], and Hussain et al. [20] have used operational methods to increase the resilience of EGNs.

Authors in Aldarajee et al. [2] introduce a new planner-disaster-risk framework for integrated planning of gas networks in major disaster events. The proposed approach takes into account the various components of vulnerable areas in EGNs as potential catastrophic events. It prioritizes diverse areas based on the associated vulnerabilities further to strengthen the resilience of the IES. Also, a new parameter (i.e., resiliency budget) is introduced to control the level of resilience in an IES based on an information-gap decision theory (IGDT) multi-incident, and multi-zone resilience is introduced.

The proposed framework using a three-level algorithm includes:

- Level 1: The problem of planning under normal circumstances.
- Level 2: Assessing resilience in the event of a severe accident.

- Level 3: Risk-free rescheduling in the event of a significant accident using the IGDT concept is resolved.

Reference Lv et al. [30] has proposed a multi-stage planning strategy to increase resilience, which as an emergency response source, heat storage is taken into account. The designed resilient planning framework consists of three steps:

- Rolling reserve optimization stage
- Day-ahead economic dispatch stage
- Fault restoration stage

In Shao et al. [49] as a method of hardening, the optimal planning of EGNs has been used. This programming is done with a two-stage robust optimization problem method:

Stage 1: Modeling the uncertainties and presenting a robust proposed model considering the constraints of the problem.
Stage 2: In the second stage, investment decisions are assessed using conditional events.

The proposed planning problem's objective is to minimize the total costs for the integrated operation and planning of EGNs.

In He et al. [18, 19], an effective hardening method is considered to increase the resilience of EGNs. For this purpose, a three-step robust optimization method is used. To solve this problem, the studied gas network's nonlinear constraints are modeled linearly by Taylor series expansion. In this paper, DGs and gas storage devices are considered as emergency response sources to supply high-priority gas/electricity demands during incidents. Reference Yan et al. [58] has used a coordinated regional-district operation model of IES with three-stage optimization and with two phases (i.e., preventive, corrective) in order to increase the resilience of EGNs. The steps of robust optimization could be summarized as follows:

- Extreme level: The extreme level minimizes the unit commitment cost (e.g., start-up and shutdown cost of thermal units).
- Middle level: The middle level finds the scenario in which the largest load curtailment occurred (worst-case scenario).
- Inner level: In this level, the committed units in the preventive stage will be employed to reduce/remove load curtailment in the worst-case scenario.

Another robust optimization method is presented in the reference Cong et al. [12]. The basis of this method and its purpose is to minimize the loads that are cut during the accident. This paper's resilience assessment is modeled as three stages or DAD (defender-attacker-defender) and is formulated as a min-max-min problem:

- Step 1: The defender optimizes the reinforcement in the transmission lines to reduce the damage caused by unexpected events.
- Step 2: Attackers create the worst type of attack by disrupting the system.
- Step 3: Defender responds to damage from attacks to optimize grid performance.

Reference Liu et al. [29] has used optimization methods to increase the resilience of EGNs. The method used in this study is the use of two-stage data-based optimization. The problem formulation consists of three levels.

- Lower level: At this level, the constraints of EGNs are presented.
- Middle level: The uncertainty set is built by combination of the data-based method with the hurricane's Markov property.
- Upper level: As a preventive action, system planners reinforce in the critical components in order to avoid outages.

In Tian et al. [53], an optimal allocation method for gas turbines and storage facilities in an IES is presented. Resilience assessment modeling in this paper is also based on defender-attacker-defender. The algorithm used to solve this problem is the progressive hedging algorithm (PHA). It should be noted that the power flow model used in this study is DC power flow.

One of the operational measures to increase resilience of EGNs is introduced in the reference Lin et al. [28]. In this study, by taking into account reconfiguration and DG islanding, a combined repair crew dispatch problem is introduced. This study assumes that EGN connection is through gas-fired DGs and electrically driven gas compressors.

The constraints considered for the power grid are:

- Nodal power balance
- Branch flow constraints
- Load shedding constraints
- Voltage constraints
- DG output constraints

Gas network constraints include the following:

- Gas well capacity constraint
- Nodal gas flow balance
- Active gas pipeline/compressor constraints
- Inactive gas pipeline constraints

It is also possible to write the connection constraints of the two electricity and gas networks as follows:

- DG gas consumption as a gas demand
- The compressor as an electric load

In Li et al. [25], for enhancing the resilience of the integrated distribution of EGNs, a minimax-regret robust resilience-constrained unit commitment model is developed. The problem of three-level robust optimization is used in this study, and to solve it the SOC (second-order conic)-based column-and-constraint generation algorithm is used.

Finally, reference Wang et al. [55] provides a method for identifying vulnerable components in EGNs and ensuring the network's resilient operation. The three-level optimization used in this study is solved using a nested column-and-constraint generation (C&CG).

5.3.3 Discussion

In the previous section, studies and methods to increase resilience of EGNs were reviewed. It is observed that the method used in previous studies is often based on optimization methods. Therefore, mastering optimization methods and algorithms in this regard seems to be inevitable. Simultaneous consideration of EGN constraints complicates the issue.

To linearize the problem's constraints and equations, each of the previous studies has used a specific method (e.g., Taylor series expansion). Therefore, it seems that the study of linearization methods and mastery of the mathematical part of the problem seems to be essential and vital. The model considered in EGN resilience studies could be DAD (defender-attacker-defender), ADA (attacker-defender-attacker), or DA (defender-attacker).

In general, it can be said that the differences in the constraints of the problem, adding the constraints of connecting the two EGNs, the method of solving the optimization problem, and the linearization of the equations are among the differences between the studies. Hence, in this chapter, after introducing the methods and algorithms used to increase the resilience of EGNs, we present a simple model used in most previous studies with basic constraints on the problem. This provides a better understanding and acquaintance with the mathematical nature of the problem.

5.3.4 Objective Function of Problem

The basis of objective function used in most studies is as follows:

$$\text{OF} = \text{Investment Cost (IC)} + \text{Operational Cost (OC)} \tag{5.1}$$

For example, IC and OC in Shao et al. [49] are modeled as (5.2) and (5.3).

$$\begin{aligned}\text{IC} = &\sum_{gi,t} c_{gi,t}\left(u_{gi,t} - u_{gi,t-1}\right) + \sum_{pl,t} c_{pl,t}\left(u_{pl,t} - u_{pl,t-1}\right)\\ &+ \sum_{gs,t} c_{gs,t}\left(u_{gs,t} - u_{gs,t-1}\right) + \sum_{gp,t} c_{gp,t}\left(u_{gp,t} - u_{gp,t-1}\right)\end{aligned} \tag{5.2}$$

where $u_{gi,\,t}$ is the expansion state of generators, $u_{pl,\,t}$ is the expansion state of power lines, $u_{gs,\,t}$ is the expansion state of gas sources, and $u_{gp,\,t}$ is the expansion state of gas pipelines.

$$\text{OC} = \sum_t c_t \sum_b \Delta_b \left(\sum_{gi \in NGFG} f_{gi}\left(p^0_{gi,b,t}\right) + \sum_{gs} f_{gs}\left(g^0_{gs,b,t}\right)\right) \tag{5.3}$$

where f_{gi} is the operation cost function of generators, f_{gs} is the operation cost function of gas sources, $p_{gi,\,b,\,t}$ is the power output of generators, $g_{gs,\,b,\,t}$ is the gas output of gas sources, and NGFG means non- gas-fired generators.

It can be said that the studies performed so far using different optimization algorithms and different methods satisfy the above objective function. In this section, the example of problem objective function formulation was simply stated. The purpose of presenting the formulation is to present the basis of the problem and to provide a mathematical perspective of resiliency in EGNs. It should be noted that in studies on the resilience of EGNs, the authors may have presented different objective functions from what is presented in this chapter.

In summary, some of the differences between previous studies can be highlighted as follows:

- Gas network constraints.
- Electricity network constraints.
- Considering or not taking into account the connection constraints between electricity and gas networks.
- Differences in objective function.
- Differences in the definition of the optimization problem.
- Differences in linearization method of network equations.
- The optimization algorithm is used to solve the problem.

5.4 Remarks

According to studies, it can be said that methods to increase resilience can be divided into two categories, namely, hardening and operational. Many articles have covered the evaluation and enhancement of power grid resilience, while this is not the case for interconnected networks (electricity and gas/hydrogen). The issue of resilience and its concept and definitions were presented in the second part. Also, the importance of simultaneous operation of EGNs and the vital role of the mentioned systems' resilience were examined in the following.

Finally, studies to increase the resilience of EGNs were reviewed and analyzed, and a simple mathematical model of the problem was expressed. It can be said that all the articles in question have solved the problem using different optimization methods and algorithms, and the main difference between them is the type of linearization of gas and power equations and the algorithm used. Attention to the model and its development, along with innovation in mathematical methods, can be a future task.

A summary of studies that investigated the possibility of increasing the resilience of electricity and gas systems is demonstrated in Table 5.3.

Table 5.3 Methods for improving the resilience of EGNs

References	Method
Aldarajee et al. [2]	A new planner-disaster-risk framework
Lv et al. [30]	A multistage planning strategy Rolling reserve optimization stage Day-ahead economic dispatch stage Fault restoration stage
Shao et al. [49]	Optimal planning of EGNs using a two-stage robust optimization problem
He et al. [18, 19]	A three-step robust optimization method
Yan et al. [58]	The coordinated regional-district operation model of IES with three-stage optimization
Cong et al. [12]	DAD (defender-attacker-defender) model with min-max-min problem formulation
Liu et al. [29]	Two-stage data-based optimization
Tian et al. [53]	An optimal allocation method for gas turbines and storage systems in integrated model for EGNs
Lin et al. [28]	A combined repair crew dispatch problem considering reconfiguration and DG islanding
Li et al. [24, 26]	A minimax-regret robust resilience-constrained unit commitment model
Wang et al. [55]	"Present a method for" identifying vulnerable components in EGNs and ensuring the network's resilient operation using the three-level optimization

References

1. Accendo Reliability. (2021). *Reliability engineering professional development.* Accessible in 2021, https://accendoreliability.com/
2. Aldarajee, A. H., Hosseinian, S. H., Vahidi, B., & Dehghan, S. (2020). A coordinated planner-disaster-risk-averse-planner investment model for enhancing the resilience of integrated electric power and natural gas networks. *International Journal of Electrical Power & Energy Systems, 119*, 105948.
3. Ameli, H., Qadrdan, M., & Strbac, G. (2019). Coordinated operation strategies for natural gas and power systems in presence of gas-related flexibilities. *IET Energy Systems Integration, 1*(1), 3–13.
4. Ameli, H., Qadrdan, M., Strbac, G., & Ameli, M. T. (2020). Investing in flexibility in an integrated planning of natural gas and power systems. *IET Energy Systems Integration, 2*(2), 101–111.
5. Arghandeh, R., Von Meier, A., Mehrmanesh, L., & Mili, L. (2016). On the definition of cyber-physical resilience in power systems. *Renewable and Sustainable Energy Reviews, 58*, 1060–1069.
6. Barnes, A., Nagarajan, H., Yamangil, E., Bent, R., & Backhaus, S. (2019). Resilient design of large-scale distribution feeders with networked microgrids. *Electric Power Systems Research, 171*, 150–157.
7. Baros, S., Shiltz, D., Jaipuria, P., Hussain, A., & Annaswamy, A. M. (2017). *Towards resilient cyber-physical energy systems*. IEEE.
8. Bhusal, N., Abdelmalak, M., Kamruzzaman, M., & Benidris, M. (2020). Power system resilience: Current practices, challenges, and future directions. *IEEE Access, 8*, 18064–18086.

9. Buldyrev, S. V., Parshani, R., Paul, G., Stanley, H. E., & Havlin, S. (2010). Catastrophic cascade of failures in interdependent networks. *Nature, 464*(7291), 1025–1028.
10. Chaudry, M., Ekins, P., Ramachandran, K., Shakoor, A., Skea, J., Strbac, G., Wang, X., & Whitaker, J. (2011). *Building a resilient UK energy system* (p. UKERC).
11. Cimellaro, G. P., Reinhorn, A. M., & Bruneau, M. (2010). Framework for analytical quantification of disaster resilience. *Engineering Structures, 32*(11), 3639–3649.
12. Cong, H., He, Y., Wang, X., & Jiang, C. (2018). Robust optimization for improving resilience of integrated energy systems with electricity and natural gas infrastructures. *Journal of Modern Power Systems and Clean Energy, 6*(5), 1066–1078.
13. Council. (2009). *Critical infrastructure resilience final report and recommendations*. National Infrastructure Advisory Council. Tech. Rep.
14. Hollnagel, E., Woods, D. D., & Leveson, N. (2006). *Resilience engineering: Concepts and precepts*. Ashgate Publishing.
15. Espinoza, S., Panteli, M., Mancarella, P., & Rudnick, H. (2016). Multi-phase assessment and adaptation of power systems resilience to natural hazards. *Electric Power Systems Research, 136*, 352–361.
16. Gholami, A., Aminifar, F., & Shahidehpour, M. (2016). Front lines against the darkness: Enhancing the resilience of the electricity grid through microgrid facilities. *IEEE Electrification Magazine, 4*(1), 18–24.
17. Gholami, A., Shekari, T., Amirioun, M. H., Aminifar, F., Amini, M. H., & Sargolzaei, A. (2018). Toward a consensus on the definition and taxonomy of power system resilience. *IEEE Access, 6*, 32035–32053.
18. He, C., Dai, C., Wu, L., & Liu, T. (2018). Robust network hardening strategy for enhancing resilience of integrated electricity and natural gas distribution systems against natural disasters. *IEEE Transactions on Power Systems, 33*(5), 5787–5798.
19. He, C., Zhang, X., Liu, T., Wu, L., & Shahidehpour, M. (2018). Coordination of interdependent electricity grid and natural gas network—A review. *Current Sustainable/Renewable Energy Reports, 5*(1), 23–36.
20. Hussain, A., Oulis Rousis, A., Konstantelos, I., Strbac, G., Jeon, J., & Kim, H. (2019, January). Impact of uncertainties on resilient operation of microgrids: A data-driven approach. *IEEE Access, 7*, 1492414937.
21. Jackson, M., & Fitzgerald, J. (2016). Resilience profiling in the model-based design of cyber-physical systems. In *14th overture workshop: Towards analytical tool chains, technical report ECE-TR-28* (pp. 1–15).
22. Jufri, F. H., Widiputra, V., & Jung, J. (2019). State-of-the-art review on power grid resilience to extreme weather events: Definitions, frameworks, quantitative assessment methodologies, and enhancement strategies. *Applied Energy, 239*, 1049–1065.
23. Keogh, M., & Cody, C. (2013, November). *Resilience in regulated utilities*. National Association of Regulatory Utility Commissioners. Accessible at: www.naruc.org/Grants/Documents/Resilience% 20in%20Regulated% 20Utilities% 20ONLINE% 2011_12.pdf
24. Li, W., Li, Y., Tan, Y., Cao, Y., Chen, C., Cai, Y., Kwang, Y. L., & Pecht, M. (2019). Maximizing network resilience against malicious attacks. *Scientific Reports, 9*(1), 1–9.
25. Li, Y., Li, Z., Wen, F., & Shahidehpour, M. (2018). Minimax-regret robust co-optimization for enhancing the resilience of integrated power distribution and natural gas systems. *IEEE Transactions on Sustainable Energy, 11*(1), 61–71.
26. Li, Y., Xie, K., Wang, L., & Xiang, Y. (2019). Exploiting network topology optimization and demand side management to improve bulk power system resilience under windstorms. *Electric Power Systems Research, 171*, 127–140.
27. Lin, Y., & Bie, Z. (2016). Study on the resilience of the integrated energy system. *Energy Procedia, 103*, 171–176.
28. Lin, Y., Chen, B., Wang, J., & Bie, Z. (2019). A combined repair crew dispatch problem for resilient electric and natural gas system considering reconfiguration and DG islanding. *IEEE Transactions on Power Systems, 34*(4), 2755–2767.

29. Liu, R. P., Lei, S., Peng, C., Sun, W., & Hou, Y. (2020). Data-based resilience enhancement strategies for electric-gas systems against sequential extreme weather events. *IEEE Transactions on Smart Grid, 11*(6), 5383–5395.
30. Lv, C., Yu, H., Li, P., Zhao, K., Li, H., & Li, S. (2020). Coordinated operation and planning of integrated electricity and gas community energy system with enhanced operational resilience. *IEEE Access, 8*, 59257–59277.
31. McGranaghan, M., Olearczyk, M., & Gellings, C. (2013). Enhancing distribution resiliency: Opportunities for applying innovative technologies. *Electricity Today, 28*(1), 46–48.
32. McJunkin, T. R., & Rieger, C. G. (2017, September). Electricity distribution system resilient control system metrics. In *2017 resilience week (RWS)* (pp. 103–112). IEEE.
33. Mezzadra, S., & Adger, W. N. (2000). Social and ecological resilience: Are they related? *Progress in Human Geography, 24*(3), 347–364.
34. Mi, J., & Khodayar, M. E. (2019). Operation of natural gas and electricity networks with line pack. *Journal of Modern Power Systems and Clean Energy, 7*(5), 1056–1070.
35. National Research Council. (2012). *Disaster resilience: A national imperative*. National Research Council.
36. Nazemi, M., Moeini-Aghtaie, M., Fotuhi-Firuzabad, M., & Dehghanian, P. (2019). Energy storage planning for enhanced resilience of power distribution networks against earthquakes. *IEEE Transactions on Sustainable Energy, 11*(2), 795–806.
37. Ouyang, M., Dueñas-Osorio, L., & Min, X. (2012). A three-stage resilience analysis framework for urban infrastructure systems. *Structural Safety, 36*, 23–31.
38. Stockton, P. (2014). *Resilience for black sky days*. National Association of Regulatory Utility Commissioners, , Tech. Rep.
39. Panteli, M., & Mancarella, P. (2015). Modeling and evaluating the resilience of critical electrical power infrastructure to extreme weather events. *IEEE Systems Journal, 11*(3), 1733–1742.
40. Panteli, M., & Mancarella, P. (2015). The grid: Stronger, bigger, smarter?: Presenting a conceptual framework of power system resilience. *IEEE Power and Energy Magazine, 13*(3), 58–66.
41. Panteli, M., Mancarella, P., Trakas, D. N., Kyriakides, E., & Hatziargyriou, N. D. (2017). Metrics and quantification of operational and infrastructure resilience in power systems. *IEEE Transactions on Power Systems, 32*(6), 4732–4742.
42. Panteli, M., Trakas, D. N., Mancarella, P., & Hatziargyriou, N. D. (2017). Power systems resilience assessment: Hardening and smart operational enhancement strategies. *Proceedings of the IEEE, 105*(7), 1202–1213.
43. Panwar, M., Chanda, S., Mohanpurkar, M., Luo, Y., Dias, F., Hovsapian, R., & Srivastava, A. K. (2019). Integration of flow battery for resilience enhancement of advanced distribution grids. *International Journal of Electrical Power & Energy Systems, 109*, 314–324.
44. Poudel, S., & Dubey, A. (2018). Critical load restoration using distributed energy resources for resilient power distribution system. *IEEE Transactions on Power Systems, 34*(1), 52–63.
45. Preston, B. L., Backhaus, S. N., Ewers, M., Phillips, J. A., Silva-Monroy, C. A., Dagle, J. E., Tarditi, A. G., Looney, J., & King, T. J., Jr. (2016). *Resilience of the US electricity system: A multi-hazard perspective*. US Department of Energy Office of Policy.
46. Qadrdan, M., Ameli, H., Strbac, G., & Jenkins, N. (2017). Efficacy of options to address balancing challenges: Integrated gas and electricity perspectives. *Applied Energy, 190*, 181–190.
47. Rieger, C. G. (2014, August). Resilient control systems practical metrics basis for defining mission impact. In *2014 7th international symposium on resilient control systems (ISRCS)* (pp. 1–10). IEEE.
48. Shabazbegian, V., Ameli, H., Ameli, M. T., & Strbac, G. (2020). Stochastic optimization model for coordinated operation of natural gas and electricity networks. *Computers & Chemical Engineering, 142*, 107060.

49. Shao, C., Shahidehpour, M., Wang, X., Wang, X., & Wang, B. (2017). Integrated planning of electricity and natural gas transportation systems for enhancing the power grid resilience. *IEEE Transactions on Power Systems, 32*(6), 4418–4429.
50. Shi, Q., Li, F., Kuruganti, T., Olama, M., Dong, J., Wang, X., & Winstead, C. (2020). Resilience-oriented DG siting and sizing considering stochastic scenario reduction. *IEEE Transactions on Power Systems, 36*(4), 3715–3727.
51. Stevens, L. M. (2010). *HTGR resilient control system strategy* (No. INL/EXT–10-19645). Idaho National Laboratory (United States). Funding organisation: DOE-NE (United States).
52. Thompson, M. A., Ryan, M. J., Slay, J., & McLucas, A. C. (2016, July). A new resilience taxonomy. In *INCOSE international symposium* (Vol. 26, No. 1, pp. 1318–1330).
53. Tian, H., Huang, G., Chen, Y., Li, G., Li, X., & Bie, Z. (2020, November). Resilience-based optimal placement method for integrated electricity and gas energy system. In *2020 4th international conference on HVDC (HVDC)* (pp. 1188–1194). IEEE.
54. UNISDR. (2005). *Building the resilience of nations and communities to disasters: Hyogo framework for action 2005–2015*. UNISDR.
55. Wang, C., Wei, W., Wang, J., Liu, F., Qiu, F., Correa-Posada, C. M., & Mei, S. (2016). Robust defense strategy for gas–electric systems against malicious attacks. *IEEE Transactions on Power Systems, 32*(4), 2953–2965.
56. Wang, Y., Chen, C., Wang, J., & Baldick, R. (2015). Research on resilience of power systems under natural disasters—A review. *IEEE Transactions on Power Systems, 31*(2), 1604–1613.
57. Haimes, Y. Y. (2018). On the definition of resilience in systems. *Risk Analysis, 29*(4), 498–501.
58. Yan, M., He, Y., Shahidehpour, M., Ai, X., Li, Z., & Wen, J. (2018). Coordinated regional-district operation of integrated energy systems for resilience enhancement in natural disasters. *IEEE Transactions on Smart Grid, 10*(5), 4881–4892.

Chapter 6
Optimal Placement of Combined Heat and Power (CHP) Systems Considering the Cost of Environmental Pollutants

Sasan Azad, Mohammad Mehdi Amiri, and Mohammad Taghi Ameli

6.1 Introduction

Advances in technology have led to many changes in energy production, and countries worldwide are looking to use more efficient and environmentally friendly technologies to produce energy. The use of distributed generation sources, especially (CHP), is one of the technologies and methods that has attracted more attention in recent years. At this regard, cogeneration of energy system has been so appealing to the researchers, scholars, and operators. In smart energy systems, a couple of cogeneration approaches have been considered including gas and electricity [1–5], cool and electricity [6], heat and electricity [7], and so on.

Combined heat and power (CHP) systems are a part of smart energy systems and have been enjoyed in the hybrid systems so far. Hence the CHPs are so effective, but there are several difficulties to use, including technical, environmental, operational, and location problems, and it needs a proper procedure. Here the CHP owners prospective and grid point of view should be considered as well. Planning at the first place and operational problems then should be considered as well. Planning for the use of CHP in the network should be considered from various technical, environmental, economic, etc. aspects, and the benefits of the owners of CHP and the network should be considered simultaneously [8]. The profit from installing CHP in the power and heating network depends on various parameters such as type of technology used, location and size of CHP unit, heat-to-power ratio (HPR), planning horizon, and physical, environmental, and operational constraints [8–11]. Today,

S. Azad · M. M. Amiri
Shahid Beheshti University, Tehran, Iran
e-mail: sa_azad@sbu.ac.ir; moh_amiri@sbu.ac.ir

M. T. Ameli (✉)
Faculty of Power Engineering, Shahid Beheshti University, Tehran, Iran
e-mail: m_ameli@sbu.ac.ir

© The Author(s), under exclusive license to Springer Nature Switzerland AG 2022

V. Vahidinasab, B. Mohammadi-Ivatloo (eds.), *Whole Energy Systems*, Power Systems, https://doi.org/10.1007/978-3-030-87653-1_6

fuel cells (FC) [12–17], microturbines (MTs) [18, 19], and internal combustion engines (ICEs) [19, 20] are mostly used as CHP. ICE and MT technologies are more efficient than FC, but FC has better conditions for CO2 production and environmental pollutants [21]. In locating and determining the optimal size of CHP, in addition to the desired technical goals, economic goals for the optimal sale of heat and heat are also considered. Therefore, university centers and hospitals, etc., which have constant heat and heat consumption, can be suitable places to install CHP.

There are several methods to locate and determine the optimal CHP size. These methods can be divided into two categories: analytical methods and random search methods [22]. In the following, several research types focusing on the optimal placement and sizing of CHP will be briefly reviewed.

The basis for using analytical methods is based on mathematical methods. In [23], the authors have investigated the optimal capacity of CHP of the fuel cell type for installation in the heating network using mathematical methods. In this article, the aim is to reduce operating costs and environmental pollutants. Also, the issue for the 24-hour time horizon has been investigated. In [8], the FC, ICE, and MT location and optimal size for CHP networks in the 5-year time horizon have been studied. This paper's objective function is a multi-objective function and considers the problem from CHP owners and network operators perspective. In this paper, the effect of voltage stability index on placement planning and determining the optimal CHP size is investigated in this paper. The model presented in this paper is an MINLP model. In [24], the authors have examined the optimal size of CHP for different technologies in the power and heating distribution network, regardless of the optimal location. In this study, environmental and economic constraints have been considered. The model presented in this paper for problem optimization is an integer linear programming model. The goal is to minimize the annual costs of operating the system, maintenance, and resource owners' costs. In [25], an optimal programming method is proposed to determine the optimal CHP size. In this paper, based on a mathematical theory, the optimal CHP size is determined to minimize the system's annual cost. The problem studied in this paper is modeled as a mixed non-integer programming problem. In [11], the optimal size of CHP for home use is determined. This paper uses a nonlinear mixed-integer model that aims to minimize the annual cost to a customer. Next, the optimal placement and sizing of CHP is examined using random search methods.

Authors [26] have presented a method that uses particle swarm optimization to select the optimal allocation for CHP systems considering maximum allowable capacity with power loss reduction, voltage profile, and reliability improvement of microgrids considering network loading conditions. This problem presents as a multi-objective problem based on the weight factors method. The decision variables are the location and capacity of the CHP systems that determine optimally. The results in this paper confirm that the proposed method can select the optimal location and size for CHP systems by improving the customer's reliability and voltage profile and reducing power losses in an effective and rapid approach. In this study, the PSO algorithm's superiority is confirmed compared to the genetic algorithm (GA). In [15], the location of fuel cell units has been investigated by considering environmental pollutants and excluding the heating network. In this paper, a new fuzzy adaptive modified particle swarm optimization (FAMPSO) algorithm is used. This paper also uses Pareto optimization to obtain a set of optimal solutions to a multi-objective problem. In [21], the optimal placement and sizing of CHP units in power and heating networks for a 5-year time horizon has been developed. In this study, the planning of various renewable and nonrenewable technologies with an improved particle swarm algorithm is solved. In [27] for determining the optimal size of CHP, a possible method for the long-term time horizon is presented. In this paper, a two-step stochastic optimization algorithm is developed using Monte Carlo to solve the optimization problem. A hospital in Italy has been used as a case study. Authors [28] developed a model to optimize the capacity and long-term operations of CHP. This planning approach is solved using particle swarm optimization (PSO). The long-term optimization results show that there are changes in the installed capacity and operation. In [20], regardless of environmental pollutants, the optimal size of CHP units in microgrids has been investigated using the PSO algorithm. In this article, the aim is to determine the ratio of profit to cost. The technologies used in this paper are gas turbines, MTs, and ICEs. The results show that the highest profit ratio to cost is obtained in MT technology. To select the optimal CHP location, the loss sensitivity index in each bus was used.

In the optimal placement of CHP, taking into account the effects of environmental pollutants and using different types of CHP technologies can play an essential role in the planning and implementing of these systems. In fact, the release of environmental pollutants can change their owners' profits and the type of technology they use. In this chapter, the optimal placement of CHP systems considering environmental pollutants is studied. Features are presented in this chapter, providing a comprehensive model for CHP optimal placement systems, considering nonrenewable sources, including electrical and heat energy networks, annual planning for horizon which is 5 years, and the cost of environmental pollutants.

The structure of the chapter is as follows: in Sect. 6.2 to present and formulate the problem, in Sect. 6.3 to present the study network and information needed to solve the problem, in Sect. 6.4 to discuss and simulate the results, and finally in Sect. 6.5 to discuss the general result overview of the chapter.

6.2 Problem Formulation

In this section, the objective function and constraints used in the problem of optimal placement of different CHP technologies for a 5-year time horizon are stated.

6.2.1 Objective Function

The objective function of the problem, which is the profit of the network operator from the optimal allocation of CHP, consists of two terms of revenue and cost, which are mentioned in Eq. (6.1):

$$\text{Benefit}_{\text{CHP}} = \text{Revenue}_{\text{CHP}} - \text{Cost}_{\text{CHP}} \tag{6.1}$$

where $\text{Revenue}_{\text{CHP}}$ and Cost_{CHP} are obtained using Eqs. (6.2) and (6.3), respectively:

$$\begin{aligned}\text{Revenue}_{\text{CHP}} = \sum_{n=1}^{N_h}\sum_{y=1}^{N_y}\sum_{s=1}^{N_s}\sum_{t=1}^{N_T} \beta_{y,s}^{H} S_s H_{n,y,s}^{t} + \\ \sum_{i=1}^{N_b}\sum_{y=1}^{N_y}\sum_{s=1}^{N_s}\sum_{t=1}^{N_T} \beta_{y,s}^{E} S_s P_{i,y,s}^{t} + \sum_{y=1}^{N_y}\sum_{s=1}^{N_s} \beta_{y,s}^{E} S_s \left(P_{\text{loss},y,s} - P_{\text{loss},y,s}^{\text{CHP}}\right) \\ + \sum_{y=1}^{N_y} \text{URC}_y\end{aligned} \tag{6.2}$$

In which the first term is the CHP units revenue from the sale of heat energy, the second term is the CHP units revenue from the sale of electrical energy, the third term is the network operator's revenue from the reduction of electrical energy losses, and the fourth semester is the revenue from the reduction of investment costs for equipment upgrades. The amount of investment reduction depends on CHP networks' structure, which is considered a fixed annual number here [20]:

$$\begin{aligned}\text{Cost}_{\text{CHP}} = \sum_{i=1}^{N_b}\sum_{y=1}^{N_y}\sum_{s=1}^{N_s}\sum_{t=1}^{N_T} P_{i,y,s}^{t} \text{MOC}_t S_s + \\ \sum_{i=1}^{N_b}\sum_{y=1}^{N_y}\sum_{s=1}^{N_s}\sum_{t=1}^{N_T} P_{i,y,s}^{t} \text{ER}_t^{\text{CHP}} S_s E_t^{\text{CO}_2} \\ + \sum_{i=1}^{N_b}\sum_{y=1}^{N_y}\sum_{t=1}^{N_T} IC_t^{\text{CHP}} I_{i,y,t}^{\text{CHP}} - IC_t^{\text{CHP}} I_{i,y,t-1}^{\text{CHP}}\end{aligned} \tag{6.3}$$

In which the first term shows the cost of maintenance of CHP units. Maintenance cost is expressed as a linear function of generated electrical power [29]. The second term shows the cost of carbon dioxide emissions as environmental pollutants [30]. The third term shows the cost of installing different CHP units in the electricity and heat network.

6.2.2 Problem Constraints

In this section, the constraints related to the location and optimal size of different CHP units in electrical and heat networks are expressed to maximize the problem's objective function, which is the network operator's profit from different CHP technologies.

6.2.2.1 Power Flow Constraints in the Electrical Network

The objective function of the problem is optimized with equality and inequality constraints as follows:

$$\sum_{i=1}^{N_G} P_{i,y,s}^{G} + \sum_{t=1}^{N_t} P_{i,y,s}^{t} - P_{i,y,s}^{L} = \sum_{j=1}^{N_b} V_{i,y,s} V_{j,y,s} Y_{\text{ij}} \cos\left(\varphi_{\text{ij}} + \theta_{j,y,s} - \theta_{i,y,s}\right) \tag{6.4}$$

$$\sum_{i=1}^{N_G} Q_{i,y,s}^{G} - Q_{i,y,s}^{L} = -\sum_{j=1}^{N_b} V_{i,y,s} V_{j,y,s} Y_{\text{ij}} \sin\left(\varphi_{\text{ij}} + \theta_{j,y,s} - \theta_{i,y,s}\right) \tag{6.5}$$

$$V_i^{\min} \le V_i \le V_i^{\max} \tag{6.6}$$

$$P_{\min}^{\text{Gi}} \le P^{\text{Gi}} \le P_{\max}^{\text{Gi}} \tag{6.7}$$

$$Q_{\min}^{\text{Gi}} \le Q^{\text{Gi}} \le Q_{\max}^{\text{Gi}} \tag{6.8}$$

$$S_{\text{ij},y,s} \le S_{\text{ij},y,s}^{\max} \tag{6.9}$$

In which Eqs. (6.4) and (6.5) are the power flow equality constraints, Eq. (6.6) is allowable voltage limits for the operation of the electrical network, Eqs. (6.7) and (6.8) are the minimum and maximum limits of active and reactive power of generators, and Eq. (6.9) is apparent allowable power passing through the lines.

6.2.2.2 Heat Network Constraints

In this section, the relations related to the thermal network to solve the problem are modeled as follows:

$$\sum_{t=1}^{N_t} H_{m,y,s,t}^{\text{CHP}} + H_{m,y,s}^{\text{boil}} + \sum_{k \in N_k} f_{\text{kn}} = \sum_{k \in N_k} f_{\text{nk}} + H_{m,y,s}^{L} \tag{6.10}$$

$$f_{\text{nk}} = \text{sng}\left(p_{n,y,s} - p_{k,y,s}\right) C_{\text{nk}}^{t} \sqrt{\text{sng}\left(p_{n,y,s} - p_{k,y,s}\right)\left({p^2}_{n,y,s} - {p^2}_{k,y,s}\right)} \tag{6.11}$$

$$p_n^{\min} \leq p_n \leq p_n^{\max} \tag{6.12}$$

$$f_{\mathrm{km}} \leq f_{\mathrm{km}}^{\max} \tag{6.13}$$

In which Eq. (6.10) shows the thermal equilibrium in the heat network [31]. Eq. (6.11) shows the heat flow through the lines. Equations (6.12) and (6.13) show the pressure limits at the heating network's buses and the limits of the thermal power passing through the heating network, respectively.

6.2.2.3 Conversion of Electrical Power to Heat

According to the following equation, electrical power is converted to heat in the proposed problem [31]:

$$\mathrm{HPR}_{i,y,s,t}^{\mathrm{CHP}} = \frac{H_{i,y,s,t}^{\mathrm{CHP}}}{P_{i,y,s,t}^{\mathrm{CHP}}} \tag{6.14}$$

6.2.2.4 CHP Various Technologies Investment Constraint

The binary variable determining the CHP unit's status will not change by installing different CHP units in the power and heat network during the planning period. It means that this unit will be operated in the years after the installation time:

$$I_{i,y,t-1}^{\mathrm{CHP}} \leq I_{i,y,t}^{\mathrm{CHP}} \tag{6.15}$$

6.2.2.5 Limitation of Electrical and Heat Power of Various CHP Technologies

$$P_{i,y,s}^{t,\min} \times I_{i,y,t}^{\mathrm{CHP}} \leq P_{i,y,s}^{t} \leq P_{i,y,s}^{t,\max} \times I_{i,y,t}^{\mathrm{CHP}} \tag{6.16}$$

$$H_{i,y,s}^{t,\min} \times I_{i,y,t}^{\mathrm{CHP}} \leq H_{i,y,s}^{t} \leq H_{i,y,s}^{t,\max} \times I_{i,y,t}^{\mathrm{CHP}} \tag{6.17}$$

6.2.3 *Prices and Loads of Electrical and Heat Energy*

In the issue of placement and determining the optimal size of CHP in the long run, attention to price and the electrical load is of particular importance. Also, since the price and amount of electric and heat load are different for different seasons of the year (spring and autumn are assumed to be the same), so in this issue, three patterns of load and price for the year are considered. In this modeling, increasing the load and price of electrical and heat energy are also considered:

$$P^L_{i,y,s} = (1+\alpha)^{y-1} \times \text{coef}^E_s \times P^L \tag{6.18}$$

$$Q^L_{i,y,s} = (1+\alpha)^{y-1} \times \text{coef}^E_s \times Q^L \tag{6.19}$$

$$H^L_{n,y,s} = (1+\alpha)^{y-1} \times \text{coef}^H_s \times H^L \tag{6.20}$$

$$\beta^E_{y,s} = (1+\beta)^{y-1} \times \text{coef}^E_s \times \beta^E_{y=1} \tag{6.21}$$

$$\beta^H_{y,s} = (1+\beta)^{y-1} \times \text{coef}^H_s \times \beta^H_{y=1} \tag{6.22}$$

6.3 Study Network and Problem Required Data

The electrical and heat network studied in Fig. 6.1 is considered based on the standard 14-bus IEEE network [20]. As shown in Fig. 6.1, electrical and heat networks have a similar structure, and energy is transmitted to the place of consumption through electrical and heat lines. In the studied network, each bus has different electrical and heat loads [20]. The CHP units are placed in the network to supply the heat loads in the best technical and economical way. Also, electrical loads are supplied by preinstalled units, and the electrical power of CHP units is provided in a better technical and economic condition. Figure 6.2 shows a general schematic of the system under study. The electrical and heat loads of the various bus networks under study in the first year of operation are shown in Fig. 6.3.

In the following, the data is required for CHP placement with environmental pollutants. The required information is presented in Tables 6.1 and 6.2, respectively. Table 6.1 provides information on CHP systems for fuel cell technologies, microturbines, and internal combustion engines as nonrenewable systems. This information includes installation, maintenance costs, HPR value, unit capacity, and emission rate of environmental pollutants [21]. The efficiency of FC units is 98%. Other parameters required to solve the optimization problem are presented in Table 6.2.

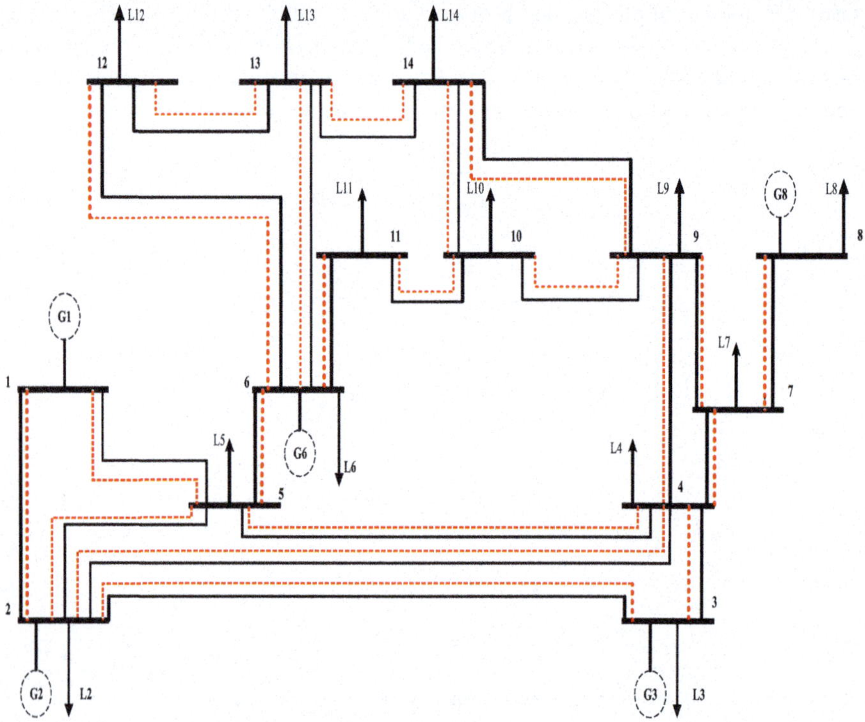

Fig. 6.1 Electrical and heat network studied

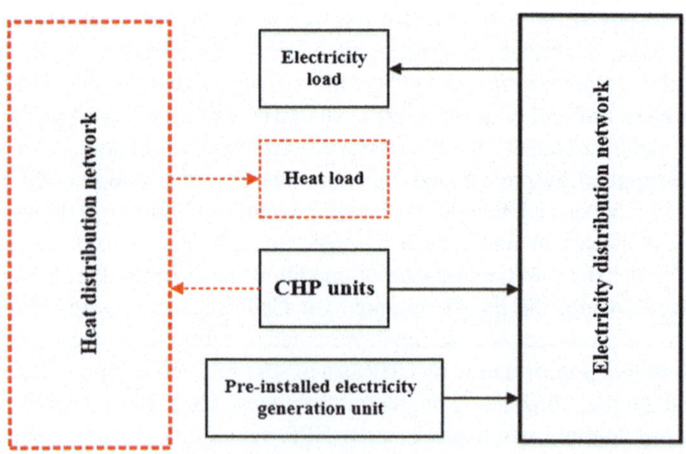

Fig. 6.2 General schematic of the studied network

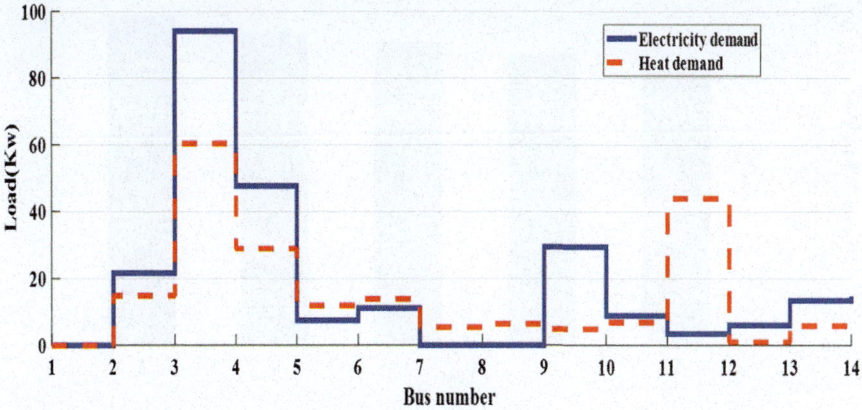

Fig. 6.3 Electrical and heat loads of the studied network in the first year of operation

Table 6.1 CHP data [8, 26]

Technology	Capacity (MW)	HPR	IC_t^{CHP}	MOC_t	ER_t^{CHP}	$H^{\max}$	$P^{\min}$	$P^{\max}$
ICE	5	0.95	2900	0.025	790	1.1	0	1.1
FC	20	1.5	6500	0.038	20	1.1	0	1.1
MT	1	0.6	4300	0.013	150	1.1	0.2	1.1

Table 6.2 Different parameters [8, 26]

Parameter	Value	Parameter	Value
$coef_{s=1}^{E}$	1	α	0.1
$coef_{s=2}^{E}$	1.3	β	0.1
$coef_{s=3}^{E}$	0.8	$V^{\min}$	0.9
$coef_{s=1}^{H}$	1	$V_i^{\max}$	1.05
$coef_{s=2}^{H}$	0.7	E^{CO_2}	0.0029 $\$/kg$
$coef_{s=3}^{H}$	1.4	URC	120 $\$/MWh$
β_y^H	50 $\$/MWh$	β_y^E	80$\$/MWh$

6.4 Simulation and Analysis of Results

As mentioned, the network in question is based on the IEEE 14-bus network. This network has five preinstalled units to supply electrical charges. With the addition of heat loads, CHP units that are optimally installed in the network are used to supply them. Electric and heat loads and the price of energy increase each year of operation according to coefficients α and β. It should be noted that during the operation period, there is always a balance in the electrical and heat networks with the optimal operation of the CHP units, while the goal is the maximum profit of the network operator. Loads and the price of electrical and heat energy change in each season of the operation period based on coefficients $coef_s^E$ and $coef_s^h$. In the following, the optimal placement of CHP systems based on fuel cell, microturbine,

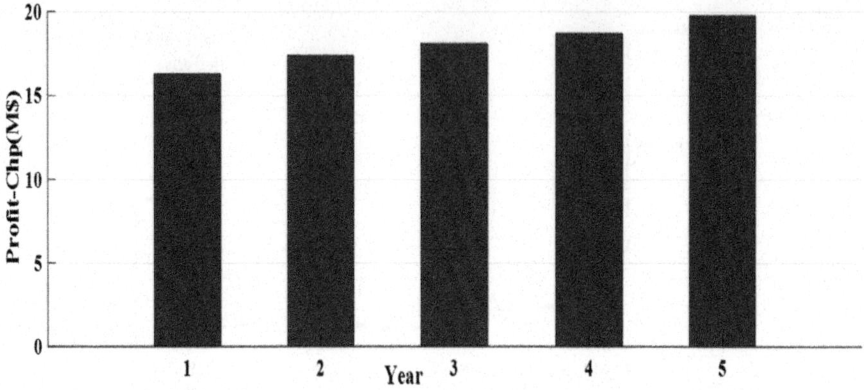

Fig. 6.4 Profit changes on the planning horizon

Table 6.3 Optimal CHP allocation *with* considering environmental pollution

Type	Bus number	Year of unit installation	Electricity capacity (MW)	Heat capacity (MW)	Profit (k$)	Cost of environmental Pollution ($)
FC	4	1	20	30	69,358	10,232
	14	1	20	30		
MT	9	1	1	0.6	3467.9	3875.2
	10	1	1	0.6		
ICE	9	1	5	4.75	17339	100,264
	10	1	5	4.75		

and internal combustion engine technologies is examined considering the cost of environmental pollution.

First, the results of the implementation of the problem are examined according to the input data. Figure 6.4 shows the changes in network operator profits from different CHP technologies over a 5-year horizon. Examination of Fig. 6.4 shows that the operating profit increases over the operating time. The simulation results show that the highest revenue is related to the sale of renewable energy, and the highest cost is related to the installation of these technologies. In total, the profit from the installation of CHP technologies in the network is $ 90,165,000.

The results of optimal location and size of nonrenewable resources are shown in Table 6.3. The results show that the profit ratio to environmental pollutants' cost due to fuel cell is more than two technologies. Therefore, this technology can be an excellent option to increase the profit of the network operator. Also, the emission of environmental pollutants from fuel cells is less than the other two technologies, so it is also environmentally friendly. The cost of environmental pollution and the amount of pollution produced by the internal combustion engine is higher than the other two technologies. Therefore, due to increasing environmental concerns, this technology's installation is not suitable in terms of environment and pollution.

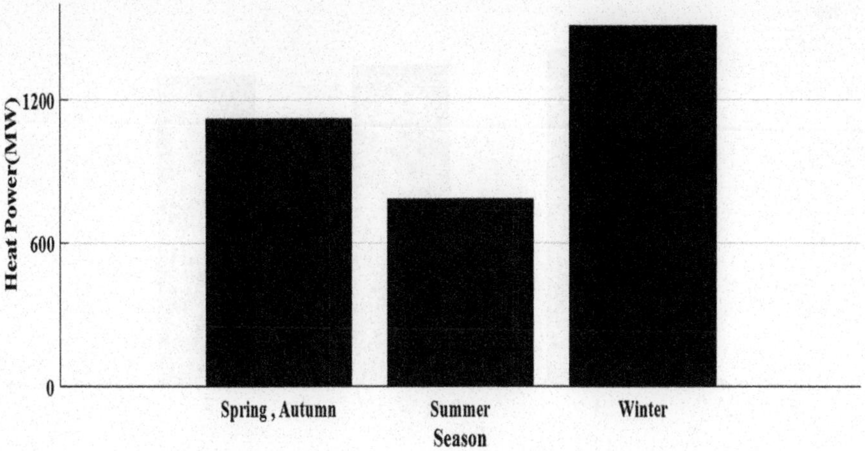

Fig. 6.5 CHP units heat power

However, suppose the cost of environmental pollutants is excluded, considering that the cost of maintenance and installation of fuel cell technology is more than the other two technologies. In that case, the internal combustion engine and microturbine are economical for the network operator and their owners. One of the issues that shows the importance of long-term planning is the cost of installing various CHP technologies. The greater the planning horizon, the more economical the installation of technologies will be with the higher installation costs. In general, as the planning horizon increases, the installation of technologies is more appropriate with less environmental emissions and lower maintenance costs.

The amount of heat power output of the three technologies discussed in this chapter increases with decreasing winter temperature. In the cold season, the amount of heat load increases, and the price coefficient of electrical energy is 0.8, and the price coefficient of thermal energy is 1.3. Therefore, increasing the heat output of the units in the cold season can be justified. Figure 6.5 shows the heat output of all three technologies for the planning horizon. As it is known, the amount of heat power in the cold season is higher due to profitability.

In the following, the sensitivity of the operator's final profit of the CHP installation to HPR is examined. To do this, the HPR value is the same for all three technologies and varies from 0.5 to 1.5. The total profit for different HPRs is shown in Fig. 6.6. Examination of Fig. 6.6 shows that with increasing HPR, the amount of profit decreases. That is because the price of thermal energy is lower than that of electricity. For example, a 50% increase in the amount of HPR (from 0.5 to 1) will result in a 6% decrease in profit. Therefore, it can be said that determining the appropriate value for HPR is of particular importance.

In order to investigate the effect of heat loads on the operating profit, the electrical network is omitted. In this way, CHP units and other sources already installed in the network supply electrical loads. Figure 6.7 shows the operating

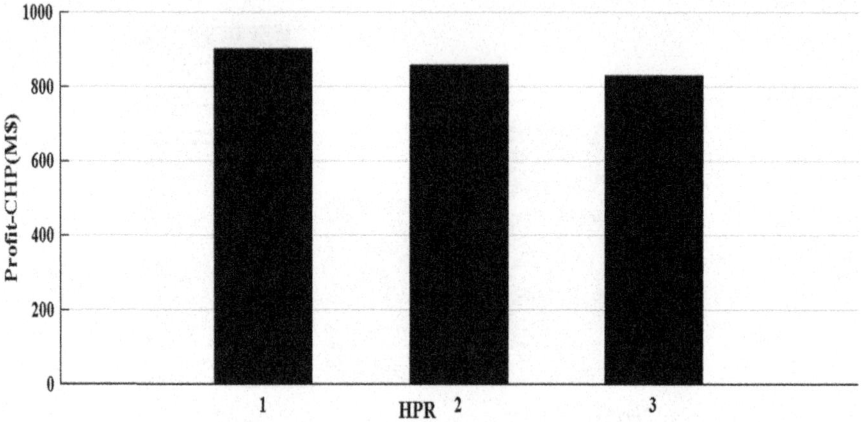

Fig. 6.6 Investigating the effect of HPR value on operating profit due to CHP installation

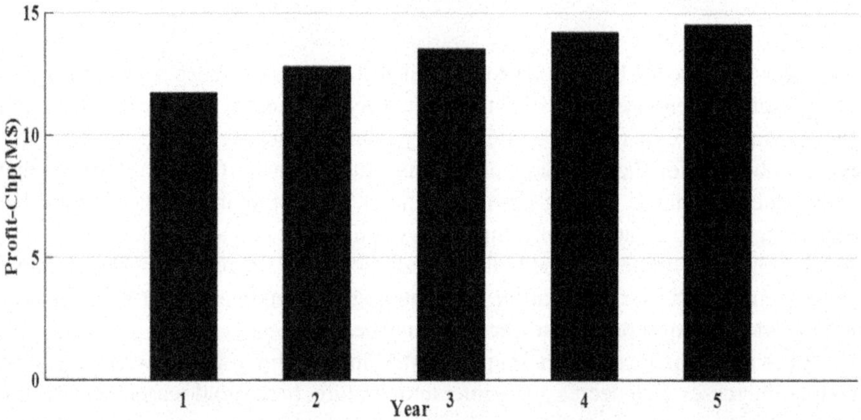

Fig. 6.7 Profit changes in the planning horizon without heat network

profit from the installation of CHP units. Examination of the figure shows that the operator benefited significantly from the supply of thermal loads, and this shows the importance of the heating network in increasing the operator's profit.

In this case, by removing the electrical network, the CHP units are placed in bus 11 with the maximum possible value. It can be seen that there is only one unit of each technology, which is less than the case with the electrical network because of the elimination of the heat network.

Finally, the effect of eliminating environmental pollutants on the optimal allocation of CHP technologies is investigated. Given the importance of environmental issues in long-term and medium-term planning, this issue must be considered. The results for CHP systems' placement based on nonrenewable technologies without

Table 6.4 Optimal CHP placement

Type	Bus number	Year of unit installation
FC	4	1
	14	1
MT	4	1
	12	1
ICE	9	1
	14	1

considering environmental pollutants are shown in Table 6.4. According to Table 6.4, due to eliminating the cost of environmental pollutants, the profit of network operators and resource owners will increase. The profit margin is $ 576,624. Also, the installation location of different technologies has been affected by this issue, as shown in Table 6.4.

6.5 Conclusion

The increasing population of the world and, naturally, the increasing need for electrical and heat energy have caused more attention and expansion of CHP systems globally. Given the importance of environmental issues, environmental pollutants as one of the crucial factors in the long-term planning of various CHP technologies should always be considered. In this chapter, a comprehensive model for the optimal planning and allocation of CHP technologies such as microturbines, fuel cells, and internal combustion engines with environmental pollutants was presented. The amount of load and price of electrical and heat energy during the planning period have had seasonal and annual changes in the studied model. The proposed model was implemented and tested on an IEEE 14-bus network with electric and heat loads over a 5-year time horizon. The simulation results showed that increasing the HPR reduces the operating profit. Also, fuel cell technology was the most suitable, and the internal combustion engine the worst technology in terms of environmental pollutants. In addition to affecting the profits of network operators and resource owners, environmental pollutants caused changes in the installation location of various CHP technologies. On the other hand, it was found that the presence of thermal loads has increased the profit of the network operator. One of the things that can be considered in the optimal allocation planning of CHP is the planning horizon's effect on the optimal allocation and profitability of the network operator. The use of electrical and heat storage can also be considered to increase operating profits.

Nomenclature

Sets

N_b	Number of network buses	V_i^{min}/V_i^{maxn}	Minimum/maximum bus voltage
N_T	Number of CHP technology	$P_{loss,y,s}$	Network loss before CHP allocation (MW)
N_S	Number of seasons	$P_{\text{loss},y,s}^{\text{CHP}}$	Network loss after CHP allocation (MW)
N_y	Number of time for the planning	S_s	The total number of hours in season s
N_g	Number of generating units	$\beta_{y,s}^{H}$, $\beta_{y,s}^{E}$	Heat/electricity price in year y at season s
Indices		$H_{n,y,s}^{t}$, $P_{i,y,s}^{t}$	Heat/electricity produced by CHP unit type t (MW)
i,j	Electric network buses	MOC_t	Maintenance and operation cost of CHP units ($/MWh)
n,k	Heat network buses	ER_t^{CHP}	Rate of CO2 emissions by CHP unit type t (kg/MWh)
y	Index of years	$E_t^{CO_2}$	Cost for each kg of CO2 ($/kg)
T	Types of CHP technologies	IC_t^{CHP}	Investment cost of CHP units ($/kW)
s	Index of seasons	$I_{i,y,t-1}^{\text{CHP}}$	A binary variable for the place of CHP(0,1)
Variables and parameters		$\text{HPR}_{i,y,s,t}^{\text{CHP}}$	Heat-to-power ratio of CHP units
UCR_y	Upgrade cost reduction in year y	$H_{m,y,s}^{\text{boil}}$	Heat power of boiler
$P^{l}{}_{i,y,s}$	Active load of electricity system	f_{nk}	Heat power flow
α	Load growth rate	p_n	Heat network buses pressure
β	Price growth rate	$p_n^{\max}$, $p_n^{\min}$	Maximum/minimum heat network buses pressure
coef_s^E, coef_s^H	Electricity/heat load and price coefficient in each season	C_{nk}^{t}	Coefficient of heat network's branches loss
$P_{\max}^{Gi}/P_{\min}^{Gi}$, $Q_{\max}^{Gi}/Q_{\min}^{Gi}$	Maximum/minimum active/reactive power generation of preinstalled generation units (MW/MVAr)	$P_{i,y,s}^{G}$, $Q_{i,y,s}^{G}$	Active/reactive power generation of preinstalled generation units (MW/MVAr)

(continued)

H^{max}, H^{min}	Maximum/minimum acceptable of heat produced by CHP unit	P^{max}, P^{min}	Maximum/minimum acceptable of electricity produced by CHP unit
$P^{L}_{i,y,s}$, $Q^{L}_{i,y,s}$	Active/reactive load of electricity network (MW/MVAr)	$H^{L}_{m,y,s}$	Heat load of heat network (MW)
$V_{i,y,s} \angle \theta_{i,y,s}$	Voltage phasor at bus i at year y and season s	$Y_{ij} \angle \varphi_{ij}$	Polar form of ijth element of admittance matrix
$S_{i,j,y,s}$	Apparent power flow in feeder i–j at year y and season s	$S^{max}_{ij,y,s}$	Maximum apparent power flow in feeder *i*–*j* at year *y* and season *s*

References

1. Shabazbegian, V., Ameli, H., Ameli, M. T., Strbac, G., & Qadrdan, M. (2021). Co-optimization of resilient gas and electricity networks; a novel possibilistic chance-constrained programming approach. *Applied Energy, 284*, 116284. https://doi.org/10.1016/j.apenergy.2020.116284
2. Shabazbegian, V., Ameli, H., Ameli, M. T., & Strbac, G. (2020). Stochastic optimization model for coordinated operation of natural gas and electricity networks. *Computers Chemical Engineering, 142*, 107060. https://doi.org/10.1016/j.compchemeng.2020.107060
3. Beigvand, S. D., Abdi, H., & La Scala, M. (2017). Economic dispatch of multiple energy carriers. *Energy, 138*.
4. Beigvand, S. D., Abdi, H., & La Scala, M. (2017). A general model for energy hub economic dispatch. *Applied Energy, 190*, 1090–1111.
5. Sanjani, K., et al. (2019). A robust-stochastic approach for energy transaction in energy hub under uncertainty. In *Robust optimal planning and operation of electrical energy systems* (pp. 219–232). Springer.
6. Dargahi, A., et al. (2020). Scheduling of air conditioning and thermal energy storage systems considering demand response programs. *Sustainability, 12*(18), 7311.
7. Aunedi, M., et al. Modelling of national and local interactions between heat and electricity networks in low-carbon energy systems. *Applied Energy, 276*(2020), 115522.
8. Yadegari, S., Abdi, H., & Nikkhah, S. (2020). Risk-averse multi-objective optimal combined heat and power planning considering voltage security constraints. *Energy, 212*, 118754.
9. Boljevic, S., & Conlon, M. F. (2011). *Optimal sizing of combined heat & power (CHP) generation in urban distribution network (UDN)*. In 46th International Universities' Power Engineering Conference (UPEC). VDE.
10. Sadeghian, H. R., & Ardehali, M. M. (2016). A novel approach for optimal economic dispatch scheduling of integrated combined heat and power systems for maximum economic profit and minimum environmental emissions based on Benders decomposition. *Energy, 102*, 10–23.
11. Ren, H., Gao, W., & Ruan, Y. (2008). Optimal sizing for residential CHP system. *Applied Thermal Engineering, 28*(5–6), 514–523.
12. Briguglio, N., et al. (2011). Evaluation of a low temperature fuel cell system for residential CHP. *International Journal of Hydrogen Energy, 36*(13), 8023–8029.
13. Asl, S. M. S., Rowshanzamir, S., & Eikani, M. H. (2010). Modelling and simulation of the steady-state and dynamic behaviour of a PEM fuel cell. *Energy, 35*(4), 1633–1646.

14. Moradi, M. H., & Abedini, M. (2012). A combination of genetic algorithm and particle swarm optimization for optimal DG location and sizing in distribution systems. *International Journal of Electrical Power & Energy Systems, 34*(1), 66–74.
15. Farjah, E., et al. (2012). Placement of combined heat, power and hydrogen production fuel cell power plants in a distribution network. *Energies, 5*(3), 790–814.
16. El-Sharkh, M. Y., et al. (2010). Economics of hydrogen production and utilization strategies for the optimal operation of a grid-parallel PEM fuel cell power plant. *International Journal of Hydrogen Energy, 35*(16), 8804–8814.
17. Adam, A., Fraga, E. S., & Brett, D. J. L. (2015). Options for residential building services design using fuel cell based micro-CHP and the potential for heat integration. *Applied Energy, 138*, 685–694.
18. Tichi, S. G., Ardehali, M. M., & Nazari, M. E. (2010). Examination of energy price policies in Iran for optimal configuration of CHP and CCHP systems based on particle swarm optimization algorithm. *Energy Policy, 38*(10), 6240–6250.
19. Buoro, D., Pinamonti, P., & Reini, M. (2014). Optimization of a DISTRIBUTED COGENERATION SYSTEM with solar district heating. *Applied Energy, 124*, 298–308.
20. Basu, A. K., Chowdhury, S., & Chowdhury, S. P. (2010). Impact of strategic deployment of CHP-based DERs on microgrid reliability. *IEEE Transactions on Power Delivery, 25*(3), 1697–1705.
21. Arandian, B., & Ardehali, M. M. (2017). Effects of environmental emissions on optimal combination and allocation of renewable and non-renewable CHP technologies in heat and electricity distribution networks based on improved particle swarm optimization algorithm. *Energy, 140*, 466–480.
22. Arandian, B., & Ardehali, M. M. (2017). Renewable photovoltaic-thermal combined heat and power allocation optimization in radial and meshed integrated heat and electricity distribution networks with storages based on newly developed hybrid shuffled frog leaping algorithm. *Journal of Renewable and Sustainable Energy, 9*(3), 033503.
23. Adam, A., Fraga, E. S., & Brett, D. J. L. (2015). Options for residential building services design using fuel cell based micro-CHP and the potential for heat integration. *Applied Energy, 138*, 685–694.
24. Buoro, D., Pinamonti, P., & Reini, M. (2014). Optimization of a distributed cogeneration system with solar district heating. *Applied Energy, 124*, 298–308.
25. Beihong, Z., & Weiding, L. (2006). An optimal sizing method for cogeneration plants. *Energy and Buildings, 38*(3), 189–195.
26. Naderipour, A., et al. Optimal allocation for combined heat and power system with respect to maximum allowable capacity for reduced losses and improved voltage profile and reliability of microgrids considering loading condition. *Energy, 196*(2020), 117124.
27. Urbanucci, L., & Testi, D. (2018). Optimal integrated sizing and operation of a CHP system with Monte Carlo risk analysis for long-term uncertainty in energy demands. *Energy Conversion and Management, 157*, 307–316.
28. Takada, K., & Ikegami, T. (2018). *Long-term Planning model for a CHP system using particle swarm optimization algorithm and mixed integer linear programming*. In 2018 International conference on Smart Energy Systems and Technologies (SEST). IEEE.
29. Wang, H., et al. (2015). Modelling and optimization of CHP based district heating system with renewable energy production and energy storage. *Applied Energy, 159*, 401–421.
30. Gharavi, H., Ardehali, M. M., & Tichi, S. G. (2015). Imperial competitive algorithm optimization of fuzzy multi-objective design of a hybrid green power system with considerations for economics, reliability, and environmental emissions. *Renewable Energy, 78*, 427–437.
31. Amirnekooei, K., Ardehali, M. M., & Sadri, A. (2017). Optimal energy pricing for integrated natural gas and electric power network with considerations for techno-economic constraints. *Energy, 123*, 693–709.

Chapter 7
Optimal Coalition Operation of Interconnected Hybrid Energy Systems Containing Local Energy Conversion Technologies, Renewable Energy Resources, and Energy Storage Systems

Behzad Motallebi Azar, Amir Mirzapour-Kamanaj, Rasool Kazemzadeh, Behnam Mohammadi-Ivatloo, and Kazem Zare

7.1 Introduction

The issue of energy is one of the main challenges of human beings in the present century. Nowadays, due to the advent of various technologies and human-life dependency on energy, these challenges have become more apparent. These challenges appear in multiple forms, such as energy supply, trade, and consumption. For decades fossil fuels have been the main primary fuel for power plants, which was not cost-effective due to the low efficiency of primary energy production. In recent years, due to increasing energy demand, economic and environmental challenges, energy security, depleting fossil fuels, and high penetration of renewable energy resources (RERs), the utilization of hybrid energy systems (HESs) has become more regarded by stakeholders. HES can use multiple types of energy as input, and after converting, storing, and conditioning, the various types of energy demands can be supplied [1]. The use of HESs highlights the concept of whole energy

B. Motallebi Azar (✉) · R. Kazemzadeh
Faculty of Electrical Engineering, Sahand University of Technology, Tabriz, Iran
e-mail: b_motallebiazar98@sut.ac.ir; r.kazemzadeh@sut.ac.ir

A. Mirzapour-Kamanaj · K. Zare
Faculty of Electrical and Computer Engineering, University of Tabriz, Tabriz, Iran
e-mail: a.mirzapour@tabrizu.ac.ir; kazem.zare@tabrizu.ac.ir

B. Mohammadi-Ivatloo
Department of Electrical and Electronics Engineering, Muğla Sıtkı Koçman University, Muğla, Turkey
e-mail: bmohammadi@tabrizu.ac.ir

Faculty of Electrical and Computer Engineering, University of Tabriz, Tabriz, Iran
e-mail: bmohammadi@tabrizu.ac.ir

© The Author(s), under exclusive license to Springer Nature Switzerland AG 2022

V. Vahidinasab, B. Mohammadi-Ivatloo (eds.), *Whole Energy Systems*, Power Systems, https://doi.org/10.1007/978-3-030-87653-1_7

systems, which means the coordination of electricity, gas, heat, traffic, and water systems [2] in order to transform clean and techno-economic energy to customers. Also, exchanging energy between interconnected hybrid energy systems (IHESs) can be an effective way to improve the energy systems' performance. Therefore, some research needs to investigate the optimal operation of IHESs to overcome the mentioned challenges.

In 2005, researchers at the Zurich Laboratory in Switzerland came up with a new definition of vision of future energy networks (VOFEN) as a novel move toward more efficient and optimized energy systems [3]. In 2007, authors of [4] used the energy hub for the first time, which was a continuation of the VOFEN project. Mathematical equations of the energy hub concept are also present in [5], which are utilized for optimizing energy carriers' integration. The energy hub concept expansion is a promising option for optimal management of multiple energy systems and attaining a comprehensive sustainable energy system model. The energy hub concept is defined as the linkage between consumers, manufacturers, storage, and transmission devices that manage one or more carriers directly or with converters' aid. A comprehensive review regarding different concepts and models for energy hubs is investigated in [1]. In this paper, strengths, weaknesses, and challenges toward dominant structures of energy hubs are identified, and sufficient explanations are expressed.

In general, energy hub studies can be classified into two groups: The first group includes the financial aspects of energy hubs development. The second group includes various methods of control and optimal operation of energy hubs [6]. In this study, the optimal operation of energy hubs is discussed. In reference [7], the optimal performance of an energy hub is investigated utilizing nonlinear programming. A robust optimization approach for managing the energy hub's operation is also presented in [8]. The authors of [4] provided a framework for networked energy systems with multiple energy carriers within a single energy hub. In [9], a probabilistic economic dispatch optimization model for the optimal performance of multiple energy systems is developed by the multi-agent genetic algorithm optimization method. The authors of [10] also presented a multi-period optimization method to solve an energy hub's power flow by considering biomass energy for regional heating with multiple energy carriers. Developing a cluster of energy hubs is an effective strategy to mitigate system costs and greenhouse gas emissions. In [11], a complex energy hub network is modeled and optimized under different scenarios, and both financial variables and the potential reduction of greenhouse gas emissions are tested. In this research, two case studies, including two and three energy hubs in a network, have been considered, and the multi-objective optimization method has been used to obtain the Pareto curve. Despite most researchers' interest in studying energy resources' optimal operation, some studies aim to determine the optimal size of energy resources' internal components. The authors of [12] presented an optimal planning model based on reliability for energy hubs connections. The proposed model calculates the minimum cost of transmission lines network and natural gas pipelines. To obtain the coupling matrix and mitigate the computational burden, a smart modeling and optimization method is illustrated for the energy hub in [13], in which the complex energy hub model separates into several simple energy hubs. The authors of [14] presented an integrated energy hub system based on power-to-gas and compressed air energy storage technologies. The main contribution of this paper focuses on the conditional value-at-risk-based stochastic model and a demand response programming to determine day-ahead optimal scheduling and load shifting in the energy hub system with introduced technologies. A multi-objective stochastic unit commitment problem with battery-based energy storage and demand response programming is also presented in [15] to minimize the system's operating costs and decrease CO_2 emissions of the electricity grid. The authors of [16] presented a combined methodology of decomposed particle swarm optimization (PSO) and the interior point method to solve multiple energy hubs' optimal programming problems considering battery lifetime. The authors of [17] proposed a micro-energy grid (MEG) architecture based on the energy hub. An optimal operation model for MEG is also presented to minimize the cost of daily operation. The results show that the proposed MEG model reduces operating costs, reduces CO_2 emissions, and improves energy efficiency. The authors of [18] presented an efficient model for the energy hub concept based on graph theory. In the current research, a proposed multistage modeling method is used to mitigate the computational burden to develop the coupling matrix, which is obtained by converting nonlinear model equations to linear model equations without losing any information. The results show a significant reduction in the operating costs of the multiple energy system (MES). The authors of [19] presented a strategic optimization framework for the optimal operation of interconnected energy hub systems in the distribution network. The bi-level optimization problem is converted to a MILP problem using the KKT terms and then investigated in the 33-bus network. The results indicate that the system's total cost, which is proportional to the whole power purchased from the upstream network and energy exchanged with energy hubs, has decreased.

The electrical energy supply of power grids, buildings, vehicle charging stations, water desalination, and concerns about increased CO_2 emissions are among the reasons for renewable energy resource utilization. Hydrogen as a green energy carrier is an essential factor in supplying energy for different energy sectors due to its high energy density and ease of storage and transportation. Although hydrogen is known as clean energy, hydrogen-based products have rarely been used in energy systems. Therefore, some studies have attempted to investigate hydrogen storage systems' coordination and renewable sources for hydrogen production for use in hydrogen-based industries, supplying the fuel of hydrogen vehicles, and converting to electricity if necessary [20]. Power-to-gas (P2G) technology has provided a new route for the optimized utilization of distributed renewable energies. With the aid of renewable energies (wind and solar energies in this study), water is converted to hydrogen and oxygen using electrolyzer devices. In this process, the wind turbine's electrical power is utilized to electrolyze water to O_2 and H_2. Water electrolysis is an environmentally friendly reaction through the $H_2O + 2F \rightarrow H_2 + \frac{1}{2}O_2$. F represents the Faraday constant that measures one mole of electricity. The generated hydrogen is either stored in the hydrogen tank or used to generate electrical power in fuel cells, which is presented as $H_2 + \frac{1}{2}O_2 \rightarrow H_2O + 2F$, to compensate electricity in case of power generation shortage [21]. In [22], a planning model for the joint optimization of the wind power transmission project and the hydrogen system is investigated. The transmission project's optimal values and the hydrogen system are obtained under different hydrogen production modes, hydrogen exchange, and hydrogen demand levels. The results show that it is challenging to supply hydrogen in real time due to wind power's intermittent nature. In such cases, the power grid can be used to eliminate hydrogen deficiency. The optimal size of hydrogen systems and wind farms in excessive wind power to produce hydrogen in an isolated power system based on different wind power penetration and wind curtailment level is investigated in [23]. In [24], a robust technique is proposed for the day-ahead scheduling of a smart micro-energy hub (SMEH) integrated with hydrogen-to-power conversion technology, along with an incentive-based integrated demand response to control the load pattern of consumers. The proposed model was formulated as a mix-integer linear programming (MILP) problem. Wind energy is converted to hydrogen and stored as clean energy at low-cost hours. Stored hydrogen can be converted to electricity at high price hours. The robust optimization method was also implemented to consider the electricity price uncertainty. The presented results ensure that the proposed procedure for SMEH planning is optimal, economical, and robust. The authors of [25] evaluated the impact of P2G technology on improving wind energy regulation capability and illustrate that P2G technology improves wind flexibility. In [26], energy management of an energy hub is proposed to minimize operation and emission costs in the presence of wind turbines and demand response. This problem is solved as a dual-objective MILP optimization problem. In some cases, produced hydrogen is utilized in refueling stations to provide energy to the hydrogen/electric-based vehicles charging process. The authors in [27] have determined the optimal performance of a network of hydrogen refueling stations supplied by renewable electricity, based on population density and renewable energy supply. A significant advantage of hydrogen refueling stations is that they can separate refueling demand from electrolyzer operation. By doing so, they can help integrate renewable electricity and take advantage of the electricity market's functioning. Therefore, in [28], an optimization method is applied to determine an electrolyzer-based hydrogen refueling station's optimal performance in an advanced manufacturing park in Rotherham, England. The authors of [29] also established a stochastic model for integrating the Markov chain process into refueling hydrogen fuel vehicle and driver behavior patterns in the South African transport sector. The authors of [30] presented an energy management approach for determining retailers' selling prices in the presence of renewable energy sources and plug-in electric vehicles (PEVs). The outcomes indicate that flexible energy management of PEVs and hydrogen storage systems leads to an increase in expected profit and reduced selling price for the end-user customers. In [31], an optimal algorithm for designing and adjusting the charging station of off-grid electric and hydrogen vehicles by the solar system, diesel generator, and fuel cell is presented. The obtained results represent that increasing solar energy profile and diesel generator optimal performance reduces the cost of the whole system significantly.

According to the previous literature review, there is no focus on the optimal collation operation of IHESs with considering P2X technologies, RERs, and energy storage systems (EESs), so some research gaps can be expressed for previous works as follows:

- In some of the studies such as [4, 7, 8, 19, 32, 33, 34], and [35], the cooling and hydrogen demands are not taken into account, while the coordination of state-of-the-art P2X technologies with other facilities in HES or multi-carrier system to supplying cooling and hydrogen demands is essential.
- The interconnected operation of HES is not considered in many studies such as [7, 8, 17, 24, 36], and [37], while IHESs can be an effective way to improve the energy systems' performance.
- In [19, 24, 31, 32, 35], and [36], co-optimization of energy and emission cost is not investigated, where due to environmental challenges, considering emission cost in the optimal operation of IHESs is imperative.

According to mentioned research gaps, in this chapter, the optimal coalition operation of IHESs with considering P2X technologies, renewable and nonrenewable energy rcsources, and EESs to supplying various types of energy demands is investigated. The aim is to improve the whole system's optimal performance with different facilities' interactions inside each HES and deplete CO_2 emissions to minimize this coalition's operational costs. The main contributions of this work are summarized as follows:

- As shown in Fig. 7.1, optimal coalition operation is a novel framework in the integration of interconnected hybrid energy systems. So, along with the optimal operation of each subsystem, minimizing the whole system costs is considered in this method. In this case, the main priority is to meet the demands of each subsystem by adjacent subsystems.
- Due to the intermittent nature of wind and solar energies, the presence of power-to-hydrogen (P2Hy) technology helps convert electrical power to hydrogen energy for consumption by loads or storage in the hydrogen tank. Also, during peak hours, hydrogen can be converted into electrical energy utilizing the fuel cell.

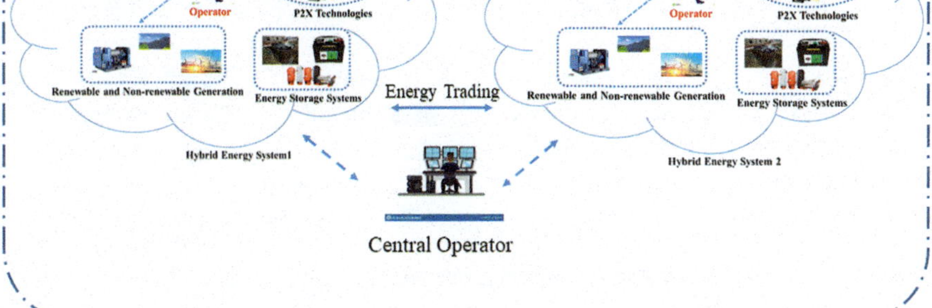

Fig. 7.1 Optimal coalition operation framework

In the following, in Sect. 7.2, the general modeling of IHESs is introduced. Section 7.3 also presents the mathematical equations for the optimization constraints. Section 7.4 deals with the results of the simulations. Section 7.5 is a brief discussion about the status quo, challenges, and outlook. Finally, Sect. 7.6 is the general summary.

7.2 Modeling

As shown in Fig. 7.2, the proposed energy system includes two IHESs that exchange electrical, thermal, and hydrogen energies together. System energy inputs include electricity and natural gas grids. The IHESs outputs contain electricity, heating, cooling, and hydrogen demands. Electricity demand is supplying by the CHP unit's

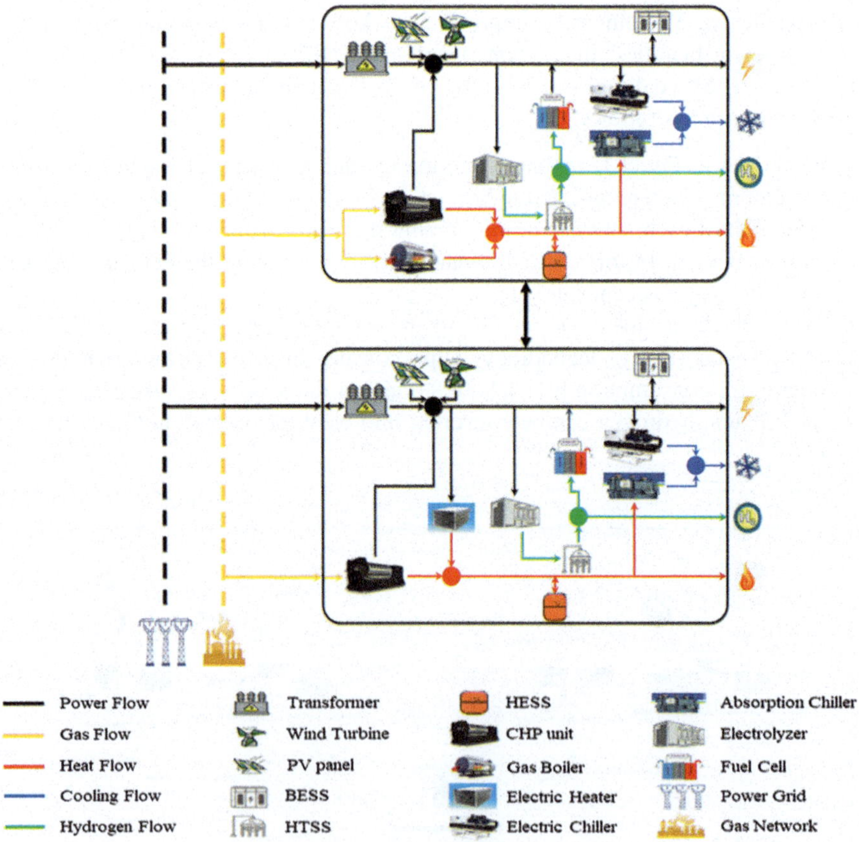

Fig. 7.2 Proposed interconnected hybrid energy system

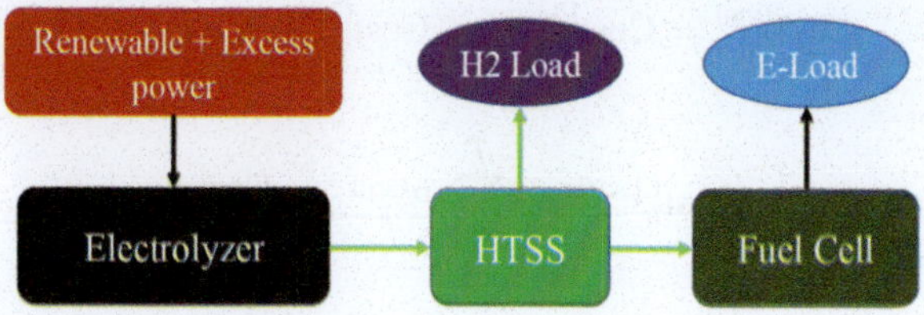

Fig. 7.3 General scheme of hydrogen system

electrical output, renewable energies, fuel cell, and electrical energy imported from the power grid. Heating demands are also provided by the CHP unit's thermal output, gas boiler, and electric heater (P2H technology), which supplies through natural gas and electrical energy in the hybrid system. For cooling applications, absorption chiller and electric chiller (P2C technology) are exploited to increase the system's flexibility in supplying cooling energy at different hours of the day. Hydrogen demands are obtained by the electrolyzer, which generates hydrogen by water electrolysis. In this regard, renewable energy outputs (wind and solar in this paper), due to their intermittent nature along with system excess power, are participating in generating hydrogen by the electrolyzer. The general scheme of the hydrogen system is shown in Fig. 7.3. The storage systems utilized in this study also contain the battery electrical storage system (BESS), heat energy storage system (HESS), and hydrogen tank storage system (HTSS), in which by storing energy at low-load periods, they can use the stored energy at peak-load periods.

7.3 Problem Formulation

In this section, linear equations of the proposed model optimal operation are investigated in singular and coalition operation modes, which are as follows.

7.3.1 Individual Operation of Each HES

In individual operation, each hybrid system operates individually in interaction with the main grid and trades no energy with each other.

7.3.1.1 Objective Function

The considered objective function in this study is to minimize the total cost of HES, which includes expenses of purchased energy from power network, natural gas grid, and costs related to CO_2 emissions:

$$\mathrm{Cost}_h^{\mathrm{Total}} = \mathrm{Cost}_h^{\mathrm{Electric}} + \mathrm{Cost}_h^{\mathrm{Gas}} + \mathrm{Cost}_h^{\mathrm{Emission}} \tag{7.1}$$

$$\mathrm{Cost}_h^{\mathrm{Electric}} = \sum_t^T P_{t,h}^{\mathrm{Grid}} \times \gamma_t^{\mathrm{Electric}} \tag{7.2}$$

$$\mathrm{Cost}_h^{\mathrm{Gas}} = \sum_t^T G_{t,h}^{\mathrm{Grid}} \times \gamma^{\mathrm{Gas}} \tag{7.3}$$

$$\mathrm{Cost}_h^{\mathrm{Emission}} = \kappa \times \sum_t^T \left(P_{t,h}^{\mathrm{Grid}} \times \mu^{\mathrm{Electric}} + G_{t,h}^{\mathrm{Grid}} \times \mu^{\mathrm{Gas}} \right) \tag{7.4}$$

In which $\mathrm{Cost}_h^{\mathrm{Total}}$ is the total cost of each HES. $\mathrm{Cost}_h^{\mathrm{Electric}}$, $\mathrm{Cost}_h^{\mathrm{Gas}}$, and$\mathrm{Cost}_h^{\mathrm{Emission}}$ are the HES's electric, gas, and emission costs, respectively. Equations (7.2) and (7.3) are problem formulation of electricity and natural gas costs, which are equal to the purchased electric power ($P_{t,h}^{\mathrm{Grid}}$) and natural gas ($G_{t,h}^{\mathrm{Grid}}$) from the primary grid multiplied by electricity price ($\gamma_t^{\mathrm{Electric}}$) and natural gas price (γ^{Gas}), respectively. Equation (7.4) also indicates emission costs, in which the first term is the amount of released emission by purchasing power from the grid and the second term is related to the amount of released emission by gas consumption of devices inside the HES. Here κ is the factor of emission cost. μ^{Electric} and μ^{Gas} are emission factors for electricity and natural gas, respectively.

7.3.1.2 Energy Balance Equations

Energy balance equations for electrical, heating, and cooling loads include the coequality of the total energy of the supply side with the demand side, which should be guaranteed at any scheduled time horizon:

$$\begin{aligned} \eta_h^T \times P_{t,h}^{\mathrm{Grid}} + P_{t,h}^{\mathrm{CHP}} + P_{t,h}^{\mathrm{BESS,dch}} + P_{t,h}^{\mathrm{FC}} + P_{t,h}^{\mathrm{PV}} + P_{t,h}^{\mathrm{Wind}} &= P_{t,h}^{\mathrm{Load}} \\ &+ P_{t,h}^{\mathrm{BESS,ch}} + P_{t,h}^{\mathrm{EC}} + P_{t,h=2}^{\mathrm{EH}} + P_{t,h}^{\mathrm{EL}} \end{aligned} \tag{7.5}$$

$$H_{t,h}^{\mathrm{CHP}} + H_{t,h=1}^{\mathrm{Bo}} + H_{t,h=2}^{\mathrm{EH}} + H_{t,h}^{\mathrm{HESS,dch}} = H_{t,h}^{\mathrm{Load}} + H_{t,h}^{\mathrm{AC}} + H_{t,h}^{\mathrm{HESS,ch}} \tag{7.6}$$

$$C_{t,h}^{\mathrm{EC}} + C_{t,h}^{\mathrm{AC}} = C_{t,h}^{\mathrm{Load}} \tag{7.7}$$

According to the Eq. (7.5), to meet consumers' electrical load ($P_{t,h}^{\mathrm{Load}}$), BESS charging ($P_{t,h}^{\mathrm{BESS,ch}}$), and electrical demands of P2X devices, including electric chiller ($P^{EC}_{t,h}$), electrolyzer ($P^{EL}_{t,h}$), and electric heater ($P^{EH}_{t,h}$), a combination of generated power by renewable energy sources ($P^{Wind}_{t,h}$ and $P^{PV}_{t,h}$), CHP unit ($P^{CHP}_{t,h}$), and BESS discharge ($P^{BESS,dch}_{t,h}$) is used along with purchased electric power from the grid with efficiencyη^{T}_{h} at the time t and HES h. The produced heat power by CHP unit ($H^{CHP}_{t,h}$), boiler ($H^{Bo}_{t,h}$), electric heater ($H^{EH}_{t,h}$), and discharging of HESS ($H^{HESS,dch}_{t,h}$) illustrated in Eq. (7.6) supplies the heating load ($H^{Load}_{t,h}$), heating requirements of the absorption chiller ($H^{AC}_{t,h}$), and HESS charging ($H^{HESS,ch}_{t,h}$) at the time t and HES h. Equation (7.7) also indicates that the cooling load ($C^{Load}_{t,h}$) can be provided by absorption chiller ($C^{AC}_{t,h}$) and electric chiller ($C^{EC}_{t,h}$) cooling outputs at the time t and HES h.

The purchased natural gas flows into both the CHP units and the gas boiler. Let $G^{CHP}_{t,h}$ and $G^{Bo}_{t,h}$ be the natural gas consumed by the CHP unit and gas boiler at the time t and HES h, respectively. So consumed natural gas amount can be described as follows:

$$G^{Grid}_{t,h} = G^{CHP}_{t,h} + G^{Bo}_{t,h} \tag{7.8}$$

7.3.1.3 Facilities Equations

Equations (7.9, 7.10, 7.11, 7.12, 7.13, 7.14, 7.15, and 7.16) represent the output energy of installed facilities in the HESs, which is utilized to supply electrical, heating, and cooling energies:

$$P^{CHP}_{t,h} = G^{CHP}_{t,h} \times \eta^{E,CHP}_{h} \tag{7.9}$$

$$H^{CHP}_{t,h} = G^{CHP}_{t,h} \times \eta^{H,CHP}_{h} \tag{7.10}$$

$$H^{Bo}_{t,h} = G^{Bo}_{t,h} \times \eta^{Bo}_{h} \tag{7.11}$$

$$H^{EH}_{t,h} = P^{EH}_{t,h} \times \eta^{EH}_{h} \tag{7.12}$$

$$C^{EC}_{t,h} = P^{EC}_{t,h} \times \mathrm{COP}^{EC}_{h} \tag{7.13}$$

$$C^{AC}_{t,h} = H^{AC}_{t,h} \times \mathrm{COP}^{AC}_{h} \tag{7.14}$$

$$\mathrm{Hy}_{t,h}^{\mathrm{EL}} = \eta_h^{\mathrm{EL}} \times P_{t,h}^{\mathrm{EL}} \tag{7.15}$$

$$P_{t,h}^{\mathrm{FC}} = \eta_h^{\mathrm{FC}} \times \mathrm{Hy}_{t,h}^{\mathrm{FC}} \tag{7.16}$$

Equations (7.9) and (7.10) describe CHP unit outputs, including electrical energy and heating energy contributions. $\eta_h^{\mathrm{E,CHP}}$ and $\eta_h^{\mathrm{H,CHP}}$ are electrical and heating efficiencies of the CHP unit in HES h, respectively. Equations (7.11) and (7.12) demonstrate the heat generated by the combustion of natural gas in the gas boiler and electricity consumption in the electric heater, along with multiplying their efficiency (η_h^{Bo} and η_h^{EH}), respectively. Equations (7.13) and (7.14) are related to meet the cooling demands of the electric/absorption chillers. Here $\mathrm{COP}_h^{\mathrm{EC}}$ and $\mathrm{COP}_h^{\mathrm{AC}}$ are coefficients of performance of the electric/absorption chillers in HES h, respectively. In the P2Hy process, in Eq. (7.15), $\mathrm{Hy}_{t,h}^{\mathrm{EL}}$ is the hydrogen produced by the electrolyzer at the time t and HES h, and η_h^{EL} is the efficiency of the electrolyzer. Equation (7.16) also illustrates the fuel cell output power. In this equation, the produced hydrogen by electrolyzer or stored in the hydrogen tank is utilized to generate electricity with the efficiency $\eta_{\mathrm{h}}^{\mathrm{FC}}$.

7.3.1.4 The Energy Level of Storages

Energy storages are vital facilities of the HESs, which elevate the flexibility and reliability of such systems. The energy level of each storage system can be formulated as follows:

$$E_{t,h}^{\mathrm{BESS}} = E_{t-1,h}^{\mathrm{BESS}} \times \left(1 - \varphi_h^{\mathrm{BESS}}\right) + \left(P_{t,h}^{\mathrm{BESS,ch}} \times \eta^{\mathrm{BESS,ch}} - \frac{P_{t,h}^{\mathrm{BESS,dch}}}{\eta^{\mathrm{BESS,dch}}}\right) \tag{7.17}$$

$$E_{t,h}^{\mathrm{HESS}} = E_{t-1,h}^{\mathrm{HESS}} \times \left(1 - \varphi_h^{\mathrm{HESS}}\right) + \left(P_{t,h}^{\mathrm{HESS,ch}} \times \eta^{\mathrm{HESS,ch}} - \frac{P_{t,h}^{\mathrm{HESS,dch}}}{\eta^{\mathrm{HESS,dch}}}\right) \tag{7.18}$$

$$E_{t,h}^{\mathrm{HTSS}} = E_{t-1,h}^{\mathrm{HTSS}} \times \left(1 - \varphi_h^{\mathrm{HTSS}}\right) + \left(\mathrm{Hy}_{t,h}^{\mathrm{EL}} - \mathrm{Hy}_{t,h}^{\mathrm{FC}}\right) - \mathrm{Hy}_{t,h}^{\mathrm{Load}} \tag{7.19}$$

In which $\mathrm{E}_{t,h}^{\mathrm{BESS}}$, $\mathrm{E}_{t,h}^{\mathrm{HESS}}$, and $\mathrm{E}_{t,h}^{\mathrm{HTSS}}$ are energy levels of BESS, HESS, and HTSS at time t and the HES h, respectively. $E_{t-1,h}$ in each equation represents the energy level of storages at the time $t - 1$ and HES h. φ_{h} is the energy loss ratio of the energy storage in HES h.η^{ch} and η^{ch} are also charging and discharging efficiencies of energy storages, respectively. Furthermore, in Eq. (7.19), we assumed no loss in the HTSS charge/discharge process.

To prevent possible errors in the state of stored energy at the end of the scheduled period in storage systems, we assumed that end of the dispatch period equal to the initial values. These assumptions are shown in Eqs. (7.20, 7.21, 7.22):

$$E_{t=24,h}^{\text{BESS}} = E_h^{\text{BESS,Initial}} \tag{7.20}$$

$$E_{t=24,h}^{\text{HESS}} = E_h^{\text{HESS,Initial}} \tag{7.21}$$

$$E_{t=24,h}^{\text{HTSS}} = E_h^{\text{HTSS,Initial}} \tag{7.22}$$

7.3.1.5 Constraints

The main power and natural gas grids' upper and lower constraints for the whole system are expressed as:

$$0 \leq P_{t,h}^{\text{Grid}} \leq P_h^{\text{Grid,Max}} \tag{7.23}$$

$$0 \leq G_{t,h}^{\text{Grid}} \leq G_h^{\text{Grid,Max}} \tag{7.24}$$

Constraint terms related to Eqs. (7.9, 7.10, 7.11, 7.12, 7.13, 7.14, 7.15, and 7.16) for installed facilities in the hybrid system are listed below, respectively, which demonstrate the upper and lower limitations:

$$0 \leq P_{t,h}^{\text{CHP}} \leq P_h^{\text{CHP,Max}} \tag{7.25}$$

$$0 \leq H_{t,h}^{\text{Bo}} \leq H_h^{\text{Bo,Max}} \tag{7.26}$$

$$0 \leq H_{t,h}^{\text{EH}} \leq H_h^{\text{EH,Max}} \tag{7.27}$$

$$0 \leq C_{t,h}^{\text{EC}} \leq C_h^{\text{EC,Max}} \tag{7.28}$$

$$0 \leq C_{t,h}^{\text{AC}} \leq C_h^{\text{AC,Max}} \tag{7.29}$$

$$0 \leq P_{t,h}^{\text{EL}} \leq P_h^{\text{EL,Max}} \tag{7.30}$$

$$0 \leq P_{t,h}^{\mathrm{FC}} \leq P_{h}^{\mathrm{FC,Max}} \tag{7.31}$$

The following equations indicate the charge/discharge ranges and upper and lower limitations of the stored energy of BESS:

$$0 \leq P_{t,h}^{\mathrm{BESS,dch}} \leq P_{h}^{\mathrm{BESS,dch,Max}} \tag{7.32}$$

$$0 \leq P_{t,h}^{\mathrm{BESS,ch}} \leq P_{h}^{\mathrm{BESS,ch,Max}} \tag{7.33}$$

$$E_{h}^{\mathrm{BESS,Min}} \leq E_{t,h}^{\mathrm{BESS}} \leq E_{h}^{\mathrm{BESS,Max}} \tag{7.34}$$

Similar to BESS, the HESS constraints are defined as follows:

$$0 \leq H_{t,h}^{\mathrm{HESS,dch}} \leq H_{h}^{\mathrm{HESS,dch,Max}} \tag{7.35}$$

$$0 \leq H_{t,h}^{\mathrm{HESS,dch}} \leq H_{h}^{\mathrm{HESS,dch,Max}} \tag{7.36}$$

$$E_{h}^{\mathrm{HESS,Min}} \leq E_{t,h}^{\mathrm{HESS}} \leq E_{h}^{\mathrm{HESS,Max}} \tag{7.37}$$

Upper and lower limitations for HTSS also are given in Eq. (7.38):

$$E_{h}^{\mathrm{HTSS,Min}} \leq E_{t,h}^{\mathrm{HTSS}} \leq E_{h}^{\mathrm{HTSS,Max}} \tag{7.38}$$

7.3.2 Coalition Operation of IHES

In coalition operation, hybrid energy systems are interconnected and can trade electricity, heat, and hydrogen. The aim is to consider the whole IHESs as a unique system to attain global optimization.

7.3.2.1 Objective Function

The whole system objective function to obtain global optimization is obtained by summation of all new costs in coalition operation status ($\mathrm{Cost}_{h,\mathrm{new}}^{\mathrm{Total}}$), which is described as follows:

$$\mathrm{Cost}^{\mathrm{Interconnected}} = \sum_{h} \mathrm{Cost}_{h,\mathrm{new}}^{\mathrm{Total}} \tag{7.39}$$

7.3.2.2 Interconnection Equations

Equations (7.40, 7.41, and 7.42) are applied to calculate the quantity of energy exchange and establish the trade balance between energy import and export in the IHES:

$$\sum_t \sum_g P_{t,h,g}^{\mathrm{Exc}} = 0 \tag{7.40}$$

$$\sum_t \sum_g H_{t,h,g}^{\mathrm{Exc}} = 0 \tag{7.41}$$

$$\sum_t \sum_g \mathrm{Hy}_{t,h,g}^{\mathrm{Exc}} = 0 \tag{7.42}$$

$P_{t,h,g}^{\mathrm{Exc}}$, $H_{t,h,g}^{\mathrm{Exc}}$, and $\mathrm{Hy}_{t,h,g}^{\mathrm{Exc}}$ are traded energy (electricity, heat, and hydrogen, respectively) at the time t between HESs h and g. These constraints represent that all imported energy at the time t and HES h must be equal to exported energy at that time and HES g.

7.3.2.3 Energy Balance Equations

In interconnected coalition operation, with inserting trade parameters, Eqs. (7.5) and (7.6) are changed as follows:

$$\begin{aligned} \eta_h^T \times P_{t,h}^{\mathrm{Grid}} + P_{t,h}^{\mathrm{CHP}} + P_{t,h}^{\mathrm{BESS,dch}} + P_{t,h}^{\mathrm{FC}} + P_{t,h}^{\mathrm{PV}} + P_{t,h}^{\mathrm{Wind}} + \sum_g P_{g,h}^{\mathrm{Exc}} = P_{t,h}^{\mathrm{Load}} \\ + P_{t,h}^{\mathrm{BESS,ch}} + P_{t,h}^{\mathrm{EC}} + P_{t,h=2}^{\mathrm{EH}} + \sum_g P_{h,g}^{\mathrm{Exc}} + P_{t,h}^{\mathrm{EL}} \end{aligned} \tag{7.43}$$

$$\begin{aligned} H_{t,h}^{\mathrm{CHP}} + H_{t,h=1}^{\mathrm{Bo}} + H_{t,h=2}^{\mathrm{EH}} + H_{t,h}^{\mathrm{HESS,dch}} + \sum_g H_{t,g,h}^{\mathrm{Exc}} = H_{t,h}^{\mathrm{Load}} + H_{t,h}^{\mathrm{AC}} + H_{t,h}^{\mathrm{HESS,ch}} \\ + \sum_g H_{t,h,g}^{\mathrm{Exc}} \end{aligned} \tag{7.44}$$

7.3.2.4 Constraints

To achieve optimal solutions in coalition operation, we assume coalition operation costs of the IHESs should be less than singular operation costs:

$$\mathrm{Cost}_{h,\mathrm{new}}^{\mathrm{Total}} \leq \mathrm{Cost}_{h,\mathrm{old}}^{\mathrm{Total}} \tag{7.45}$$

In addition, the energy-level equation of HTSS in the interconnected coalition mode can be written as below by adding exchange summations:

$$E_{t,h}^{\mathrm{HTSS}} = E_{t-1,h}^{\mathrm{HTSS}} \times \left(1 - \varphi_h^{\mathrm{HTSS}}\right) + \left(\mathrm{Hy}_{t,h}^{\mathrm{EL}} - \mathrm{Hy}_{t,h}^{\mathrm{FC}}\right) - \mathrm{Hy}_{t,h}^{\mathrm{Load}} - \sum_g \mathrm{Hy}_{t,h,g}^{\mathrm{Exc}} + \sum_g \mathrm{Hy}_{t,g,h}^{\mathrm{Exc}} \tag{7.46}$$

The energy exchange between IHESs depends on each energy transaction state, which is illustrated by the following equations:

$$-P_{t,h,g}^{\mathrm{Exc,Max}} \leq P_{t,h,g}^{\mathrm{Exc}} \leq P_{t,h,g}^{\mathrm{Exc,Max}} \tag{7.47}$$

$$-H_{t,h,g}^{\mathrm{Exc,Max}} \leq H_{t,h,g}^{\mathrm{Exc}} \leq H_{t,h,g}^{\mathrm{Exc,Max}} \tag{7.48}$$

$$-\mathrm{Hy}_{t,h,g}^{\mathrm{Exc,Max}} \leq \mathrm{Hy}_{t,h,g}^{\mathrm{Exc}} \leq \mathrm{Hy}_{t,h,g}^{\mathrm{Exc,Max}} \tag{7.49}$$

7.4 Simulation and Numerical Results

In this section, in the first part, the input data used in this chapter are introduced. Then, in the second subsection, the obtained results of the simulations are displayed in the form of tables and figures. And the obligatory explanations are expressed.

7.4.1 Input Data

This study's proposed methodology contains two interconnected hybrid energy systems with coalition correlation in electrical, thermal, and hydrogen loads to demonstrate the proficiency of the presented model. Considered renewable energies

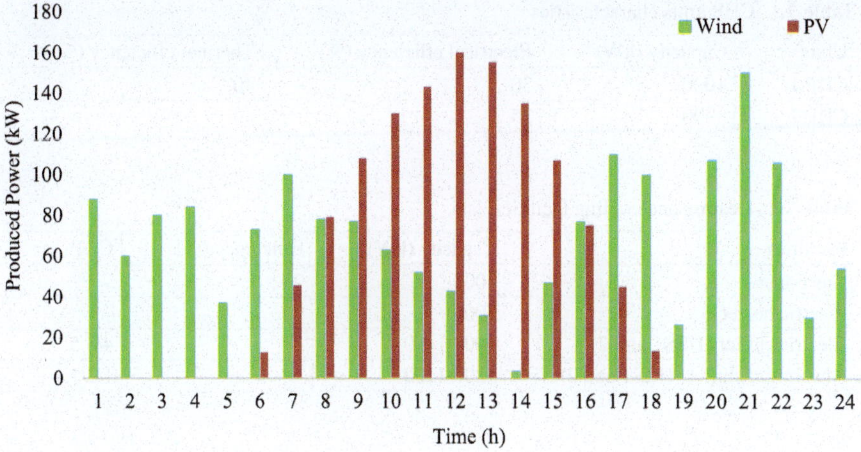

Fig. 7.4 Wind turbine and PV panel output power

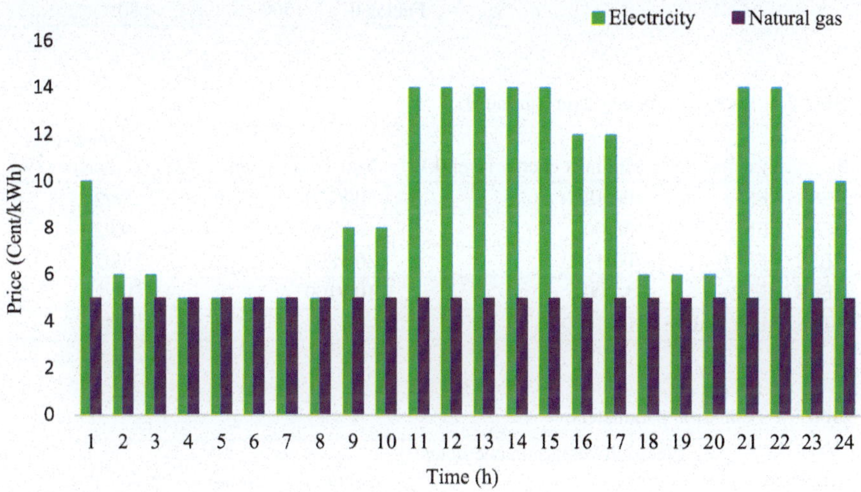

Fig. 7.5 The electricity and natural gas pricing diagram

include wind turbine and PV panel, in which their output power is shown in Fig. 7.4. Real-time electricity pricing and constant natural gas pricing diagrams are illustrated in Fig. 7.5. Parameters of CHP units are indicated in Table 7.1. Heating and cooling facilities data are given in Table 7.2. Characteristics of hydrogen systems are represented in Table 7.3, and energy storage system parameters are demonstrated in Table 7.4. Furthermore, CO_2 emission parameters are added in Table 7.5. Figures 7.6 and 7.7 are related to the electrical, thermal, and cooling demands of HES1 and HES2, respectively. Hydrogen demand information is also shown in Fig. 7.8. All renewable outputs and HES1 parameters are adopted from [17], and energy prices

Table 7.1 CHP units characteristics

Units	Capacity (kW)	Electrical efficiency (%)	Thermal efficiency (%)
CHP 1	1000	30	40
CHP 2	900	35	40

Table 7.2 Heating and cooling facilities data

Facilities	Capacity (kW)	Efficiency (%)	COP
Gas boiler	1000	90	–
Electric heater	1000	90	–
Electric chiller (HES1 and 2)	2000/1400	–	4/3.5
Absorption chiller (HES1 and 2)	1200/1100	–	1.2/1.1

Table 7.3 Hydrogen system characteristics

Devices	Capacity (kW)	Efficiency (%)
Electrolyzer	500	70
Fuel cell	500	80

Table 7.4 Energy storage system parameters

Storage systems	Min/max energy level (kW)	Max Ch/Dch range (kW)	Range/stdby efficiency (%)
BESS (HES1)	400/1800	500/700	96/2
BESS (HES2)	35/700	100/100	92/2
HESS (HES1)	400/1800	800/800	98/2
HESS (HES2)	35/700	100/100	92/2
HTSS (HES1 and 2)	100/1000	–	–/2

Table 7.5 Emission parameters

Penalty cost (κ)	Electricity emission coefficient ($\mu^{Electric}$)	Natural gas emission coefficient (μ^{Gas})
0.5 (cent/kWh)	0.972 (kg/kWh)	0.23 (kg/kWh)

and HES2 parameters are taken from [38]. HES1 and HES2 hydrogen demands are adopted from [24] and [36], respectively.

In order to solve the proposed problems, under the CPLEX solver in GAMS software, the mixed-integer linear programming (MILP) optimization method is accomplished to minimize the objective function value. All of the simulations have been done on a computer equipped with Intel(R) Core(TM) i7-7700k CPU 4.2GHZ and 16GB RAM. Also, the computational time for all simulations were less than 0.5 seconds.

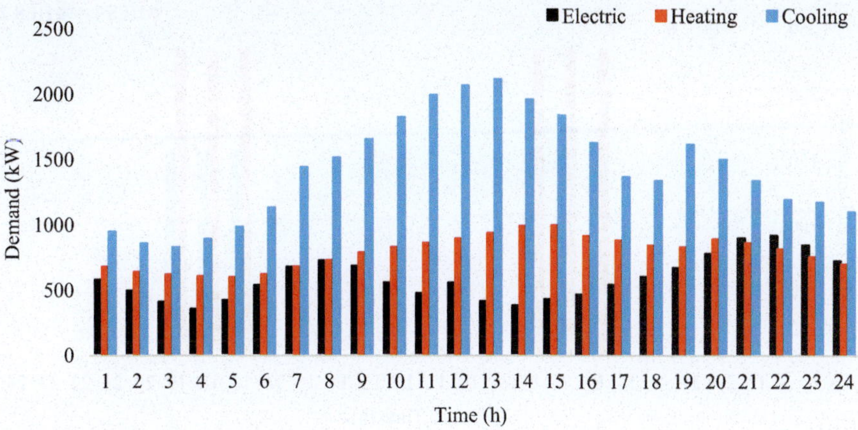

Fig. 7.6 The HES1 demand loads

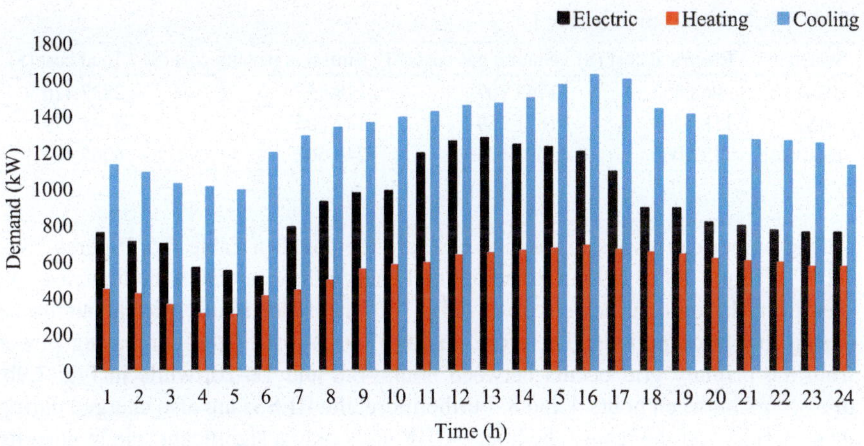

Fig. 7.7 The HES2 demand loads

7.4.2 Simulation Results and Discussion

In this subsection, in order to validate the effectiveness of the proposed model in the optimal operation of hybrid energy systems, two test cases are introduced.

For the first case, we assume that hybrid energy systems are isolated and only have interaction with the primary grid. Each of the hybrid system components has a different generation pattern according to its energy demands. In the second case, hybrid systems are interconnected and trade electricity, thermal energy, and hydrogen. Indeed, in this case, trade operation type is considered as the coalition. This means that each hybrid system has prior responsibility in meeting the shortage of energy demands.

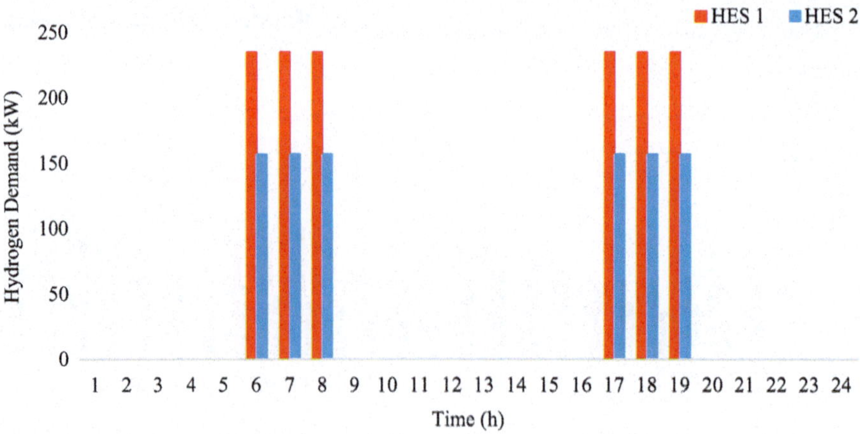

Fig. 7.8 Hydrogen demand loads

Table 7.6 Numeral results in case 1

System	Electrical cost ($)	Natural gas cost ($)	Emission penalty cost ($)	Total cost ($)
HES1	968.862	1749.707	118.857	2837.426
HES2	1188.204	1871.969	127.805	3187.978
Total costs	2157.066	3621.676	246.662	6065.404

The numerical results of the first case are tabulated in Table 7.6. Figures 7.9a, 7.10, 7.11, and 7.12b belong to the optimal energy flow curves in the first case. As seen from the optimal power curves (Fig. 7.9), most electrical power is purchased during low-price periods. Therefore, in Fig. 7.9a, most of the purchasing power from the primary grid occurs between hours 4–8 and 18–20, while in Fig. 7.9b, this occurs between hours 4 and 8. Furthermore, the BESSs are also charged during these periods. At the rest of the hours, CHP units play a significant role in meeting the electrical demands in both HESs.

Gas boiler, electric heater, and CHP units are responsible for supplying load demands for optimal heating energy flow, as demonstrated in Fig. 7.10a, b. Because CHP units do not have output during low-price electricity hours, gas boiler and electrical heater fulfill heat demands during these periods. For optimal cooling flow, the majority of demands are provided by electric chiller in both hybrid systems (Fig. 7.11a, b).

In the hydrogen system shown in Fig. 7.3, the electrolyzer converts renewable and excess powers to hydrogen. So, the optimal hydrogen flow indicated in Fig. 7.12 includes generation and consumption of hydrogen and the HTSS state of charge. As formerly explained, electrolyzer generates hydrogen in low-price electricity periods in both HESs. In Fig. 7.12a, the fuel cell has no power generation due to the high demand for hydrogen and desirable electrical power supply in this subsystem.

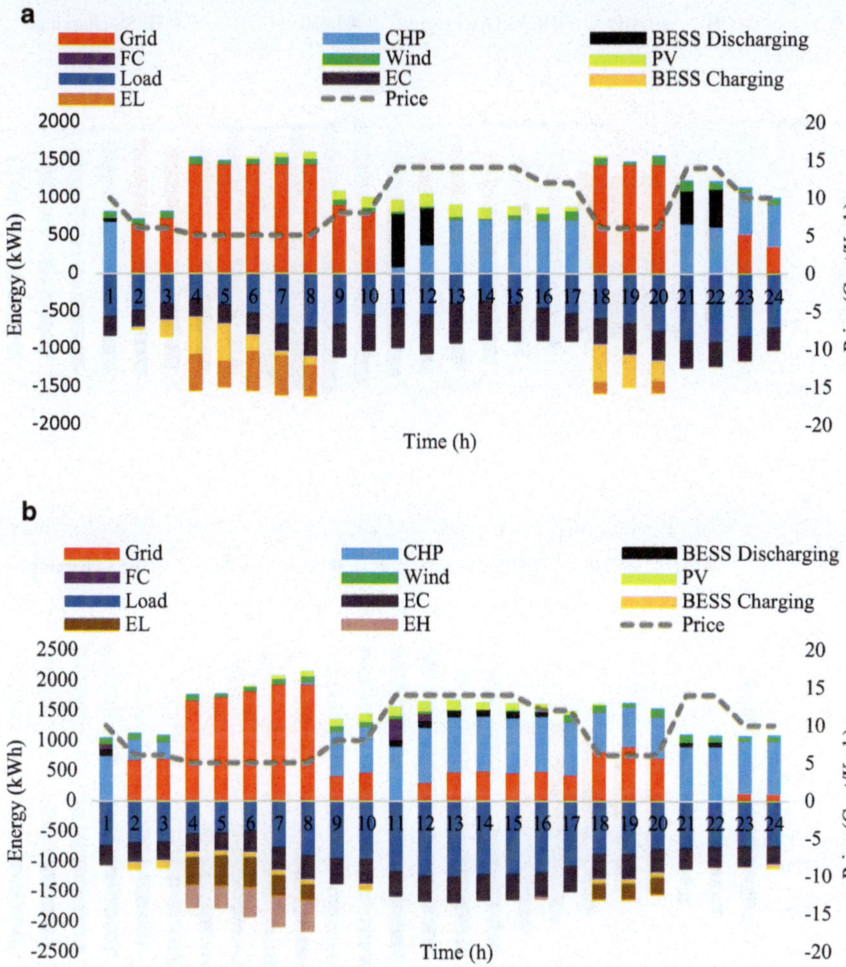

Fig. 7.9 (**a**) The HES1 electrical optimal power flow in case 1. (**b**) The HES2 electrical optimal power flow in case 1

However, the HES2 fuel cell generates electrical power in three different hours (Fig. 7.12b).

Table 7.7 also tabulated the second case numerical results. As it can be seen from the obtained results comparison in this table and Table 7.6, the costs related to electricity have decreased from \$ 2157.066 to \$ 2022.773, although this cost has increased for HES2 in the second case. In the natural gas cost section, we see a decrease in the total cost from \$ 3621.676 in the first case to \$ 3590.759 in the second case, while the HES1 natural gas cost in the second case (due to more use of gas burn facilities) is more than in the first case. The cost of emissions in HES1 for the first case is \$ 118.805, while in the second case, it is \$ 114,249. However, the

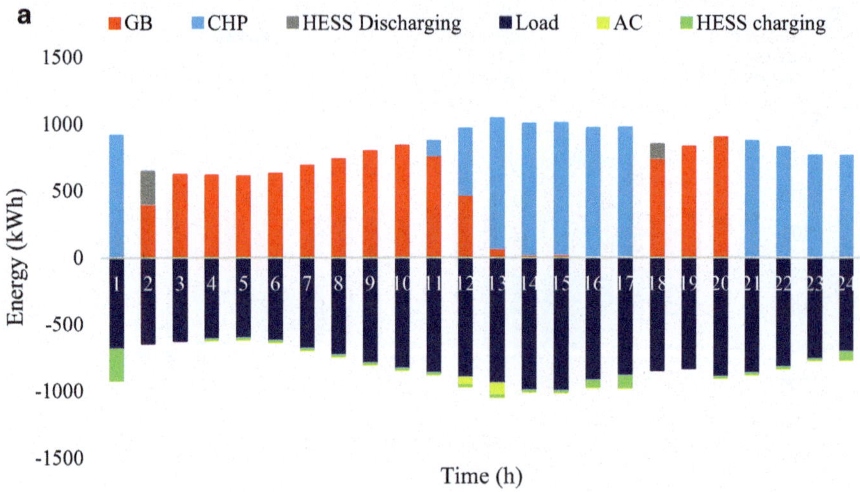

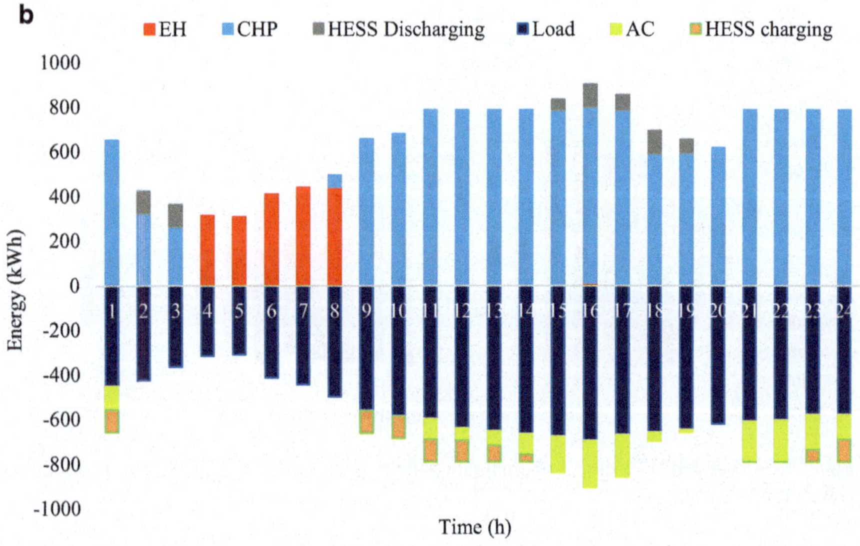

Fig. 7.10 (**a**) The HES1 optimal heating flow in case 1. (**b**) The HES2 optimal heating flow in case 1

natural gas consumed in the second case is higher. In the first case, the amount of electricity purchased from the primary grid is more than in the second case, which is the reason for the increase in emission costs. The same conditions are valid for HES2 (see Eq. 7.4). The total energy costs in HES1 for both cases are $ 2837.426 due to the existence of Eq. (7.45) (see Subsect. 7.3.2.4 for more details). Finally, in the second test case, the total value of the objective function compared with the first test case has a 2.7% decrease.

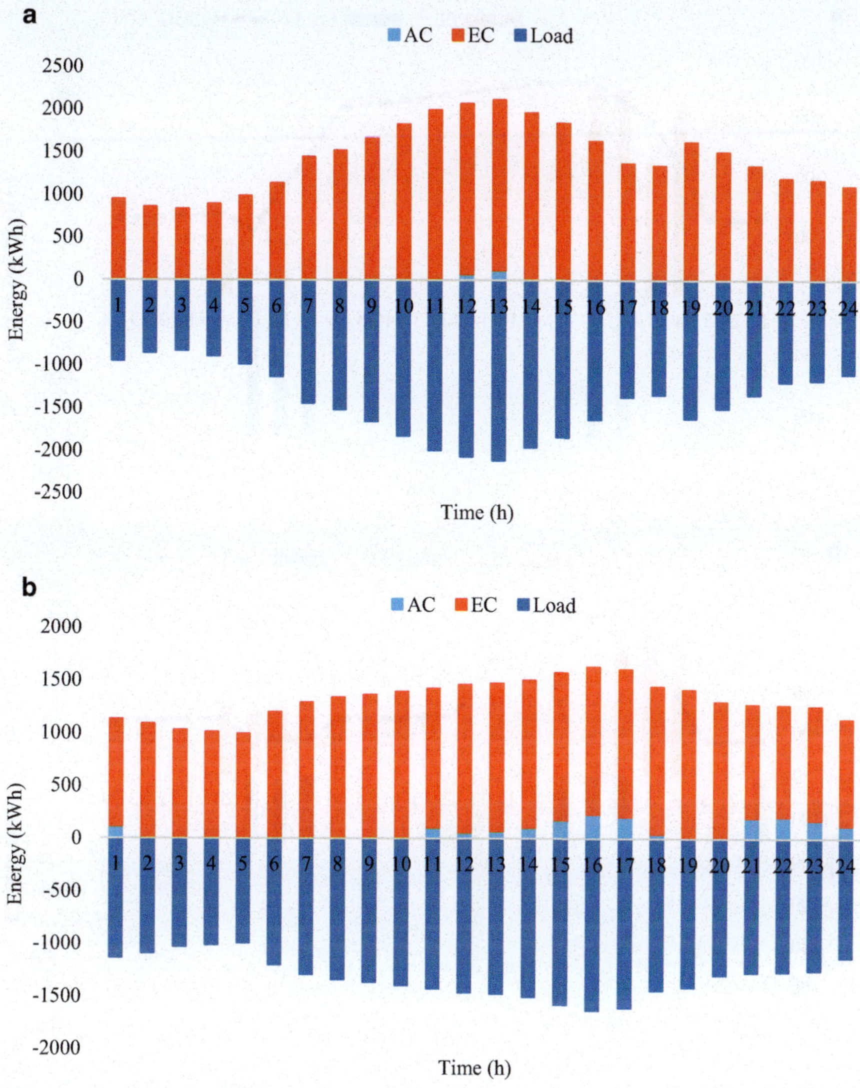

Fig. 7.11 (**a**) The HES1 optimal cooling flow in case 1. (**b**) The HES2 optimal cooling flow in case 1

Figure 7.13a, b demonstrates optimal dispatch of electrical power under interconnected coalition operation. In Fig. 7.13a, b, similar to Fig. 7.9a, b, grid power purchasing accrue in low-price periods, and the CHP units are off in these intervals either. Besides, from the power trading point of view, during the daytime, IHES2 imports more electrical power, while during the night, IHES1 is the one that receives more electrical power.

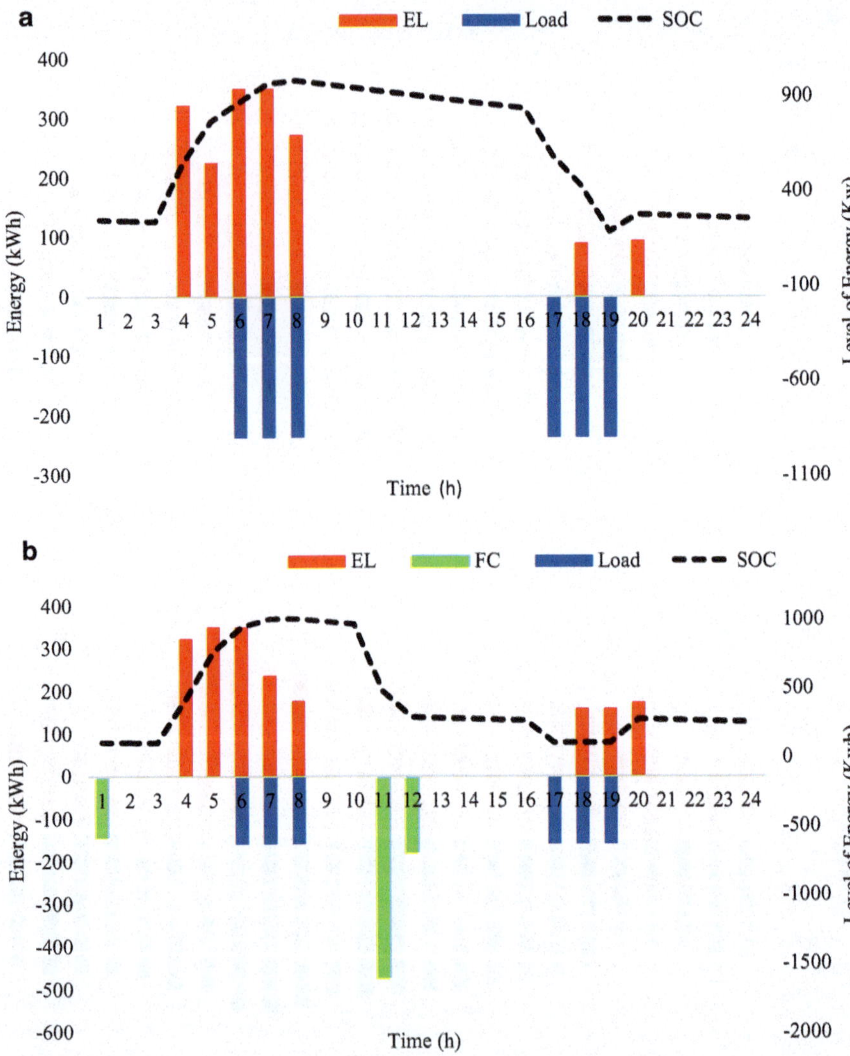

Fig. 7.12 (**a**) The HES1 optimal hydrogen flow in case 1. (**b**) The HES2 optimal hydrogen flow in case 1

Table 7.7 Numeral results in case 2

System	Electrical cost ($)	Natural gas cost ($)	Emission penalty cost ($)	Total cost ($)
HES1	822.234	1900.943	114.249	2837.426
HES2	1200.539	1689.816	135.852	3026.208
Total costs	2022.773	3590.759	250.101	5863.634

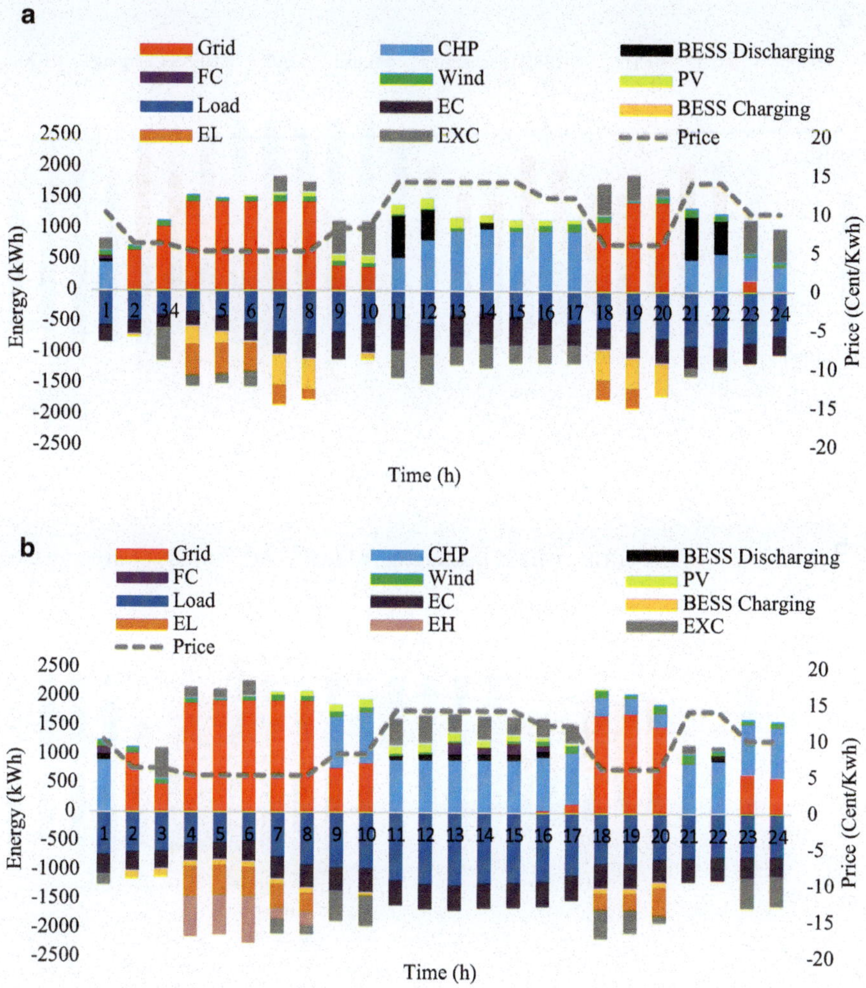

Fig. 7.13 (**a**) Optimal dispatch of electrical power in IHES1 in case 2. (**b**) Optimal dispatch of electrical power in IHES2 in case 2

As it is evident from Fig. 7.14a, b, heat generation devices' outputs (especially gas boiler) are diminished due to thermal energy exchange. Despite the fact that there are significant variations in thermal balance values in interconnected coalition operation compared to the individual operation, there is little change in the optimal dispatch of cooling energy (Fig. 7.15a, b).

The optimal hydrogen dispatch of the IHESs presented in Fig. 7.16a, b represents the reduction in the hourly rate of hydrogen production compared to the first case, mainly due to hydrogen exchanges. In this system, IHES1 acts more as a hydrogen

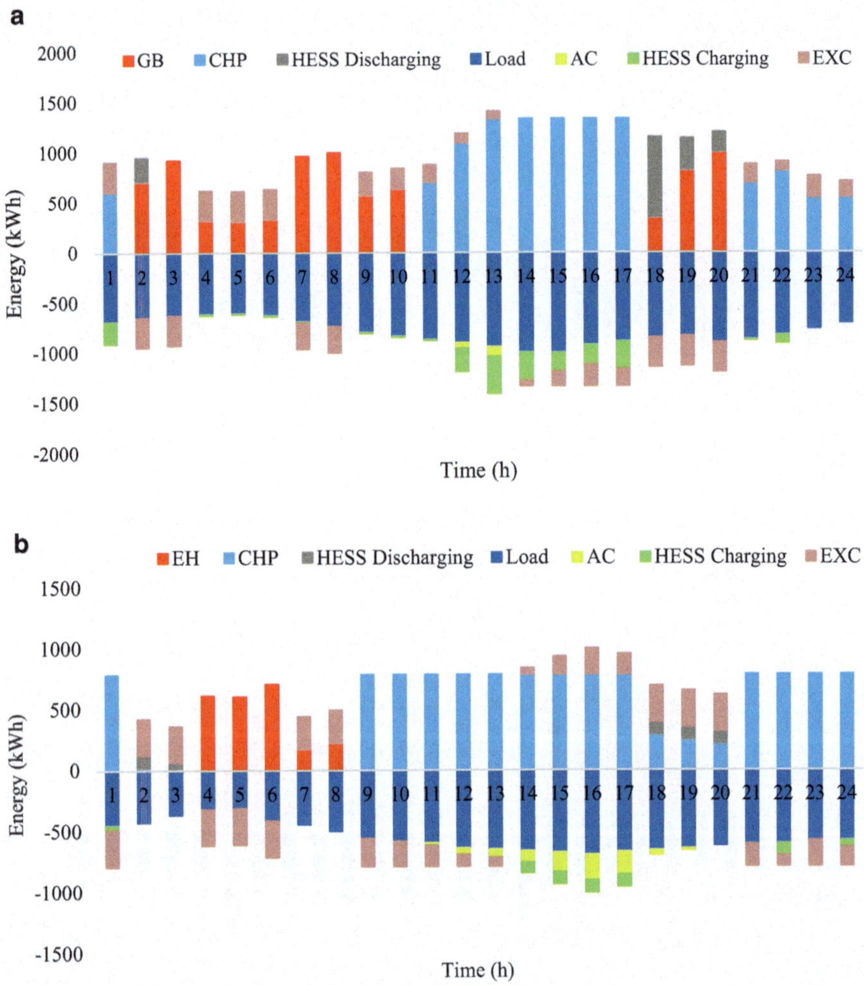

Fig. 7.14 (**a**) Optimal dispatch of heating energy in IHES1 in case 2. (**b**) Optimal dispatch of heating energy in IHES2 in case 2

transporter most of the time. Likewise, the amount of electricity generated by the fuel cell in IHES2 has increased.

7.5 Status Quo, Challenges, and Outlook

This chapter focuses on the optimal coalition operation of interconnected multi-carrier energy systems and P2X technologies. Currently, plenty of research work has

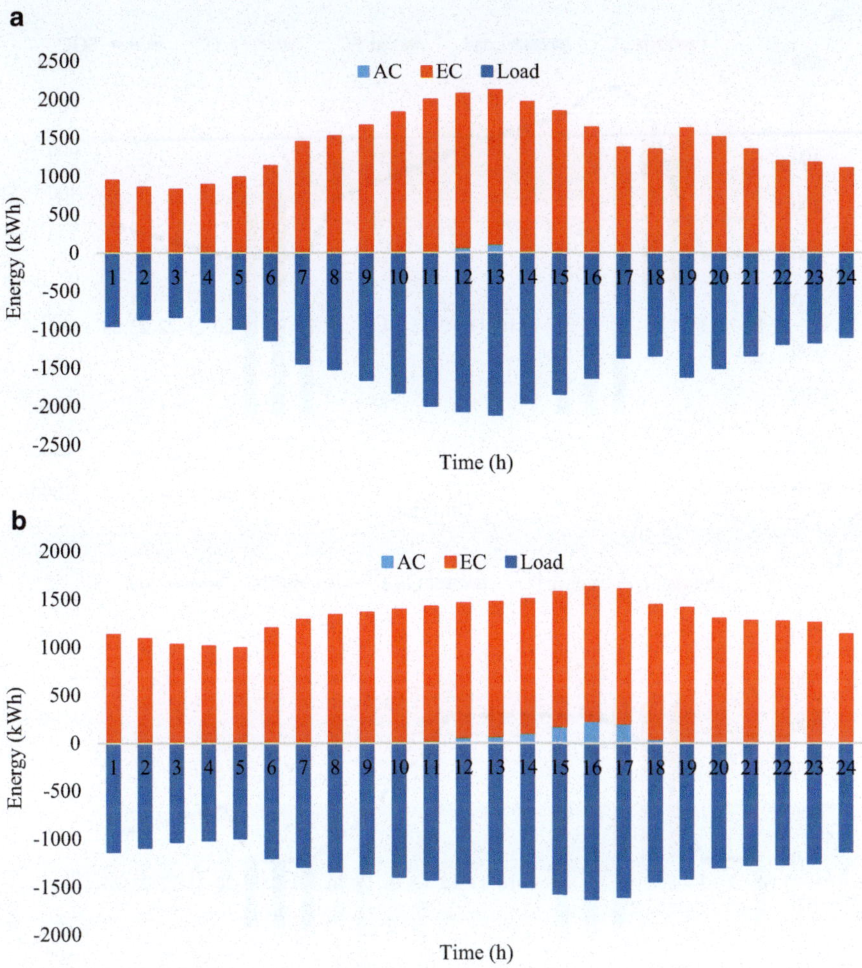

Fig. 7.15 (**a**) Optimal dispatch of cooling energy in IHES1 in case 2. (**b**) Optimal dispatch of cooling energy in IHES2 in case 2

been performed in these fields. The overall objective is to achieve a comprehensive situation of future energy systems. In the case of interconnected operation of hybrid energy systems, the majority of the methods are conceptual, and less consideration has been paid to practical schemas and prototypes. The overall structure of these systems is a set of generation, storage, and consumption devices along with energy converters to achieve optimal performance. For P2X technologies, which are considered almost new technologies, the focus is more on using these technologies to achieve more optimal operation in modern energy systems. A brief explanation of the conducted researches on these topics is provided in the Introduction Sect. (7.1).

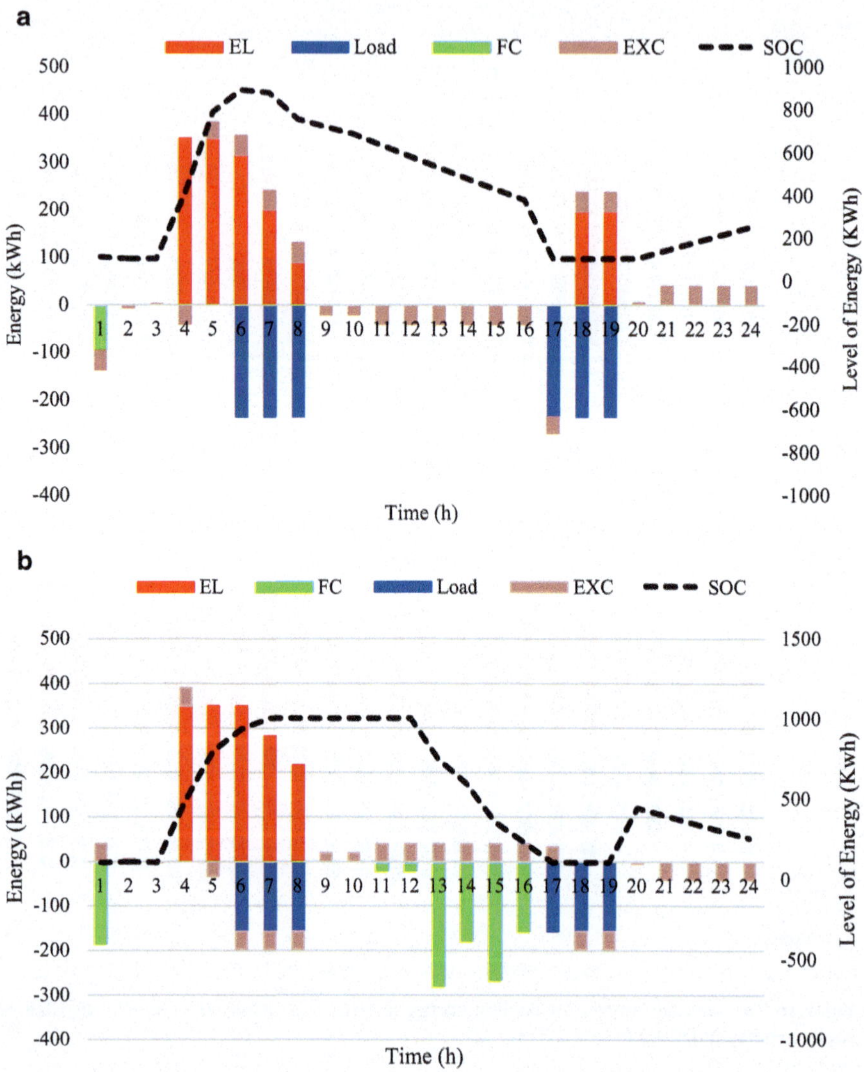

Fig. 7.16 (**a**) Optimal dispatch of hydrogen energy in IHES1 in case 2. (**b**) Optimal dispatch of hydrogen energy in IHES2 in case 2

The biggest challenges, which are related to all research in the field of modern energy systems, are the convergence and adaptability of these systems with conventional infrastructures. Therefore, most research efforts are to resolving these challenges. In the field of interconnected operation of multi-carrier energy systems, most of the challenges affiliate with modeling, optimization algorithms, and economic issues. In this regard, this chapter discusses the optimal coalition operation of hybrid energy systems. In the subject of P2X technologies, in which converters play a crucial role, these devices' efficiency and optimal performance are the main challenges for researchers in this field. In most projects, researchers have tried to significantly reduce energy loss by using local energy converters. But the main question is whether energy converters in all forms of energy can work optimally or not. In the present study, the main focus is on the power-to-hydrogen technology, which has been introduced in recent years and still faces many challenges. However, most of the proposed projects are currently more conceptual but are expected to be operational in the near future. In the context of the interconnected operation of local energy systems, it is necessary to focus more on economic issues, and new business plans need to be adopted to ensure the participants' profitability according to different circumstances. In P2X technologies, it is essential to design a comprehensive pattern to help manage the energy allocation for conversions and whether such conversion technologies can be used efficiently in practice.

7.6 Conclusion

This study developed a methodology about the optimal coalition operation of interconnected hybrid energy systems containing local energy conversion (power-to-X) technologies, renewable energies, and energy storage systems. This problem is considered centralized, and the project aims to mitigate the operation and emission costs of the whole system in the presented approach, even though individual benefits of subsystems may be ignored. In this regard, to validate the proposed model, two test cases are considered. In the first case, hybrid energy systems are isolated and only interact with the primary grid. In this case study, the total operation cost of the HES1 and HES2 obtained \$ 2737.42 and \$ 3187.97, respectively. In the second case, system operation is based on the interconnected coalition, and hybrid energy systems trade electricity, thermal energy, and hydrogen. In this case study, the total operation costs of the HES1 and HES2 are, respectively, \$ 2737.42 and \$ 3026.2. The obtained numerical results indicate that in the central coalition operation of IHESs, the cost reduction for HES2 is \$ 161.7, while the operation cost of HES1 is equal to case 1. It means that HES2 has more opportunities to reduce its operation cost under the collation operation of IHESs. However, the optimal energy dispatch cost in the IHES operation compared to the individual HES operation is more beneficial and decreases the entire system operational cost by up to 2.7%. This outcome proves the expediency and effectiveness of the proposed model. In future works, the uncertainty of some parameters such as renewable generation, electricity price, and energy demands can be modeled by different uncertainty approaches to evaluate their impacts on the optimal operation of IHESs. Also, demand-side management (DSM) services can be implemented to reduce operation costs and change customers' consumption patterns in the HES studies.

Acknowledgements This publication was partially supported by award NPRP12S-0125-190013 from the QNRF-Qatar National Research Fund, a member of The Qatar Foundation. The information and views set out in this publication are those of the authors and do not necessarily reflect the official opinion of the QNRF.

References

1. Mohammadi, M., Noorollahi, Y., Mohammadi-Ivatloo, B., & Yousefi, H. (2017). Energy hub: From a model to a concept–a review. *Renewable and Sustainable Energy Reviews, 80*, 1512–1527.
2. Berjawi, A., Walker, S., Patsios, C., & Hosseini, S. (2021). An evaluation framework for future integrated energy systems: A whole energy systems approach. *Renewable and Sustainable Energy Reviews, 145*, 111163.
3. Favre-Perrod, P. (2005). *A vision of future energy networks*. In 2005 IEEE power engineering society inaugural conference and exposition in Africa, IEEE, pp. 13–17.
4. Geidl, M., Koeppel, G., Favre-Perrod, P., Klockl, B., Andersson, G., & Frohlich, K. (2006). Energy hubs for the future. *IEEE Power and Energy Magazine, 5*(1), 24–30.
5. Geidl, M. (2007). *Integrated modeling and optimization of multi-carrier energy systems*. ETH Zurich.
6. Sheikhi, A., Rayati, M., Bahrami, S., Ranjbar, A. M., & Sattari, S. (2015). A cloud computing framework on demand side management game in smart energy hubs. *International Journal of Electrical Power & Energy Systems, 64*, 1007–1016.
7. Sheikhi, A., Mozafari, B., & Ranjbar, A. M. (2011). *CHP optimized selection methodology for a multi-carrier energy system*. In 2011 IEEE Trondheim PowerTech, IEEE, pp. 1–7.
8. Parisio, A., Del Vecchio, C., & Vaccaro, A. (2012). A robust optimization approach to energy hub management. *International Journal of Electrical Power & Energy Systems, 42*(1), 98–104.
9. Moeini-Aghtaie, M., Dehghanian, P., Fotuhi-Firuzabad, M., & Abbaspour, A. (2013). Multi-agent genetic algorithm: an online probabilistic view on economic dispatch of energy hubs constrained by wind availability. *IEEE Transactions on Sustainable Energy, 5*(2), 699–708.
10. Schulze, M., & Del Granado, P. C. (2009). *Multi-period optimization of cogeneration systems: Considering biomass energy for district heating*. In 2nd Power systems modeling conference (vol. 126).
11. Maroufmashat, A., et al. (2015). Modeling and optimization of a network of energy hubs to improve economic and emission considerations. *Energy, 93*, 2546–2558.
12. Tian, M.-W., et al. (2019). Risk-based stochastic scheduling of energy hub system in the presence of heating network and thermal energy management. *Applied Thermal Engineering, 159*, 113825.
13. Liu, T., Zhang, D., Dai, H., & Wu, T. (2019). Intelligent modeling and optimization for smart energy hub. *IEEE Transactions on Industrial Electronics, 66*(12), 9898–9908.
14. Mirzaei, M. A., et al. (2020). Integrated energy hub system based on power-to-gas and compressed air energy storage technologies in the presence of multiple shiftable loads. *IET Generation, Transmission & Distribution, 14*(13), 2510–2519.
15. Ebadi, R., Yazdankhah, A. S., Mohammadi-Ivatloo, B., & Kazemzadeh, R. (2020). Coordinated power and train transportation system with transportable battery-based energy storage and demand response: A multi-objective stochastic approach. *Journal of Cleaner Production, 275*, 123923.
16. Chamandoust, H., Derakhshan, G., Hakimi, S. M., & Bahramara, S. (2019). Tri-objective optimal scheduling of smart energy hub system with schedulable loads. *Journal of Cleaner Production, 236*, 117584.
17. Ma, T., Wu, J., & Hao, L. (2017). Energy flow modeling and optimal operation analysis of the micro energy grid based on energy hub. *Energy Conversion and Management, 133*, 292–306.

18. Liu, T., Zhang, D., Wang, S., & Wu, T. (2019). Standardized modelling and economic optimization of multi-carrier energy systems considering energy storage and demand response. *Energy Conversion and Management, 182*, 126–142.
19. Mirzapour-Kamanaj, A., Majidi, M., Zare, K., & Kazemzadeh, R. (2020). Optimal strategic coordination of distribution networks and interconnected energy hubs: A linear multi-follower bi-level optimization model. *International Journal of Electrical Power & Energy Systems, 119*, 105925.
20. Fasihi, M., & Breyer, C. (2020). Baseload electricity and hydrogen supply based on hybrid PV-wind power plants. *Journal of Cleaner Production, 243*, 118466.
21. Bianchini, C., & Barbaro, P. (2009). *Catalysis for sustainable energy production*. Wiley.
22. Deng, Z., & Jiang, Y. (2020). Optimal sizing of wind-hydrogen system considering hydrogen demand and trading modes. *International Journal of Hydrogen Energy, 45*(20), 11527–11537.
23. Troncoso, E., & Newborough, M. (2011). Electrolyzers for mitigating wind curtailment and producing 'green' merchant hydrogen. *International Journal of Hydrogen Energy, 36*(1), 120–134.
24. Mansour-Saatloo, A., Agabalaye-Rahvar, M., Mirzaei, M. A., Mohammadi-Ivatloo, B., Abapour, M., & Zare, K. (2020). Robust scheduling of hydrogen based smart micro energy hub with integrated demand response. *Journal of Cleaner Production, 267*, 122041.
25. Syed, F. (2011). *Analysis of a clean energy hub interfaced with a fleet of plug-in fuel cell vehicles*. University of Waterloo.
26. Kholardi, F., Assili, M., Lasemi, M. A., & Hajizadeh, A. (2018). *Optimal management of energy hub with considering hydrogen network*. In 2018 International conference on Smart Energy Systems and Technologies (SEST), IEEE, pp. 1–6.
27. Dagdougui, H., Ouammi, A., & Sacile, R. (2012). Modelling and control of hydrogen and energy flows in a network of green hydrogen refuelling stations powered by mixed renewable energy systems. *International Journal of Hydrogen Energy, 37*(6), 5360–5371.
28. Carr, S., Zhang, F., Liu, F., Du, Z., & Maddy, J. (2016). Optimal operation of a hydrogen refuelling station combined with wind power in the electricity market. *International Journal of Hydrogen Energy, 41*(46), 21057–21066.
29. Isaac, N., & Saha, A. (2021). Analysis of refueling behavior of hydrogen fuel vehicles through a stochastic model using Markov Chain Process. *Renewable and Sustainable Energy Reviews, 141*, 110761.
30. Nojavan, S., Zare, K., & Mohammadi-Ivatloo, B. (2017). Application of fuel cell and electrolyzer as hydrogen energy storage system in energy management of electricity energy retailer in the presence of the renewable energy sources and plug-in electric vehicles. *Energy Conversion and Management, 136*, 404–417.
31. Mehrjerdi, H. (2019). Off-grid solar powered charging station for electric and hydrogen vehicles including fuel cell and hydrogen storage. *International Journal of Hydrogen Energy, 44*(23), 11574–11583.
32. Majidi, M., & Zare, K. (2018). Integration of smart energy hubs in distribution networks under uncertainties and demand response concept. *IEEE Transactions on Power Systems, 34*(1), 566–574.
33. Mohammadi, M., Noorollahi, Y., & Mohammadi-Ivatloo, B. (2020). Fuzzy-based scheduling of wind integrated multi-energy systems under multiple uncertainties. *Sustainable Energy Technologies and Assessments, 37*, 100602.
34. Jin, X., Mu, Y., Jia, H., Wu, J., Xu, X., & Yu, X. (2016). Optimal day-ahead scheduling of integrated urban energy systems. *Applied Energy, 180*, 1–13.

35. Jadidbonab, M., Madadi, S., & Mohammadi-Ivatloo, B. (2018). Hybrid strategy for optimal scheduling of renewable integrated energy hub based on stochastic/robust approach. *Journal of Energy Management and Technology, 2*(4), 29–38.
36. Mansour-Saatloo, A., Mirzaei, M. A., Mohammadi-Ivatloo, B., & Zare, K. (2020). A risk-averse hybrid approach for optimal participation of power-to-hydrogen technology-based multi-energy microgrid in multi-energy markets. *Sustainable Cities and Society, 63*, 102421.
37. Najafi-Ghalelou, A., Nojavan, S., Zare, K., & Mohammadi-Ivatloo, B. (2019). Robust scheduling of thermal, cooling and electrical hub energy system under market price uncertainty. *Applied Thermal Engineering, 149*, 862–880.
38. Rakipour, D., & Barati, H. (2019). Probabilistic optimization in operation of energy hub with participation of renewable energy resources and demand response. *Energy, 173*, 384–399.

Chapter 8
Optimal Co-Generation of Electric and Heat Energy Systems Considering Heat Energy Storage Systems and CHP Units

Sasan Azad, Khezr Sanjani, and Mohammad Taghi Ameli

8.1 Introduction

As a solution to overcome the environmental issues, hybrid energy systems are to dominate the conventional separate systems, because of its numerous opportunities. Hybrid energy systems are also on the focus in which the energy usage is increasing so fast in the new smart grid platforms. Every energy sector is been facing huge increase in demand recently [1]. To solve the energy commitment problem considering the renewable and emission challenges, the operators have been practicing the hybrid systems. In different works different energy systems have been considered to pair and build a hybrid system, e.g., water and power [2–4], gas and electricity [5–9], heat, cool, and fresh water [10], energy and food [11, 12], heat and electricity [13, 14], and so on. Figure 8.1 represents a schematic of the co-generation of heat and electricity for a proposed user.

Electric and heat systems, as the two of the biggest and most important energy systems, have been considered to study in this chapter. Electricity is to provide a vast energy demand types directly and indirectly. Heat is one of the biggest demand of electric system, and it is crucial to be committed well. As long as the systems get smarter, the electric demand is exponentially increasing and it gets harder to deal with. This is why if we could decline a part of this demand, it would work best for the electric system. Combining electric and heat systems to decline a part of these demand is so applicable. Combined heat and power systems are playing a key role in the hybrid electric and heat system by providing the electric and heat energy

S. Azad · K. Sanjani · M. T. Ameli (✉)
Shahid Beheshti University, Tehran, Iran
e-mail: sa_azad@sbu.ac.ir; m_ameli@sbu.ac.ir

© The Author(s), under exclusive license to Springer Nature Switzerland AG 2022

V. Vahidinasab, B. Mohammadi-Ivatloo (eds.), *Whole Energy Systems*, Power Systems, https://doi.org/10.1007/978-3-030-87653-1_8

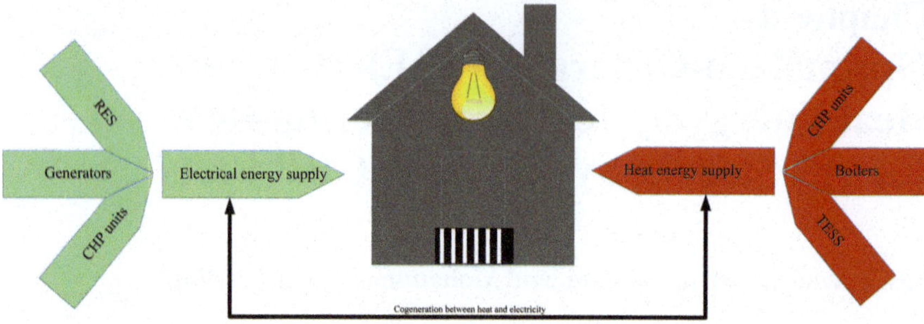

Fig. 8.1 A graphical schematic co-generation between heat and electricity networks

simultaneously. Such components are so necessary and effective in the system which makes the system operation much easier to handle. Because of that, the system operator is willing to use and enjoy such a hybrid energy sources.

Combined heat and power (CHP) systems are a part of HRESs and have been practiced, due to its features, in the heat and electric systems recently. Hence the CHPs are so useful, but there are some difficulties to use, including technical, environmental, operational, and location problems, and it needs a proper procedure. Here the CHP owners prospective and grid point of view should be considered as well. Planning at the first place and operational problems then should be considered as well. The authors in [15] have considered a multi-objective problem for planning of CHPs in their work. In addition to that, in the [16, 17] the optimal planning of CHP units has been investigated optimally. The [17] had considered a probabilistic optimization of molten carbonate fuel cell-CHP and renewable energy sources in microgrids considering hydrogen storage with point estimate method. Operation at the second place is an important and huge task which needs to be handled well. In the [18], the role of centralized combined heat and power systems has been analyzed in a European decarbonized power system. Some review papers have comprehensively analyzed CHPs [19–21]. The most common types of renewable and non-renewable CHP units integrated into power systems and their impacts on system characteristics have been investigated in [22], including the fuel cells (FCs), internal combustion engines (ICEs), and micro-turbines (MTs). For more research, detailed information of CHP systems could be found in [23–25]. The current chapter focuses on operational issues of CHPs by considering heat energy storage systems. The authors have considered three types of renewable and non-renewable CHP units including fuel cells (FCs), internal combustion engine (ICE), and micro-turbines (MTs), as considered in [22]. Hence the ICE and MT are strong to use as CHP; the emission output of such CHPs is not applicable in the environmental emission point of view which they produce 0.79 and 0.15 $kgCO_2/kWh$, respectively, compared to the FC units which produce 0.02 $kgCO_2/kWh$. FC-CHP systems been presented as useful and effective technologies which been widely employed in small/large-scale co-generation applications because of high power density, high co-generation performance, and fast startup time due to its low operation temperature.

Thermal energy storage systems, on the other hand, have been practicing in the energy systems to relax the operation problem and shift a part of heat/cool loads from the peak to the valleys. For some, energy storage systems (ESS) are economically beneficial and for the others ESSs have technical benefit to overcome the system issues. Both of these features could be obtained using ESSs in the systems, and here the authors employed heat energy storage systems to overcome heat load issues during the high winter peaks. Using PSO optimization algorithm, the authors in [26] have investigated the optimal sizing design and operation of electric/heat storage systems in smart buildings. A latent design of heat energy storage systems for large solar plants has been presented in [27]. To reduce the summer peak load using thermal energy storage systems, Dargahi et al. have utilized a thermal energy storage for air-conditioning system in [28]. In the [29, 30] planning problem of thermal energy storages had been analyzed. In [29] a cost-effective and large-scale planning design for hot water thermal energy storage in Renewable District heating systems has been presented. More planning design for TES units could be find in the following references [31–34]. At this work, we have chosen the HESSs to couple with CHPs considering the electric and heat interconnected systems to optimize the net loads and minimize the total cost of the whole system. The electric and heat systems are considered to co-optimize. This chapter's points and contributions are as follows:

- The authors addressed both electric and heat systems through a complex topology of both regarding the heat pipelines and electric transmission lines, simultaneously.
- The net cost of the combined system is minimized through the model. Considering the combined electric and heat systems, the system cost is lower than summation of both systems separately operating.
- The combined heat and power systems will be analyzed in both heat and electricity point of views. Here the operator point of view has been considered to analyze the CHP impacts on the system operation performance.
- Heat energy storage systems have been considered as a proper auxiliary unit for heat system specifically. Using HESSs, the heat system load will be much easier to commit.

The rest of the paper is organized as follows: In the second section the problem description has been presented. After that and in the third section the problem formulation has been discussed. Then in the fourth section discussion and results are described. Finally, the conclusion is been presented in the fifth section.

8.2 Challenges and Outlook

Greenhouse gas emission and energy related issues have been the top priority of the science, now, for quiet sometime since we are facing huge problems for the future of the world. The more highlighted issue here is that the situation is getting worse every year and the world is facing huge issues right now. Energy, as one of the main factors of the greenhouse and climate change challenges, plays a key role and the scientists have employed various applications to handle the energy generation, transmission, and consumption in a more green way. Hybrid energy systems are one of the mentioned applications that have been tested successfully through the energy systems all around the world. Utilizing the hybrid systems, the operator could optimize the fuel consumption and decrease the percentage of greenhouse gas emission then.

The authors proposed a hybrid energy system including heat and electricity to analyze how much it could save the total cost of generation. The different objectives could be employed in different works, though. Total cost of the system is one of the most effective factors for the operator, and if we could save a percentage of it, then, we could use the saved cost in other application to improve more of the climate problems. This cost saving has some direct and some indirect effects and as much as we save of it, the system's performance would improve more. With improving the systems performance, scientists aim to control greenhouse gas emission challenges and if we could overcome all these performance issues with new application and approaches, we could get everything in control.

8.3 Problem Description

The general problem related to the chapter is to co-optimize the electric and heat systems simultaneously in which to reduce total cost of the system. In fact, co-optimizing energy systems together has vast opportunities for the system operators including operation performance, load commitment, technical improvements, and so on. For some, hybrid energy systems are to increase the economical benefits, and for others, these systems are to enhance the system's overall performance. Both of these features could be obtained from the hybrid energy systems. In one hand, technical features like voltage stability, reliability, higher energy commitment performance, and so on could be achieved using the co-operation of the energy systems. On the other hand, co-operation of the energy systems would significantly decrease the net cost of the system. Actually, a part of heat and cool energy demand is being provided by the electric system directly and co-operation of, for instance, heat and electric systems together would alleviate the demand pressure on the heat system in which the electric system is to provide a part of the heat loads indirectly. The combined heat and power (CHP) systems are to provide both electric and heat energies at the same time and to commit both systems. This means that the operator could count on the CHP units to provide the heat system with extra heat energy. At this regard, not only the system has a higher performance in providing heat load units with a high quality demand, but also this will decrease the heat systems cost and overall system costs. In the electric system's point of view, using the renewable and non-renewable CHP units would have the same impacts on both technical and economical features of the system. As such, having the hybrid energy systems would improve the overall energy system's performance in both technical and economical point of views.

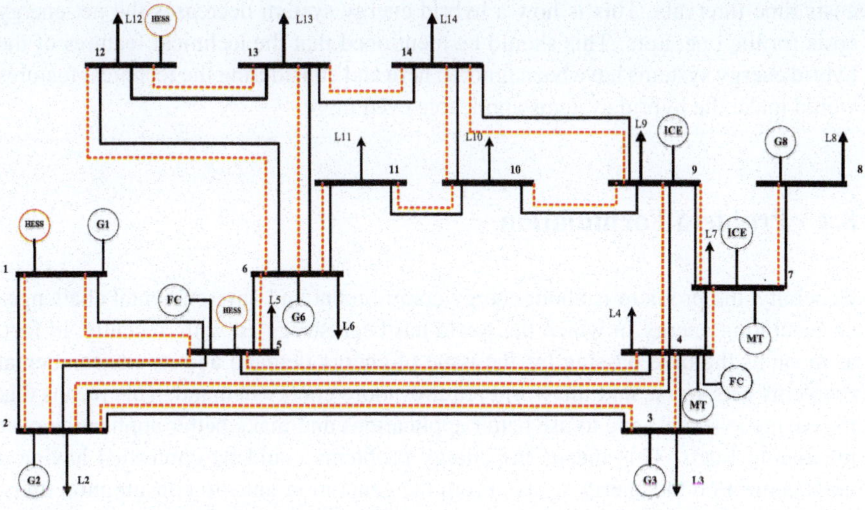

Fig. 8.2 The combined electric/heat system's schematic

Table 8.1 The HPR values for all CHP types used in the chapter [35]

CHP type	HPR
FC	1.5
MT	0.6
ICE	0.95

Generally, considering the heat system along with electric system needs the heat pipelines to be considered and modeled through the system as the authors have included heat pipelines in the chapter. The heat and electric systems have been comprehensively depicted in the Fig. 8.2. The red lines are the heat pipelines, and the black lines are the electric transmission lines. Here, the HESSs have been simulated with a red-dashed circle in node No. 1, 5, 12. Moreover, the CHP units have been also simulated in the chart with a black circle around in the bus No. 4, 5, 7, and 9. The maximum output value for the CHP units are 0.2, 0.01, and 0.05, respectively. The heat power ratio (HPR) values for the CHP types, including fuel cell, micro-turbine, and internal combustion engine used in the model have been written in the following Table 8.1.

After combining these two and creating a hybrid energy system, now, the objective of the new hybrid system, at the current work, is to minimize the net energy costs. The objective is the summation of the electric and heat system costs without any weighting factors for both. Having the CHP units in the system as an extra source of power and heat generators not only provides the two systems with extra energy but also it will decline the market prices and would lead to decline in energy costs in both systems and in the hybrid energy system compared to the separated systems. Considering the heat energy storage systems is also decreasing the heat system costs by shifting the heat loads from peak times to the other low costs time intervals. This is how a hybrid energy system decreases the net energy costs for the operators. This should be mentioned that the technical features of the hybrid energy systems have been ignored here and considering the technical features would make the hybrid systems applicable even more.

8.4 Problem Formulation

Generally, the problem is about energy commitment and environmental challenges of generating energy in which the world has been struggled with, recently. In fact, as much as the time passing by, the issue of energy demand and its environmental problems gets worse and the world requires more energy demand. This means that the energy systems have to use better applications and make better application over the energy loads. This means the energy problems could be controlled having a good control on the energy usage. From the generation units to consumption units, the system needs to have a better control over the energy [36]. One of the most effective and crucial solutions is to co-optimize the energy sectors and to minimize the total cost and emissions, simultaneously. And since a decade ago, the combined energy systems have been increased so fast and these combined systems' impact on the environmental issues is notable.

Hybrid energy systems are to couple two or more energy systems together and to co-optimize these systems altogether. These led to minimize the total system costs and greenhouse gas (GHG) emissions. Since these systems are interconnected, co-optimizing leads to a better energy management and the costs and GHG values notably decrease. Electric and heat systems are the examples of the energy systems which are interdependent and co-optimizing these systems would led to a better energy management. That is why the system operators are willing to use the hybrid energy systems.

The graphical chart of coupled system of electric and heat systems is figured in the Fig. 8.2. The black lines are the electric transmission lines and the red lines are the heat pipelines. It should be mentioned that the HESS units have been considered in the node NO.1, 5, 12.

In the hybrid energy system works, different energy sectors could be employed and the objective could be whether economy or emission or both. These different objectives can be employed based on the work's vision whether it is in the operator point of view or the facility owner's side. It is worth to mention that the operator objective is to maximize the social welfare and at the same time, the owner's objective is to maximize its own profit. At the current chapter, the authors have employed electric and heat energy sectors, and the objective is to minimize the total costs of the combined energy systems. The objective function is modeled through (8.1).

$$TC = \sum \lambda^{E} \times P_{g,t}^{G} + \sum \lambda^{H} \times h_{n,t}^{boil} + \sum \lambda^{H} \times h_{c,t}^{CHP} + \sum \lambda^{E} \times P_{c,t}^{CHP} \quad (8.1)$$

Here, the objective consists of four terms including the cost of conventional electric generators ($P_{g,t}^{G}$) at the first term, the cost of heat generators ($h_{n,t}^{boil}$) at the second term, the cost of heat generation from the CHP units ($h_{c,t}^{CHP}$) in the third term, and the cost of electric generation from CHP units ($P_{c,t}^{CHP}$) in the last term. λ^{E} and λ^{H} are the electricity and heat market price. The electric and heat systems are co-optimizing without considering weight for any. Constraints of the two electric and heat systems are described separately as follows:

8.4.1 The Electricity System's Constraints

The most important constraint in every system is about the equality constraint. The equality constraint poses the equilibrium criteria among the value of generated energy and energy demand value. Here, the (8.2) presents the electricity equality constraint.

$$\sum_{G} P_{g,t}^{G} - \sum_{L} P_{l,t}^{L} + \sum_{CHP} P_{c,t}^{CHP} = \sum_{j} P_{i,j,t} \tag{8.2}$$

The equality constraint consists of conventional generation units ($P_{g,t}^{G}$), load units ($P_{l,t}^{L}$), CHP units ($P_{c,t}^{CHP}$), and the exchanged power through lines i and j ($P_{i,j,t}$). The other constraints of the electric system are presented below:

$$0 \leq P_{c,t}^{CHP} \leq P_{c,t}^{CHP,max} \tag{8.3}$$

$$P_{g,t}^{G,min} \leq P_{g,t}^{G} \leq P_{g,t}^{G,max} \tag{8.4}$$

$$-P_{i,j,t}^{max} \leq P_{i,j,t} \leq P_{i,j,t}^{max} \tag{8.5}$$

The last constraints including (8.3, 8.4, 8.5) are presenting the CHP units up/down limitation, conventional generation up/down limitations, and the transmitted power between lines, respectively.

8.4.2 The Heat System's Constraints

The heat system's equality constraint has been modeled through (8.6) including all the heat energy sectors: traditional boiling units, CHP units and, charge/discharge of HESS units ($H_{s,t}^{dch}$, $H_{s,t}^{ch}$). As mentioned, here, the heat pipelines have been considered and the heat transfer between pipelines has been modeled here, too. The $f_{m,n,t}$ is responsible for the heat transfer between lines (8.7, 8.9).

$$H_{n,t}^{boil} + H_{c,t}^{CHP} - H_{n,t}^{load} + H_{s,t}^{dch} - H_{s,t}^{ch} = \sum_{m} f_{m,n,t} - \sum_{m} f_{n,m,t} \tag{8.6}$$

The rest of constraints of the heat system are presented below:

$$f_{m,n} = sgn(pr_{n,t} - pr_{m,t}) \times C_{n,m}^{K} \times \sqrt{sgn(pr_{n,t} - pr_{m,t})(pr_{n,t}^{2} - pr_{m,t}^{2})} \tag{8.7}$$

$$pr_{n,t}^{min} \leq pr_{n,t} \leq pr_{n,t}^{max} \tag{8.8}$$

$$- f_{n,m}^{max} \leq f_{n,m} \leq f_{n,m}^{max} \tag{8.9}$$

$$H_{i,t,k}^{CHP} = H_{i,t,k}^{CHP} \times HPR_{k}^{CHP} \tag{8.10}$$

The pressure limit in heat system is modeled through (8.8). And the heat output is calculated using the power level of the CHP and a HPR ratio in (8.10).

8.4.3 *Heat Energy Storage System's Constraints*

The heat energy storage system is modeled as follows:

$$E_{s,t} = E_{s,t-1} + H_{s,t}^{ch} - H_{s,t}^{dch} \tag{8.11}$$

$$0 \leq H_{s,t}^{ch} \leq H_{s,t}^{ch,max} \times U_{s,t}^{ch} \tag{8.12}$$

$$0 \leq H_{s,t}^{dch} \leq H_{s,t}^{dch,max} \times U_{s,t}^{dch} \tag{8.13}$$

$$U_{s,t}^{dch} + U_{s,t}^{ch} \leq 1 \tag{8.14}$$

The (8.11) is presenting the energy level of the HESS and the (8.12, 8.13) are presenting the charge and discharge limitations of the HESS. The last constraint is to limit the storage function to whether charge or discharge at a certain time interval. It should be noted that the $U_{s,t}^{ch}$ and $U_{s,t}^{dch}$ are the storage's binary variables.

8.5 Discussion and Results

Modeling a mixed integer non-linear programming problem in GAMS, here, we are discussing the simulation results. The system considered in this chapter is a combination of electric and heat systems operating together. The proposed combined system has been graphically illustrated in the Fig. 8.2. As shown in the Fig. 8.2, the heat pipelines have been taken into account. The heat system consists of 14 nodes and the electric system is the IEEE 14-bus test system. Both systems are operating at the same time, and a co-optimization strategy is considered to minimize the total system costs. The CHP units and HESSs are also functioning through the system to decrease the system costs and provide the operator with more heat and electricity generations. The net total cost of the system is now equal to 6.0159×10^5 where the system costs for the electric and heat systems in a separate operation mode are 5.7303×10^5 and 1.1295×10^5, respectively. Here the difference between co-generation and separated modes of the system is evident. With co-generating the electric and heat systems, the operator could save almost $100,000, which is a considerable value. On the other hand, the operation reliability is higher and the system could provide the consumption units with a reliable energy source. Such a methodology is practical and this could be implemented in real systems.

Normalized and standard load curves have been considered for both electricity and heat consumption units. Table 8.2 and Fig. 8.3 show the consumption values of electricity and heat systems. In Fig. 8.3, the heat load values have been multiplied by 10 to become more comparable with electric values. From the Table 8.2, it is clear that the load curve has a high peak from time interval 10 to time interval 19. Moreover, the low peak (valley) is from time interval 3 to time interval 6.

Figure 8.3 is demonstrating the peak and valley of both heat and electricity curves.

The day-ahead electricity and heat price curves are shown in the Fig. 8.4. The red line is the heat price and black curve is for the electricity price. A part from the most published works that the heat price curve is dismissed, the authors have considered day-ahead price curves for both heat and electric systems.

The HESS units, as shown in Fig. 8.5, are relaxing the problem by declining the high peak values of heat load curve. Moreover, the energy level of the first HESS, which is installed in node No.1, has been shown in the Fig. 8.6. Here, three different

Table 8.2 Electricity and heat day-ahead demand values

Hour	Electricity	Heat	Hour	Electricity	Heat
1	1.66	0.124	13	2.57	0.193
2	1.65	0.123	14	2.60	0.195
3	1.50	0.113	15	2.60	0.195
4	1.45	0.109	16	2.52	0.189
5	1.45	0.109	17	2.49	0.187
6	1.50	0.113	18	2.49	0.187
7	1.66	0.124	19	2.41	0.181
8	1.98	0.148	20	2.39	0.179
9	2.26	0.169	21	2.39	0.179
10	2.47	0.185	22	2.41	0.181
11	2.57	0.193	23	2.26	0.169
12	2.60	0.195	24	1.87	0.140

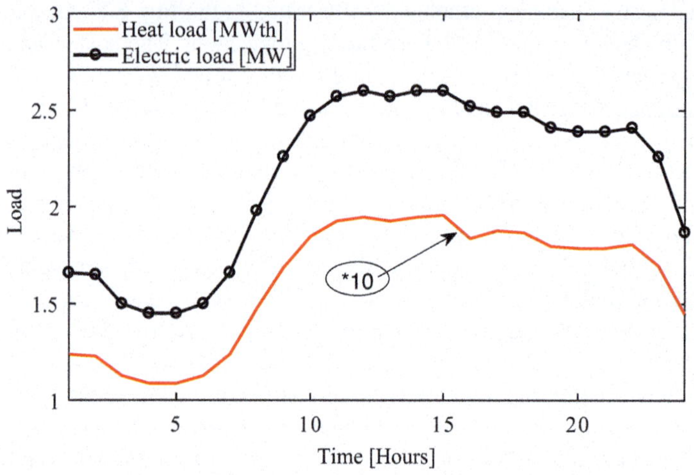

Fig. 8.3 Electric and heat loads (heat load values have been multiplied by 10 to become comparable to electric load)

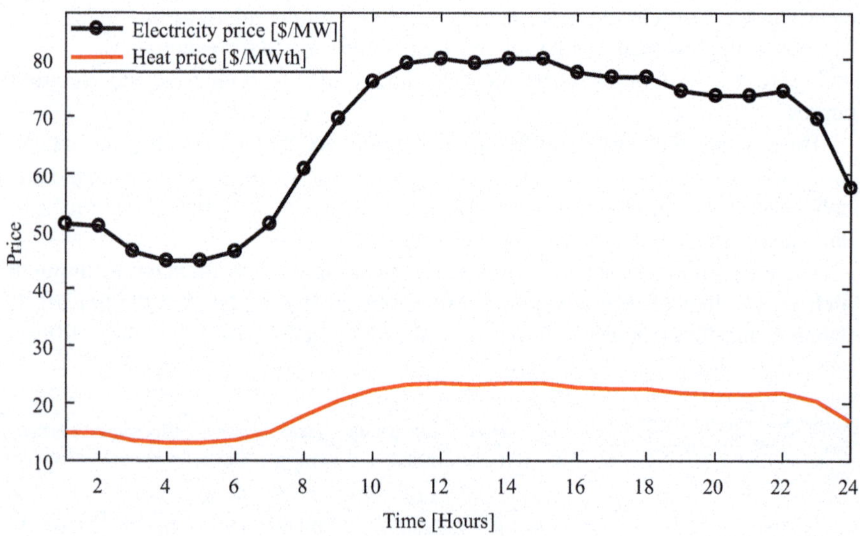

Fig. 8.4 Day-ahead electricity and heat market prices

HESSs have been considered with different maximum of charge/discharge power. It should be mentioned in all three HESSs, the energy level at the start and at the end of operation are the same and equal to 0.25 p.u.

Figure 8.5 is presenting the charge and discharge spots of the HESS along with the heat load curve. Here, we plot just the first HESS as a test case and the other HESS units could be plotted just the same. The heat load curve has two peak times

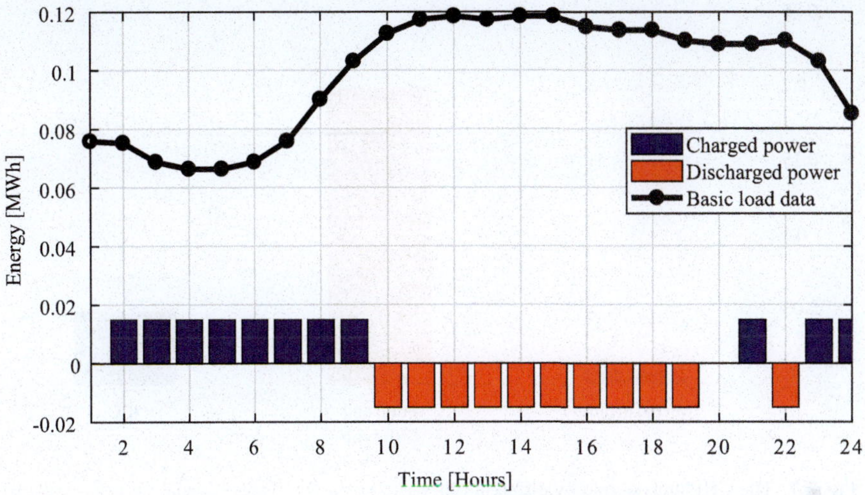

Fig. 8.5 Heat load curve with charge/discharge of the HESS figure

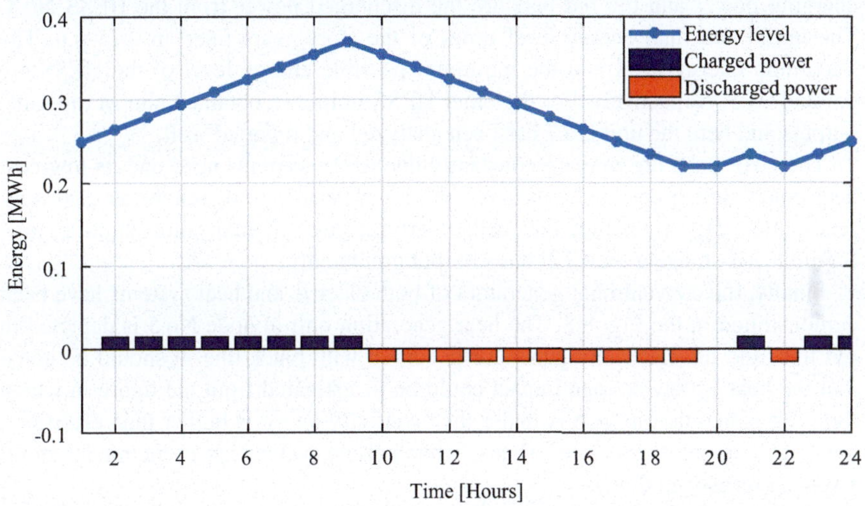

Fig. 8.6 HESS No.1 energy level with charge and discharge time spots and values

one from 10 to 18 and the other is in 22 time intervals. Further, there is a valley at the beginning of the day from 0 to almost 9 time intervals. The HESS schedule is to relax the heat load curve by charging in the valley and discharging the power in the high peak time intervals.

Like Fig. 8.5, Fig. 8.6 presents the day-ahead charge and discharge power time spots of the HESS. This figure is also for the HESS unit No.1 along with its energy level. The blue line is for its day-ahead energy level and the blue bars are the

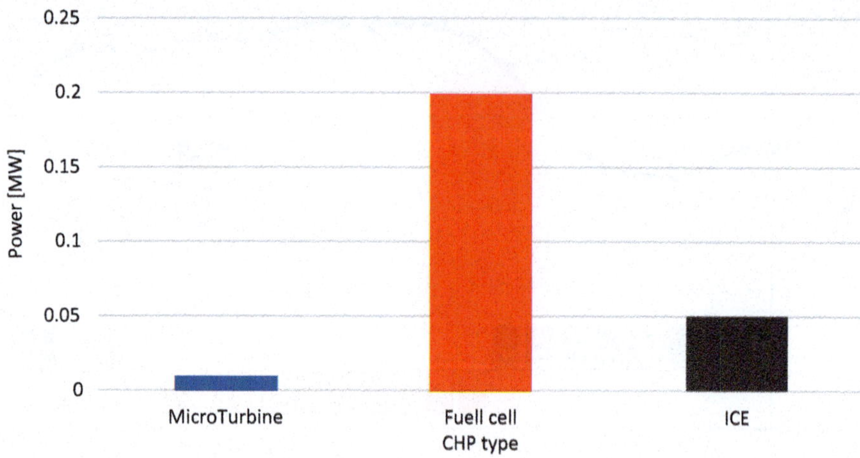

Fig. 8.7 The CHP units output by different types

charging power and the red bars are the discharged power from the HESS No.1. The initial and final energy level times of the storage are fixed in 0.25 p.u. The maximum energy level and the minimum possible energy level of the HESS are 0.5 and 0.05, respectively. For the other HESS units, we obtained almost the same outputs and here the unit No.1 has been analyzed and reported well.

The CHP units are running constantly due to the systems need and its low-cost heat generation. Here, we considered three different type of renewable and non-renewable CHPs including fuel cell, micro-turbine, internal combustion engine. Figure 8.7 is showing the CHP units output graphically.

Finally, the conventional generators of both electric and heat systems have been demonstrated in the Fig. 8.8. The heat generation unit in node No.5 is depicted in red line, and the electricity generator is curved with black line. It should be noted that the heat energy storage impact could be investigated from the figure in which the peak generation in heat is under the electricity and it is higher than electricity generation in valley intervals. That is what storage systems do in the model and it has been investigated well.

8.6 Conclusion

This chapter presented a co-optimization technique for electric and heat systems together. Co-operation of the energy systems has some positive impacts on the whole energy systems including enhancing the reliability of load commitment level, reducing the total operation costs, and so on. Here, the authors have analyzed the co-optimization of the electric and heat systems, simultaneously. In addition to that, the CHP units, both renewable and non-renewables including fuel cell, micro-turbine,

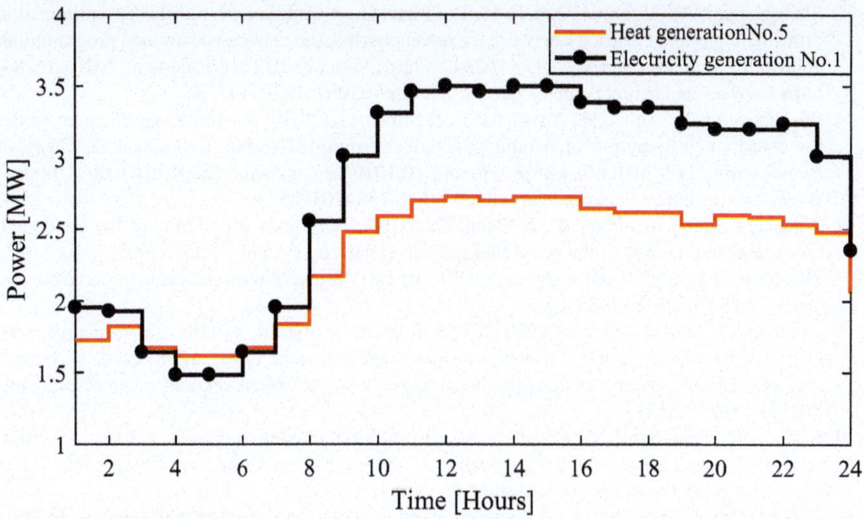

Fig. 8.8 The heat and electricity conventional generators

and internal combustion engine, and HESSs have been considered to the model to enhance the operation output in the model. The HESSs are considered to improve the heat operation system and the CHP units are considered in both systems in which they are generating both heat and electricity at the same time. Three different heat storage systems and three different CHP units have been included through the model. The simulation results have proved the mentioned performance of the combined electric and heat system. Co-generating the electric and heat systems had reduced the operation net cost by approximately $ 100,000 which is over 15% and it is quite huge. Having such effective impacts on the energy systems, it is practical to implement these co-generation systems in real energy systems.

References

1. International Energy Agency. [Online] Available: https://www.iea.org/.
2. Jurasz, J., & Ciapała, B. (2018). Solar—hydro hybrid power station as a way to smooth power output and increase water retention. *Solar Energy, 173*, 675–690. https://doi.org/10.1016/j.solener.2018.07.087. https://www.sciencedirect.com/science/article/pii/S0038092X18307527.
3. Al-Nimr, M. A., & Qananba, K. S. (2018). A solar hybrid system for power generation and water distillation. *Solar Energy, 171*, 92–105. https://doi.org/10.1016/j.solener.2018.06.019. https://www.sciencedirect.com/science/article/pii/S0038092X18305590.
4. Valenzuela, C., Mata-Torres, C., Cardemil, J. M., & Escobar, R. A. (2017). Csp+pv hybrid solar plants for power and water cogeneration in northern Chile. *Solar Energy, 157*, 713–726. https://doi.org/10.1016/j.solener.2017.08.081. https://www.sciencedirect.com/science/article/pii/S0038092X17307624.

5. Shabazbegian, V., Ameli, H., Ameli, M. T., Strbac, G., & Qadrdan, M. (2021). Co-optimization of resilient gas and electricity networks; a novel possibilistic chance-constrained programming approach. *Applied Energy, 284*, 116284. https://doi.org/10.1016/j.apenergy.2020.116284. https://www.sciencedirect.com/science/article/pii/S030626192031672X.
6. Shabazbegian, V., Ameli, H., Ameli, M. T., & Strbac, G. (2020). Stochastic optimization model for coordinated operation of natural gas and electricity networks. *Computers & Chemical Engineering, 142*, 107060. https://doi.org/10.1016/j.compchemeng.2020.107060. https://www.sciencedirect.com/science/article/pii/S0098135420300533.
7. Zeng, Q., Zhang, B., Fang, J., & Chen, Z. (2017). A bi-level programming for multistage co-expansion planning of the integrated gas and electricity system. *Applied Energy, 200*, 192–203. https://doi.org/10.1016/j.apenergy.2017.05.022. https://www.sciencedirect.com/science/article/pii/S0306261917305123.
8. Yildirim, N., & Bilir, L. (2017). Evaluation of a hybrid system for a nearly zero energy greenhouse. *Energy Conversion and Management, 148*, 1278–1290. https://doi.org/10.1016/j.enconman.2017.06.068. https://www.sciencedirect.com/science/article/pii/S0196890417306131.
9. Sanjani, K., Vahabzad, N., Nazari-Heris, M., & Mohammadi-ivatloo, B. (2019). *A robust-stochastic approach for energy transaction in energy hub under uncertainty* (pp. 219–232). Cham: Springer International Publishing.
10. Calise, F., d'Accadia, M. D., & Vicidomini, M. (2019). Optimization and dynamic analysis of a novel polygeneration system producing heat, cool and fresh water. *Renewable Energy, 143*, 1331–1347. https://doi.org/10.1016/j.renene.2019.05.051. https://www.sciencedirect.com/science/article/pii/S0960148119307116.
11. Zeng, X., Zhang, J., Yu, L., Zhu, J., Li, Z., & Tang, L. (2019). A sustainable water-food-energy plan to confront climatic and socioeconomic changes using simulation-optimization approach. *Applied Energy, 236*, 743–759. https://doi.org/10.1016/j.apenergy.2018.11.086. https://www.sciencedirect.com/science/article/pii/S0306261918318038.
12. Xydis, G. A., Liaros, S., & Botsis, K. (2017). Energy demand analysis via small scale hydroponic systems in suburban areas—an integrated energy-food nexus solution. *Science of The Total Environment, 593-594*, 610–617. https://doi.org/10.1016/j.scitotenv.2017.03.170. https://www.sciencedirect.com/science/article/pii/S0048969717306915.
13. Liu, N., Wang, J., & Wang, L. (2019). Hybrid energy sharing for multiple microgrids in an integrated heat—electricity energy system. *IEEE Transactions on Sustainable Energy, 10*(3), 1139–1151. https://doi.org/10.1109/TSTE.2018.2861986.
14. Das, B. K., & Hasan, M. (2021). Optimal sizing of a stand-alone hybrid system for electric and thermal loads using excess energy and waste heat. *Energy, 214*, 119036. https://doi.org/10.1016/j.energy.2020.119036. https://www.sciencedirect.com/science/article/pii/S0360544220321435.
15. Yadegari, S., Hamdi Abdi, & Nikkhah, S. (2020). Risk-averse multi-objective optimal combined heat and power planning considering voltage security constraints. *Energy, 212*, 118754. [Online] Available: https://www.eia.gov/.
16. Wang, H., Gu, C., Zhang, X., & Li, F. (2020). Optimal chp planning in integrated energy systems considering network charges. *IEEE Systems Journal, 14*(2), 2684–2693. https://doi.org/10.1109/JSYST.2019.2921218.
17. Bornapour, M., Hemmati, R., Pourbehzadi, M., Dastranj, A., & Niknam, T. (2020). Probabilistic optimal coordinated planning of molten carbonate fuel cell-CHP and renewable energy sources in microgrids considering hydrogen storage with point estimate method. *Energy Conversion and Management, 206*, 112495. https://doi.org/10.1016/j.enconman.2020.112495. https://www.sciencedirect.com/science/article/pii/S0196890420300315.
18. Jimenez-Navarro, J.-P., Kavvadias, K., Filippidou, F., Pavičević, M., & Quoilin, S. (2020). Coupling the heating and power sectors: The role of centralised combined heat and power plants and district heat in a European decarbonised power system. *Applied Energy, 270*, 115134. https://doi.org/10.1016/j.apenergy.2020.115134. https://www.sciencedirect.com/science/article/pii/S0306261920306462.

19. Newborough, A. M. (2005). Impact of micro-chp systems on domestic sector co2 emissions. *Applied Thermal Engineering, 25*, 2653–2676.
20. Martinez, S., Michaux, G., Salagnac, P., & Bouvier, J.-L. (2017). Micro-combined heat and power systems (micro-CHP) based on renewable energy sources. *Energy Conversion and Management, 154*, 262–285.
21. Radenahmad, N., Azad, A. T., Saghir, M., Taweekun, J., Bakar, M. S. A., Reza, M. S., & Azad, A. K. (2020). A review on biomass derived syngas for SOFC based combined heat and power application. *Renewable and Sustainable Energy Reviews, 119*, 109560. https://doi.org/10.1016/j.rser.2019.109560. https://www.sciencedirect.com/science/article/pii/S1364032119307683.
22. Shao, Z., Ahangarnejad, A. H., Monazzah, A., Rao, Y., Rodriguez, D. (2019). Increasing of fuel cell economic benefits by optimal participation strategy with energy storages and other distributed resources and considering uncertainties and various markets. *International Journal of Hydrogen Energy, 44*, 1839–1850.
23. Horlock, J. H. (1987). Combined heat and power. https://www.osti.gov/biblio/5433836.
24. Hawkes, A., & Leach, M. (2007). Cost-effective operating strategy for residential micro-combined heat and power. *Energy, 32*(5), 711–723. https://doi.org/10.1016/j.energy.2006.06.001. https://www.sciencedirect.com/science/article/pii/S0360544206001277.
25. Alipour, M., Mohammadi-Ivatloo, B., & Zare, K. (2015). Stochastic scheduling of renewable and CHP-based microgrids. *IEEE Transactions on Industrial Informatics, 11*(5), 1049–1058. https://doi.org/10.1109/TII.2015.2462296.
26. Baniasadi, A., Habibi, D., Al-Saedi, W., Masoum, M. A., Das, C. K., & Mousavi, N. (2020). Optimal sizing design and operation of electrical and thermal energy storage systems in smart buildings. *Journal of Energy Storage, 28*, 101186. https://doi.org/10.1016/j.est.2019.101186. https://www.sciencedirect.com/science/article/pii/S2352152X19311545.
27. Liu, M., Riahi, S., Jacob, R., Belusko, M., & Bruno, F. (2020). Design of sensible and latent heat thermal energy storage systems for concentrated solar power plants: Thermal performance analysis. *Renewable Energy, 151*, 1286–1297. https://doi.org/10.1016/j.renene.2019.11.115. https://www.sciencedirect.com/science/article/pii/S0960148119318099.
28. Dargahi, A., Sanjani, K., Nazari-Heris, M., Mohammadi-Ivatloo, B., Tohidi, S., & Marzband, M. (2020). Scheduling of air conditioning and thermal energy storage systems considering demand response programs. *Sustainability, 12*(18), 7311. https://doi.org/10.3390/su12187311. https://www.mdpi.com/2071-1050/12/18/7311.
29. Ochs, F., Dahash, A., Tosatto, A., & Bianchi Janetti, M. (2020). Techno-economic planning and construction of cost-effective large-scale hot water thermal energy storage for renewable district heating systems. *Renewable Energy, 150*, 1165–1177. https://doi.org/10.1016/j.renene.2019.11.017. https://www.sciencedirect.com/science/article/pii/S0960148119316982.
30. Wang, Y., Wang, Y., Huang, Y., Li, F., Zeng, M., Li, J., Wang, X., & Zhang, F. (2019). Planning and operation method of the regional integrated energy system considering economy and environment. *Energy, 171*, 731–750. https://doi.org/10.1016/j.energy.2019.01.036. https://www.sciencedirect.com/science/article/pii/S0360544219300386.
31. Garmabdari, R., Moghimi, M., Yang, F., & Lu, J. (2020). Multi-objective optimisation and planning of grid-connected cogeneration systems in presence of grid power fluctuations and energy storage dynamics. *Energy, 212*, 118589. https://doi.org/10.1016/j.energy.2020.118589. https://www.sciencedirect.com/science/article/pii/S0360544220316972.
32. Liu, Z., Zhao, Y., & Wang, X. (2020). Long-term economic planning of combined cooling heating and power systems considering energy storage and demand response. *Applied Energy, 279*, 115819. https://doi.org/10.1016/j.apenergy.2020.115819. https://www.sciencedirect.com/science/article/pii/S0306261920312976.
33. Marczinkowski, H. M., & Østergaard, P. A. (2019). Evaluation of electricity storage versus thermal storage as part of two different energy planning approaches for the islands Samsø and Orkney. *Energy, 175*, 505–514. https://doi.org/10.1016/j.energy.2019.03.103. https://www.sciencedirect.com/science/article/pii/S0360544219305079.

34. Dahash, A., Ochs, F., Tosatto, A., & Streicher, W. (2020). Toward efficient numerical modeling and analysis of large-scale thermal energy storage for renewable district heating. *Applied Energy, 279*, 115840. https://doi.org/10.1016/j.apenergy.2020.115840. https://www.sciencedirect.com/science/article/pii/S0306261920313179.
35. Arandian, B., & Ardehali, M. (2017). Effects of environmental emissions on optimal combination and allocation of renewable and non-renewable CHP technologies in heat and electricity distribution networks based on improved particle swarm optimization algorithm. *Energy, 140*, 466–480. https://doi.org/10.1016/j.energy.2017.08.101. https://www.sciencedirect.com/science/article/pii/S0360544217314767.
36. Energy Information Aministration. [Online] Available: https://www.eia.gov/.

Chapter 9
Investigating the Role of Flexibility Options in Multi-vector Energy Systems

Vahid Shabazbegian, Hossein Ameli, and Mohammad Taghi Ameli

9.1 Introduction

Energy systems (e.g., gas and electricity systems) include all components designed to supply and deliver energy to the demand centers. Gas and electricity systems deliver a major share of energy to consumers and play a vital role in people's lives all around the world. On the other hand, these two sectors produce a considerable share of the total greenhouse gasses [1]. Therefore, there is a need to take a few corrective measures to curve this trend and decrease the potential consequences. To meet this purpose, a high number of studies introduced investing more in renewable energies as a solution to mitigate the proposed concern [2].

The share of renewable energy sources (RES) to supply energy is increasing, in order to meet the future carbon targets. The renewable resources are expanded to replace the power plants with high carbon intensity, such as coal plants. Gas-fired power plants are the main linkage between these two networks. Due to their characteristics, such as fast ramping rate, these plants complement the lack of renewables [3], and hence the intermittent nature of RES in the power system will be reflected in the gas network demand. As a solution, flexibility options, such as storage systems, bidirectional compressors, and power-to-gas (P2G) systems, are employed to cope with the imposed intermittency to the energy system. More precisely, storage systems provide both upward and downward reserve, discharging energy when there is a higher amount of demand and charging on the contrary case. In energy systems, various sort of storage systems, such as batteries and gas storage

V. Shabazbegian (✉) · M. T. Ameli (✉)
Faculty of Power Engineering, Shahid Beheshti University, Tehran, Iran
e-mail: m_ameli@sbu.ac.ir

H. Ameli
Control and Power group, Imperial College London, London, UK

© The Author(s), under exclusive license to Springer Nature Switzerland AG 2022
V. Vahidinasab, B. Mohammadi-Ivatloo (eds.), *Whole Energy Systems*, Power Systems, https://doi.org/10.1007/978-3-030-87653-1_9

systems, can play different roles depending on the time scale [4]. Multidirectional compressors redirect the gas flow and boost the gas pressure in different direction to ensure gas demand is met [5]. Besides, P2G systems are employed to prevent the curtailment of renewable electricity. These components produce hydrogen via the electrolysis process from the excess power of renewables and can be injected to the gas system or be utilized for different purposes [6].

Aside from that, in energy systems, reliability is defined as "designing, running, and maintaining energy supply resources, transmission, and distribution systems to provide an adequate, safe, and stable flow of energy" [7]. Meanwhile, resiliency is linked to the concept of reliability which is defined as "the capacity of an energy system to tolerate disturbance and to continue to deliver affordable energy services to consumers" [8]. Therefore, considering the ability of the flexibility options in order to cope with numerous issues in energy systems, such as mitigating renewable units output power fluctuations and peak load management, these components can be employed to enhance the reliability and resiliency of the energy systems.

Taking into account the proposed issues, in this chapter, different types of flexibility options are firstly introduced, including their uses and mathematical models. After that, the contribution of these components in mitigating the intermittency and variability of RES is investigated based on previous projects and studies. Moreover, some analyses are conducted to assist the decision-makers to choose the options and solve the problems more efficiently.

9.2 Flexibility Options

The electricity network delivers electricity from power plants to consumers. After producing electricity at power plants, transformers in step-up substations are employed to boost the voltage and transmit power to industrial units and step-down substations. Then, through employing the transformers in step-down substations that are located near the consumers, the power transmits to demand places, such as houses and small industries (Fig. 9.1) [9]. It is noteworthy to mention that the changes in the voltage level are conducted to reduce the loss of power.

The electricity is mainly generated in three ways, including (i) combustion of fossil fuels in power plants, such as coal and natural gas, (ii) nuclear fission in power plants, and (iii) other no fossil-fuel-based sources, such as wind and solar power. In recent years, however, the share of RES is increasing due to the emissions produced through fossil fuels, causing climate change and global warming [10]. Although these resources produce electricity without emissions, their output power is variable and depends on different parameters, mainly weather conditions. Different flexibility options in the electricity systems can be employed to cope with the intermittent nature of renewables, such as energy storage systems and flexible gas-fired power plants [11]. The energy storage systems store electricity during peak hours and recharge the electricity during peak hours to facilitate addressing the supply-demand balance. Moreover, owing to some characteristics

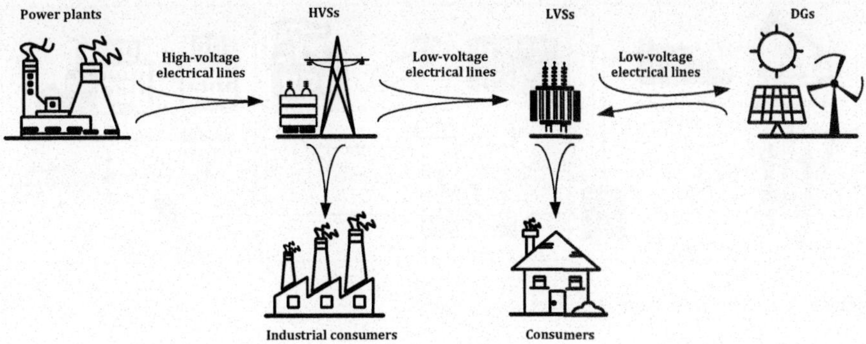

Fig. 9.1 Electricity network from suppliers to consumers

of gas-fired power plants (e.g., fast ramp up/down and short starting up and shutting down times), the power plant can be opted to deal with the variability of RES. However, it leads to transmitting the variability from the electricity system to the gas system. To mitigate the issue, some flexibility options are employed in gas networks, such as flexible compressors and gas storage facilities. The compressors are employed to boost the pressure within the pipelines, and gas storage facilities inject/withdraw gas (e.g., natural gas or hydrogen) to facilitate supply-demand balance in the gas networks. The P2G systems as a cross vector flexibility can be installed to produce hydrogen in order to utilize or inject to the gas network by using the excess of renewables. The characteristics and performance of the flexibility options will be explained more precisely in the following.

Due to the interdependency of gas and electricity network operation, it is worthwhile to explain the gas network operation in this section. The main layers of gas networks include gas terminals, gas transmission systems, and gas distributions systems [12]. In a natural gas system, the natural gas is supplied through the terminals, such as liquefied natural gas carriers, onshore/offshore gas fields, and interconnectors. In a hydrogen system, the hydrogen will be produced through fossil fuels (e.g., brown and blue hydrogen) and through electrolysis process (i.e., green hydrogen). In the transmission gas network, the gas is transmitted within the pipelines at high pressures. By using high pressures, more compressed gas can be transmitted through long distances which also reduces the cost of pipelines (i.e., due to the fact that smaller diameter is necessary to transmit more compressed gas). It should be noted that heavy industrial demands and power plants are directly supplied by the gas transmission systems. Finally, in order to deliver the gas to residential, commercial, and small industrial plants, the pressure is reduced through the regulator stations, and the consumers are supplied through the distribution systems. The performance and mathematical formulation to simulate the main components in natural gas system, including the gas pipelines, compressors, storages, and regulators, are illustrated in Fig. 9.2.

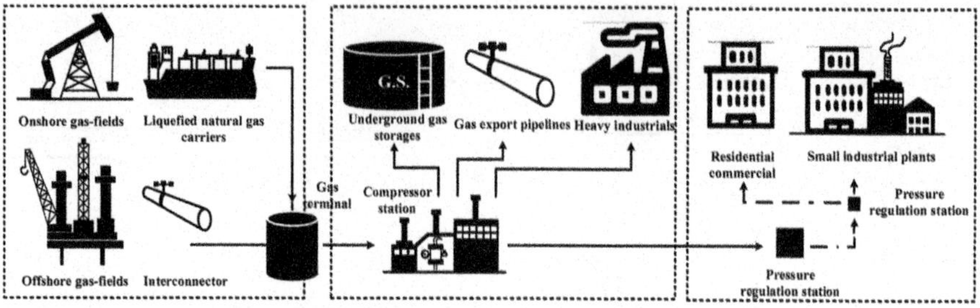

Fig. 9.2 Gas network from suppliers to consumers

9.2.1 Storage Systems

Storage system refers to devices that convert a form of energy into another form which can be stored and converted back when it is necessary. Energy storage systems, which are employed in electricity networks, can mainly categorize into four different groups, including mechanical (e.g., compressed air energy storages, pumped hydro storages, and flywheel energy storage systems), electrochemical (e.g., battery energy storage systems), and electrical (e.g., superconducting magnet energy storage system and electric double-layer capacitors or supercapacitors) [13]. However, in the gas network, gas storage systems (overground and underground) are also employed to support the supply-demand balance and provide security of supply [14].

The pumped storage systems can facilitate the storage of energy in off-peak hours and inject to the grid during peak hours [15]. This technology operates by pumping water to a higher altitude, while the compressed air energy storage systems compress the air and release it for combustion and power generation [16]. The flywheel energy storages store kinetic energy mechanically in a flywheel [17]. Furthermore, the battery storage systems store energy in the electrochemical form [18]. Aside from that, the superconducting magnet energy storages store electricity within a coil, and the capacitors store and discharge electricity benefiting from the electrolyte, placed between the conductors [19]. Moreover, in the gas network, the natural gas is stored in tankers underground, high-pressure holders, and low-pressure holders or liquefied to be stored (the characteristic of each technology is presented in Table 9.1) [20].

The model formulation for energy storage systems and gas storage systems is presented in this subsection. In Eqs. (9.1), (9.2), (9.3), and (9.4), the energy storage systems in the electricity system are formulated [21].

$$E_{r,t} = E_{r,t} + \left(P^{\mathrm{ch}}_{r,t}.\eta_{\mathrm{ch}} - P^{\mathrm{dch}}_{r,t}/\eta_{\mathrm{dch}}\right) \qquad \forall r, \forall t \qquad (9.1)$$

$$E^{\min}_{r} \leq E_{r,t} \leq E^{\max}_{r} \qquad \forall r, \forall t \qquad (9.2)$$

Table 9.1 Characteristics of different storage systems

Category	Type	Range	Response time	Efficiency
Mechanical	Pumped hydro storages	110 MW–2 GW	sec–mins	70–80%
	Compressed air energy storages	110 MW–290 MW	1–15 mins	90%
	Flywheel energy storage systems	5 kW–90 MW	sec	80–95%
Electrochemical	Battery energy storage systems	100 W–100 MW	20 ms-sec	60–80%
Electrical	Superconducting magnet energy storage system	170 kW–100 MW	<100 ms	95%
	Supercapacitors	<1 MW	8 msec	>95%

$$P_r^{\text{ch,min}} \le P_{r,t}^{\text{ch}} \le P_r^{\text{ch,max}} \qquad \forall r, \forall t \tag{9.3}$$

$$P_{r,t}^{\text{dch,min}} \le P_{r,t}^{\text{dch}} \le P_{r,t}^{\text{dch,max}} \qquad \forall r, \forall t \tag{9.4}$$

In Eqs. (9.1) and (9.2), the changes in the energy level of batteries and the limitation of the state of charge are presented, respectively. In Eqs. (9.3) and (9.4), the charging/discharging power limitation for these systems are indicated. In these equations, r represents the set of energy storage systems. The energy level of energy storage systems is also shown by $E_{r,\,t}$. Moreover, the efficiency of charge/discharge is depicted by $\eta_{\text{ch/dch}}$, and $P_{r,t}^{\text{ch/dch}}$ demonstrates the charging/discharging power of energy storage systems.

Aside from the already mentioned, in Eqs. (9.5), (9.6), (9.7), (9.8), (9.9), and ((9.10), the model formulation for gas storage systems is presented [22].

$$Gl_{q,t} = Gl_{q,t-1} + \left(Q_{q,t}^{wd}.\eta^{\text{ch}} - Q_{q,t}^{inj}/\eta^{\text{dch}}\right) \quad \forall q, \forall t \tag{9.5}$$

$$Gl_q^{\min} \le Gl_{q,t} \le Gl_q^{\max} \quad \forall q, \forall t \tag{9.6}$$

$$0 \le Q_{q,t}^{wd} \le Q_{q,t}^{\text{wd,max}} \quad \forall q, \forall t \tag{9.7}$$

$$0 \le Q_{q,t}^{inj} \le Q_{q,t}^{\text{inj,max}} \quad \forall q, \forall t \tag{9.8}$$

$$Q_{q,t}^{\text{wd,max}} = K_q.\sqrt{Gl_{q,t}} \quad \forall q, \forall t \tag{9.9}$$

$$Q_{q,t}^{\text{inj,max}} = -K_q^1 \cdot \sqrt{\frac{1}{Gl_{q,t} + Gl_{q,t}^{\text{cush}}}} + K_q^2 \quad \forall q, \forall t \tag{9.10}$$

In Eqs. (9.5) and (9.6), the changes in the gas level of the storages and the limitation of gas level are presented, respectively. In Eqs. (9.7) and (9.8), the limitation of withdrawal and injected gas of these systems are indicated. Furthermore, in Eqs. (9.9) and (9.10), it is indicated that maximum withdrawal and injected gas are proportional to the gas level in the system. In the represented model and formulation, q shows the set of gas storages, and $Gl_{q,\,t}$ indicates the gas level within the storages. The amount of withdrawal/injected natural gas is also demonstrated by $Q_{q,t}^{wd/inj}$.

9.2.2 *Flexible Power Plant*

While a low-carbon energy system has various benefits (e.g., low environmental impact and employment), the main problem with these resources is the intermittency in their output power [23]. In comparison with conventional fossil fuel-fired power plants, such as gas-fired power plants, the renewable ones cannot provide firm generation [24]. As a result, the fossil fuel-fired power plants, planned to supply baseload, should participate to deal with the changes and supply peak load. For this purpose, these power plants must turn on and off and ramp up and down although it could have some negative impacts (e.g., increase in forced outage rate). However, other options to cope with this problem, such as demand response, have still their downsides. The demand response programs need smart frameworks in order to address the challenges associated with RES to maintain the supply-demand balance.

The flexibility level of different type of power plants can be categorized into three different groups, including inflexible technologies, flexible technologies, and highly flexible technologies [25]. Relatively inflexible generation technologies include nuclear[1], coal, wind, photovoltaic, oil, and geothermal power plants. The flexible technologies include gas-fired, flexible coal-fired, combined-cycle turbine, biomass, and biogas power plants. Some technologies, such as hydropower plants, are also categorized as highly flexible power plants. In Table 9.2 (extracted from [11, 25]), the characteristics of all these technologies are represented.

According to Table 9.2, the flexible gas-fired power plants provide maximum ramping (up/down) rate and minimum start-up time. Furthermore, there is a considerable share of these power plants in the electricity networks all over the world (e.g., around 40% in the United States of America and Great Britain) [26]. As a result, the technology is one of the most commonly used options to deal with the

[1] Although nuclear plants can be flexible as well, but due to the fact that these plants usually provide the baseload, it is categorized as a relatively inflexible generation technology.

Table 9.2 Characteristics of different generating units' technologies

Technology	Minimum load (%)	Ramping rate (%)	Hot start-up time (%)
Nuclear	60	2	24
Oil	40	30	3
Geothermal	15	30	1.5
Gas-fired	40	60	0.16
Flexible coal-fired	40	60	3
Combined-cycle turbine	40	60	2
Biomass	50	40	3
Biogas	50	40	3
Hydro	5	100	0.1

intermittent nature of RES, such as wind and photovoltaic power plants. The model formulation of the gas-fired power plants is demonstrated in Eqs. (9.11), (9.12), (9.13), (9.14), (9.15), (9.16), (9.17), and (9.18) [27].

$$\mathrm{Cost}=\sum_t\sum_i b_i.P_{i,t}+\sum_t\sum_i W_{i,t}^{\mathrm{su}}+\sum_t\sum_i W_{i,t}^{\mathrm{sd}} \tag{9.11}$$

$$P_i^{\min}\leq P_{i,t}\leq P_i^{\max}\quad \forall i,\forall t \tag{9.12}$$

$$P_{i,t}-P_{i,t-1}\leq RU_i^{\max}\quad \forall i,\forall t \tag{9.13}$$

$$P_{i,t-1}-P_{i,t}\leq RD_i^{\max}\quad \forall i,\forall t \tag{9.14}$$

$$\lambda_{i,t'}-\lambda_{i,t'-1}\leq \lambda_{i,t}\quad \forall i,\forall t'\in\left[t-T^{\mathrm{on}}+1,t-1\right] \tag{9.15}$$

$$\lambda_{i,t'-1}-\lambda_{i,t'}\leq 1-\lambda_{i,t}\quad \forall i,\forall t'\in\left[t-T^{\mathrm{off}}+1,t-1\right] \tag{9.16}$$

$$W_{i,t}^{\mathrm{su}}=C_i^{\mathrm{su}}.\max\left\{\lambda_{i,t}-\lambda_{i,t-1},0\right\}\quad \forall i,\forall t \tag{9.17}$$

$$W_{i,t}^{\mathrm{sd}}=C_i^{\mathrm{sd}}.\max\left\{\lambda_{i,t-1}-\lambda_{i,t},0\right\}\quad \forall i,\forall t \tag{9.18}$$

As depicted, three terms are added into objective functions that represent the cost of generating power, start-up cost, and shutdown cost, respectively (Eq. (9.11)). Furthermore, a few constraints are added, including maximum/minimum stable output power (Eq. (9.12)), ramp up/down (Eqs. (9.13) and (9.14)), minimum up-/downtime (Eqs. (9.15) and (9.16)), and startup/shutdown costs (Eqs. (9.17) and (9.18)). In the model, i indicates the set of generating units. Furthermore, $P_{i,\,t}$,

RU_i/RD_i, and $\lambda_{i,t}$ demonstrate output power of generating units, ramp up/down, and status of generating unit (on or off).

9.2.3 Bidirectional Compressors

Compressors are deployed in the gas infrastructure to boost the pressure within the pipelines to keep the pressure of the gas in an acceptable range and avoid pressure drop over a long distance. More precisely, the transmission of gas reduces the movements as a result of the fraction. Therefore, the compressors increase the pressure by reducing the volume. This is noteworthy to mention that when the gas is compressed, its pressure rises. This is the main reason that compressor stations mainly have cooler facilities. There are different types of compressors based on prime movers, including gas turbines/engines, electric motors, and steam turbines [28]. Although a portion of gas flow is mainly used in compressor units, some countries use electrically driven compressors for security and environmental reasons. Hence, suitable type of prime movers should be employed based on economic and environmental factors. The model and formulation of compressors are presented in Eqs. (9.19), (9.20), (9.21), and (9.22) [29].

$$P_{c,t}^{\mathrm{comp}} = \frac{\beta_{comp}.\pi_{p,t}^{\mathrm{in}}.Q_{c,t}^{\mathrm{comp}}}{\eta_{\mathrm{comp}}} \cdot \left[\left(\frac{\pi_{p,t}^{\mathrm{out}}}{\pi_{p,t}^{\mathrm{in}}} \right)^{\frac{1}{\beta_{comp}}} - 1 \right] \quad \forall c, \forall t \tag{9.19}$$

$$1 \leq \frac{\pi_{p,t}^{out}}{\pi_{p,t}^{in}} \leq \mathrm{PR}^{\mathrm{max}} \quad \forall c, \forall t \tag{9.20}$$

$$Q_{c,t}^{\mathrm{comp}} \leq Q_{c,t}^{\mathrm{comp,max}} \quad \forall c, \forall t \tag{9.21}$$

$$P_{c,t}^{\mathrm{comp}} \leq P_{c,t}^{\mathrm{comp,max}} \quad \forall c, \forall t \tag{9.22}$$

In the model, the required power to drive compressors is calculated in Eq. (9.19). The ratio of output/input pressure is limited in Eq. (9.20). The limitations of gas flow through compressors and electricity consumption of the component are addressed in Eqs. (9.21) and (9.22), respectively. In the model, c stands for the set of compressor units. Moreover, $P_{c,t}^{\mathrm{comp}}$, $Q_{c,t}^{\mathrm{comp}}$, and $\pi_{p,t}^{\mathrm{in/out}}$ demonstrate electricity consumption, gas flow, and input/output pressure of compressors, respectively.

9.2.4 Power-to-Gas Systems

This technology uses electricity to produce hydrogen. Hydrogen has a high energy density and application in industries like steel making. Furthermore, the combustion of hydrogen produces water and energy ($2H_2(g) + O_2(g) \rightarrow 2H_2O + 572$kj of thermal energy where H_2: hydrogen, H_2O : water, and O_2 : oxygen) [30]. The gas can be injected into the gas system or stored to either be used for other purposes or burned to produce electricity in peak hours. It can also be used in fuel cells to produce energy without combustion in the presence of an oxidizing agent (e.g., oxygen) [31].

For a cost-effective transition to a low-/zero-carbon energy system, different methods have been invented that can couple this technology with RES [32]. Among the methods, the electrolysis process is the most popular one in which hydrogen and oxygen are produced via electrochemical conversion of water. However, based on the electrolyte, different methods have been invented that can be categorized, including (a) proton exchange membrane electrolyzer (PEM), (b) alkaline electrolyzer, and (c) solid oxide electrolyzer (SOEC). In Eqs. (9.23), (9.24), and (9.25), the model and formulation for the technology are demonstrated.

$$P_{d,t}^{\mathrm{p2g}} = \upsilon^{\mathrm{H2}}.\eta^{\mathrm{p2g}}.Q_{d,t}^{\mathrm{p2g}} \quad \forall d, \forall t \tag{9.23}$$

$$P_{d,t}^{\mathrm{p2g}} \leq P_{d,t}^{\mathrm{p2g,max}} \quad \forall d, \forall t \tag{9.24}$$

$$Q_{d,t}^{\mathrm{p2g}} \leq \varphi.Q_{d,t}^{\mathrm{p2g,max}} \quad \forall d, \forall t \tag{9.25}$$

In this model, the relation between the consumed power and the produced hydrogen is indicated in Eq. (9.23). Furthermore, the limitation of power consumption and gas production are addressed in Eqs. (9.24) and (9.25). In the equations, d demonstrates the set of P2G systems. Moreover, $P_{d,t}^{\mathrm{p2g}}$ and $Q_{d,t}^{\mathrm{p2g}}$ present the injected power into the systems and withdrawal gas, respectively. Constant for converting hydrogen energy to gas volume, efficiency of electrolyzers, and maximum allowance hydrogen injection to the natural gas system are indicated by υ^{H2}, η^{p2g}, and φ, respectively.

9.2.5 Integrated Operation of Gas and Electricity Networks

Aside from the components, the integrated operation of gas and electricity networks is a strategy that is beneficial for numerous reasons. It can be said that, in a power system with high penetration of renewable power sources, gas-fired units can be considered as a back-up option to improve the balance between generation and

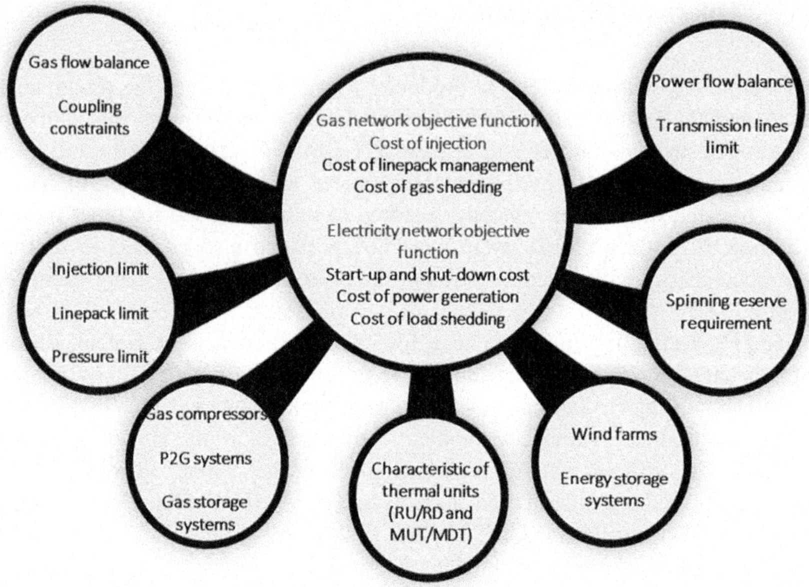

Fig. 9.3 Optimizing the integrated operation of gas and electricity networks

consumption in short-term scheduling [33]. For this purpose, in the mathematical model, the objective function is the sum of gas network operation's objective function and electricity network's objective function that is optimized subject to both networks' constraints, whereas, in an iterative strategy, initially the operation for electricity network is optimized. After that, by calculating the required amount of gas demand for the power plants, the gas network operation is optimized. The drawback of this strategy is that due to much lower speed of gas transport within the infrastructure, the required gas for power generation (i.e., output of electricity network optimization) may not be delivered on time. As a result, the gas-fired power plants should limit their output power that reduces the flexibility in the electricity network.

The integrated operation strategy facilitates the supply-demand balance in each node of the gas network, leading to an efficient operation of gas-fired power plants, which are helpful in dealing with intermittency of RES. The objective function and associated constraints of the optimization problem are depicted in Fig. 9.3.

It is noteworthy to mention that the objective function is demonstrated in the central part, and peripheral parts indicate the constraints (e.g., linepack[2] limit).

[2] Linepack refers to the amount of stored gas within the pipelines.

9.2.6 Other Flexibility Options

Other measures are available to improve the flexibility that is mainly related to the electricity network, including (a) demand response, (b) allowing RES to provide reserves, (c) reducing the minimum generation level of coal power plants, and (d) increasing the transmission capacity [34].

Demand response provides the capability to shift the demand in order to maintain the supply-demand balance. On the other hand, a minimum level of thermal generation is due to providing the reserve requirements. A large share of RES in the grid provides the capability to supply the reserve and reduce curtailment. Besides, the reduction in the minimum generation level of thermal units increases the flexibility as a result of the increase in the overall operating range of the power plants. The transmission expansion also provides the capability to import and export power (i.e., creating market opportunity) as well as scheduling the generation units by a central system that improves flexibility.

9.3 Analyses and Managerial Insights

In this section, different benefits of the options are examined considering the previous studies and projects. After that, some useful insights are provided for decision-makers and investors.

9.3.1 Flexibility Analyses in the Integrated Gas and Electricity Networks

Different projects were conducted in the world, focusing on the role of flexibility options in multi-vector energy systems. Among the projects, the research and analysis that was implemented on natural gas and electricity networks in Great Britain was a comprehensive one [35]. This study investigated the value of flexibility (flexible power plants, energy storage systems, adding interconnections, and demand response) for the United Kingdom's decarbonized gas and electricity systems for future horizons. The result of this project shows that the cost of operation with low-carbon technologies and RES highly depends on the level of flexibility in the network. In Fig. 9.4, costs and emissions of a fully flexible system in the United Kingdom are compared with an inflexible one.

As demonstrated, the increase in the level of flexibility reduces the disadvantage of wind generation versus nuclear ones from 14$/MWh greater to 1.3$/MWh lower. In comparison with the photovoltaic generation, it also reduces the cost from 2.3$/MWh higher than nuclear to 10.7$/MWh lower than nuclear. Therefore, a considerable amount of cost saving occurs as a result of the increase in the level of

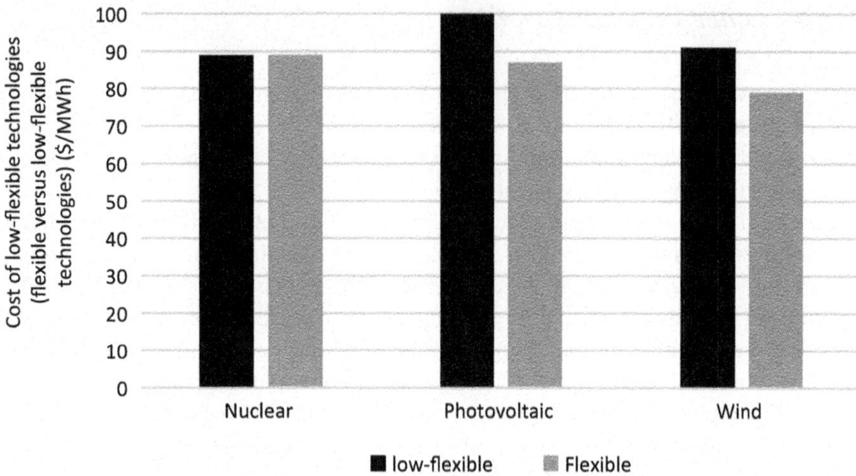

Fig. 9.4 Costs of low-flexible technologies in different flexibility level

Table 9.3 Characteristics of different generating units' technologies

Ref.	Flexibility options	Results
Qadrdan et al. (2017) [36]	Flexible gas-fired generation, P2G, electricity storage systems	Cost reduction and better performance of energy storage systems
Ameli et al. (2017) [37]	Flexible gas-fired generation and multidirectional compressor systems	Cost reduction
Strbac et al. (2020) [38]	Flexible gas-fired generations, storage systems, interconnections, and demand-side management	Cost and emission reduction
Fu et al. (2020) [39]	Different types of storage systems	Cost and emission reduction

flexibility. Aside from the already mentioned, the flexibility reduces the emissions 50% (100 g/kWh to 50 g/kWh). The main reason is the participation of RES in the generation mix as a result of flexible networks.

After this project, different studies on the role of flexibility options in the operation of Great Britain's gas and electricity networks were conducted. The main characteristics of the recent studies are indicated in Table 9.3, including the aims, the employed flexibility options, and the main results. In Fig. 9.5, the benefits of employing flexibility options were also quantified.

It is noteworthy to explain the scenarios and level of flexibility options in these studies. For instance, in [36], four cases are examined, including a reference case (i.e., no considerable flexibility), flexible gas-fired power plants, electricity storages, and P2G systems. The result shows that the integration of these options reduces

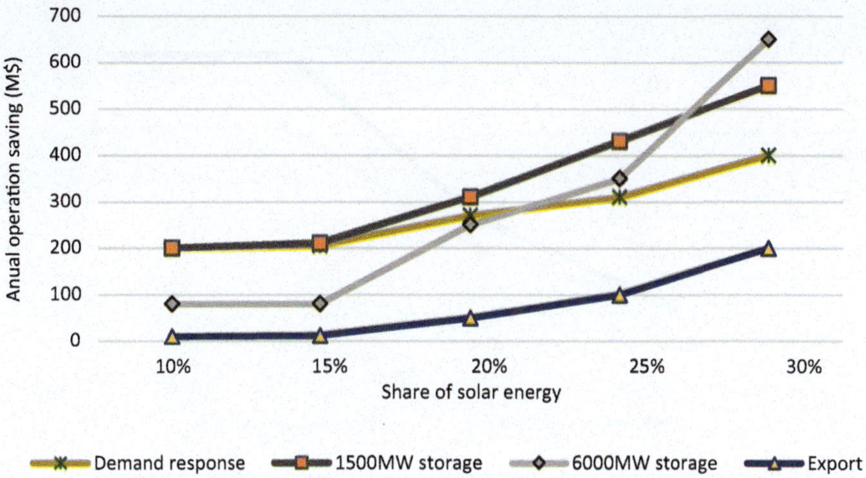

Fig. 9.5 Annual operation cost saving versus the increase in the share of photovoltaic systems in the presence of different flexibility options

the amount of curtailed wind and operating costs. Furthermore, the energy storage systems work more efficiently compared to others. In [37], however, three cases are examined, including flexible multidirectional compressors, optimizing the operation of both networks simultaneously, and combining the first and the second scenarios. The cost and emission reduction are less in the last case compared to others. Besides, in [38], flexible generation, interconnections, demand response, and energy storage systems are examined as flexibility options. In [39], the researchers focused on different types of P2G and power-to-power storage systems. The result of these studies shows that the flexibility option employment reduces the operation cost. For instance, in [36], employment of flexible gas-fired power plants, P2G systems, and electricity storage systems leads to 4% cost reduction. In [37], the potential of gas-fired power plants and multidirectional compressors are examined that indicates the cost reduction by 83%. However, in [38, 39], the decrease in the cost of operation was 10% and 21%, respectively. It should be noted that the case studies and level of flexibility option employment are different in these studies.

9.3.2 Flexibility Studies in Gas or Electricity Networks

Among the previous projects, some studies examined the impact of flexibility options only in gas or electricity network. For instance, National Renewable Energy Laboratory (NREL) focused on the benefits of different options to enhance the grid flexibility in three regions of the United States of America (California, Florida, and central parts) [40]. For this purpose, the role of different flexibility options was

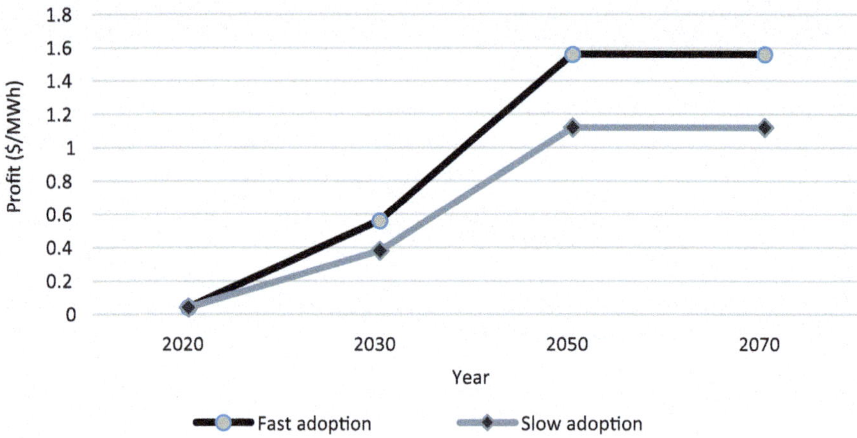

Fig. 9.6 Profit as a result of employing two scenarios for flexibility options adoption

examined considering different shares of photovoltaic generation. In Fig. 9.5, the results of this study are demonstrated in which it is evident that employing energy storage systems is the best option to reduce the costs and address supply-demand balance.

Eurogas also studied the role of flexibility options in the gas network and investigated different projects in different countries with a focus on gas storage systems and P2G systems [41]. This study employs two scenarios for the integration of the flexibility options, indicated in Fig. 9.6, where the profit of each scenario for future gas systems in a planned period is evident.

9.3.3 Managerial Insights

Flexibility refers to the ability of energy systems to respond to the variability of supply and demand. Previously used approaches for planning energy systems have supported flexibility that is sufficient to meet load reliably. However, the growing share of RES, such as photovoltaic systems and wind turbines, makes the planning methods inadequate to ensure sufficient flexibility in the future. Therefore, the flexibility of energy systems depends on its ability to accommodate renewable resources to maintain a reliable balance of supply and demand at a reasonable cost. Increasing flexibility is a low-regret option (option value), reducing the overall energy system cost (operation and reinforcement) while maintaining the security of supply. Our analyses of previous studies show the benefits. Based on the analyses, it is feasible to manage deeply decarbonized future energy systems by high penetration of RES. Therefore, achieving cost-effective deep decarbonization requires expansion of low-carbon capacity, along with increased level of system-wide flexibility (from the current levels).

9.4 Summary

This chapter addressed the role of flexibility options in future energy systems. For this purpose, at first, different types of flexibility options were demonstrated, including storage systems, P2G systems, flexible generation, integrated operation strategy, allowing RES to provide reserves, reducing minimum generation level of coal power plants, and increase in the transmission capacity. Furthermore, the model and formulation for the components, which are employed for this purpose, were presented more precisely. Afterward, results and analyses were conducted based on the previous studies, and managerial insights were provided to assist the decision-makers. All in all, it is addressed that employment of flexible components and flexible strategy (e.g., integrated operation strategy) reduce the cost of operation and could provide "option value," which could avoid/reduce additional reinforcements in the energy system infrastructure.

Based on the research and analysis that were conducted in this chapter, future research can focus on the economic analyses of each technology in the integrated operation of gas and electricity systems. The analyses can provide fundamental evidence to policy makers in regard of future energy system decarbonization. Aside from that, a couple of state-of-the-art technologies, including power-to-gas systems or electric vehicles, can be investigated in detail. The share of power-to-gas systems are increasing benefiting from storing energy in the form of gas (e.g., hydrogen). Therefore, this can be beneficial in order to facilitate a clean energy system. Moreover, using the storage capacity of electric vehicles in an optimal scheduling paradigm has a great potential that can get the attention of researchers in this field.

References

1. Hosseini, S. H. R., Allahham, A., Walker, S. L., & Taylor, P. (2020). Optimal planning and operation of multi-vector energy networks: A systematic review. *Renewable and Sustainable Energy Reviews, 133*, 110216.
2. Adedeji, A. R., Zaini, F., Mathew, S., Dagar, L., Petra, M. I., & De Silva, L. C. (2020). Sustainable energy towards air pollution and climate change mitigation. *Journal of environmental management, 260*, 109978.
3. Shabazbegian, V., Ameli, H., Ameli, M. T., & Strbac, G. (2020). Stochastic optimization model for coordinated operation of natural gas and electricity networks. *Computers & Chemical Engineering, 142*, 107060.
4. Mirzaei, M. A., Nazari-Heris, M., Zare, K., Mohammadi-Ivatloo, B., Marzband, M., Asadi, S., & Anvari-Moghaddam, A. (2020). Evaluating the impact of multi-carrier energy storage systems in optimal operation of integrated electricity, gas and district heating networks. *Applied Thermal Engineering, 176*, 115413.

5. Ameli, H., Qadrdan, M., & Strbac, G. (2020). Coordinated operation of gas and electricity systems for flexibility study. *Frontiers in Energy Research, 8*.
6. Wang, S., & Yuan, S. (2020). Interval optimization for integrated electrical and natural-gas systems with power to gas considering uncertainties. *International Journal of Electrical Power & Energy Systems, 119*, 105906.
7. Chen, F., & Wu, C. (2020). A novel methodology for forecasting gas supply reliability of natural gas pipeline systems. *Frontiers in Energy, 14*(2), 213–223.
8. Shabazbegian, V., Ameli, H., Ameli, M. T., Strbac, G., & Qadrdan, M. (2021). Co-optimization of resilient gas and electricity networks; a novel possibilistic chance-constrained programming approach. *Applied Energy, 284*, 116284.
9. Golpîra, H., Sadeghi, H., & Bahramara, S. (2021). Electricity supply chain coordination: Newsvendor model for optimal contract design. *Journal of Cleaner Production, 278*, 123368.
10. Kang, J. N., Wei, Y. M., Liu, L. C., Han, R., Yu, B. Y., & Wang, J. W. (2020). Energy systems for climate change mitigation: A systematic review. *Applied Energy, 263*, 114602.
11. Ameli, H., Qadrdan, M., Strbac, G., & Ameli, M. T. (2020). Investing in flexibility in an integrated planning of natural gas and power systems. *IET Energy Systems Integration, 2*(2), 101–111.
12. Qadrdan, M., Abeysekera, M., Wu, J., Jenkins, N., & Winter, B. (2020). The future of gas networks. In *The future of gas networks* (pp. 49–68). Springer.
13. Rahman, M. M., Oni, A. O., Gemechu, E., & Kumar, A. (2020). Assessment of energy storage technologies: A review. *Energy Conversion and Management, 223*, 113295.
14. Weihe, H. U. A. N. G., Honglong, Z. H. E. N. G., & Mingfei, L. (2019). Development history and prospect of oil & gas storage and transportation industry in China. *Oil Gas Storage Transportation, 38*(1), 1–11.
15. Smallbone, A., Jülch, V., Wardle, R., & Roskilly, A. P. (2017). Levelised Cost of Storage for Pumped Heat Energy Storage in comparison with other energy storage technologies. *Energy Conversion and Management, 152*, 221–228.
16. Bartela, Ł. (2020). A hybrid energy storage system using compressed air and hydrogen as the energy carrier. *Energy, 196*, 117088.
17. Amiryar, M. E., & Pullen, K. R. (2017). A review of flywheel energy storage system technologies and their applications. *Applied Sciences, 7*(3), 286.
18. Amirante, R., Cassone, E., Distaso, E., & Tamburrano, P. (2017). Overview on recent developments in energy storage: Mechanical, electrochemical and hydrogen technologies. *Energy Conversion and Management, 132*, 372–387.
19. Mukherjee, P., & Rao, V. V. (2019). Superconducting magnetic energy storage for stabilizing grid integrated with wind power generation systems. *Journal of Modern Power Systems and Clean Energy, 7*(2), 400–411.
20. Verma, H., Gambhir, J., & Goyal, S. (2013). Energy storage: A review. *International Journal of Innovative Technology and Exploring Engineering, 3*(1), 9.
21. Mehrjerdi, H., & Hemmati, R. (2019). Modeling and optimal scheduling of battery energy storage systems in electric power distribution networks. *Journal of Cleaner Production, 234*, 810–821.
22. Chaudry, M., Jenkins, N., & Strbac, G. (2008). Multi-time period combined gas and electricity network optimisation. *Electric power systems Research, 78*(7), 1265–1279.
23. Abujarad, S. Y., Mustafa, M. W., & Jamian, J. J. (2017). Recent approaches of unit commitment in the presence of intermittent renewable energy resources: A review. *Renewable and Sustainable Energy Reviews, 70*, 215–223.
24. Kryzia, D., Kopacz, M., & Kryzia, K. (2020). The valuation of the operational flexibility of the energy investment project based on a gas-fired power plant. *Energies, 13*(7), 1567.
25. Gonzalez-Salazar, M. A., Kirsten, T., & Prchlik, L. (2018). Review of the operational flexibility and emissions of gas-and coal-fired power plants in a future with growing renewables. *Renewable and Sustainable Energy Reviews, 82*, 1497–1513.
26. Sequeira, T. N., & Santos, M. S. (2018). Renewable energy and politics: A systematic review and new evidence. *Journal of Cleaner Production, 192*, 553–568.

27. Mirzaei, M. A., Nazari-Heris, M., Mohammadi-Ivatloo, B., Zare, K., Marzband, M., & Pourmousavi, S. A. (2020). Robust Flexible Unit Commitment in Network-Constrained Multicarrier Energy Systems. *IEEE Systems Journal*, 1–10.
28. Hendryx, M., & Luo, J. (2020). Natural gas pipeline compressor stations: VOC emissions and mortality rates. *The Extractive Industries and Society, 7*(3), 864–869.
29. Zhao, Y., Xu, X., Qadrdan, M., & Wu, J. (2021). Optimal operation of compressor units in gas networks to provide flexibility to power systems. *Applied Energy, 290*, 116740.
30. Buttler, A., & Spliethoff, H. (2018). Current status of water electrolysis for energy storage, grid balancing and sector coupling via power-to-gas and power-to-liquids: A review. *Renewable and Sustainable Energy Reviews, 82*, 2440–2454.
31. Staffell, I., Scamman, D., Abad, A. V., Balcombe, P., Dodds, P. E., Ekins, P., ... Ward, K. R. (2019). The role of hydrogen and fuel cells in the global energy system. *Energy & Environmental Science, 12*(2), 463–491.
32. Nazari-Heris, M., Mirzaei, M. A., Mohammadi-Ivatloo, B., Marzband, M., & Asadi, S. (2020). Economic-environmental effect of power to gas technology in coupled electricity and gas systems with price-responsive shiftable loads. *Journal of Cleaner Production, 244*, 118769.
33. Mirzaei, M. A., Nazari-Heris, M., Mohammadi-Ivatloo, B., Zare, K., Marzband, M., & Anvari-Moghaddam, A. (2020). A novel hybrid framework for co-optimization of power and natural gas networks integrated with emerging technologies. *IEEE Systems Journal, 14*(3), 3598–3608.
34. Sinsel, S. R., Riemke, R. L., & Hoffmann, V. H. (2020). Challenges and solution technologies for the integration of variable renewable energy sources—A review. *Renewable Energy, 145*, 2271–2285.
35. Strbac, G., Aunedi, M., Pudjianto, D., et al. (2015). *Value of flexibility in a decarbonized grid and system externalities of low-carbon generation technologies for the committee on Climate Change*. Imperial College of London.
36. Qadrdan, M., Ameli, H., Strbac, G., & Jenkins, N. (2017). Efficacy of options to address balancing challenges: Integrated gas and electricity perspectives. *Applied Energy, 190*, 181–190.
37. Ameli, H., Qadrdan, M., & Strbac, G. (2017). Value of gas network infrastructure flexibility in supporting cost effective operation of power systems. *Applied Energy, 202*, 571–580.
38. Strbac, G., Pudjianto, D., Aunedi, M., Djapic, P., Teng, F., Zhang, X., ... Brandon, N. (2020). Role and value of flexibility in facilitating cost-effective energy system decarbonisation. *Progress in Energy, 2*(4), 042001.
39. Fu, P., Pudjianto, D., Zhang, X., & Strbac, G. (2020). Integration of Hydrogen into Multi-Energy Systems Optimisation. *Energies, 13*(7), 1606. Online: https://www.theccc.org.uk/wp-content/uploads/2015/10/CCC_Externalities_report_Imperial_Final_21Oct20151.pdf
40. Denholm, P., Novacheck, J., Jorgenson, J., and O'Connell, M. (2016). *Impact of flexibility options on grid economic carrying capacity of solar and wind: three case studies*. National Renewable Energy Laboratory. : https://www.nrel.gov/docs/fy17osti/66854.pdf
41. Eurogas. (2017). *Flexibility in the energy transition: a toolbox for gas DSOs*. Online.: https://www.edsoforsmartgrids.eu/wp-content/uploads/Flexibility-in-the-energy-transition-A-tool-for-electricity-DSOs-2018-HD.pdf

Chapter 10
Impact of Demand Response Programs on the Operation of Power and Gas Systems

Mohammad Mehdi Davary, Mohammad Taghi Ameli, and Hossein Ameli

10.1 Introduction

The goal of all components designed in energy systems (e.g., power and gas systems) is to provide sustainable and economical energy to consumers. Power and gas systems have a significant share in energy consumption and play a vital role in people's lives [1]. On the other hand, with the unbalanced growth of consumption and the existence of demand peaks in limited hours of the year, a high investment cost is required to meet the demand peak. Therefore, to increase the efficiency of the energy system, consumption management is considered. Therefore, to reduce this process and its possible consequences, it is necessary to take corrective measures [2], such as demand response. The US Energy Agency defines demand response (DR) as a change in the pattern of energy consumption by consumers in response to changes in power prices over time with economic programs designed to discourage the use of power when market prices are high or when network reliability is compromised [3]. The use of demand response programs (DRPs) offers numerous benefits to consumers, the market, and the network. Consumer benefits include the continuity of power and the economics of these programs. Responsive market benefits include preventing price fluctuations and helping to stabilize the retail and wholesale markets. Also, DRP implementation saves investment and operation costs, postponing the construction of power plants

M. M. Davary
Department of Electrical Engineering, Shahid Beheshti University, Tehran, Iran

M. T. Ameli (✉)
Faculty of Power Engineering, Shahid Beheshti University, Tehran, Iran
e-mail: m_ameli@sbu.ac.ir

H. Ameli
Control and Power Group, Imperial College London, London, UK

© The Author(s), under exclusive license to Springer Nature Switzerland AG 2022

V. Vahidinasab, B. Mohammadi-Ivatloo (eds.), *Whole Energy Systems*, Power Systems, https://doi.org/10.1007/978-3-030-87653-1_10

and new lines and smoothing the network load curve [4]. DR is a developed model of demand management. Changing prices and different rewards encourage consumers to improve their consumption patterns to increase network reliability and efficiency. Restructuring and deregulation of the power industry have created new challenges for consumption management. Sufficient and accurate information is required for proper DR implementation. Due to lack of equipment, it is sometimes not possible to access enough information. In [44], to cover the lack of information about the local demand on the network, random periodic multiplicative method is used for the design of distributed algorithms.

With the increase in the power market's competitiveness, the importance of consumption management programs is felt more than ever. DRPs are divided into two main categories: (1) incentive-based programs and (2) price-based programs. Each program includes subprograms which are described in the third section. Modeling these programs to plan production and load forecasting is of great importance for power and gas companies. Modeling DR programs using demand elasticity matrices is one of the most robust and most common methods in this field. In this chapter, various modeling of DRPs based on incentives and price are modeled. On the other hand, as shown below, sometimes considering only a linear DR model can lead to load forecasting errors and impose huge costs that make planning challenging. Therefore, in this chapter, in addition to the linear model, nonlinear models (power, exponential, and logarithmic) are also considered [5].

Given the nature of the problem of economic distribution and demand response, which are related to the producer and consumer, respectively, the simultaneous study of these two issues is very beneficial. One of the most critical issues in DRP is determining the optimal incentive or price rate. If incentive or price is inappropriate set in DRP, then it may, in addition to incurring additional costs, create a new peak at other times. A suitable solution to determine the optimal rate in this research is the simultaneous implementation of the economic dispatch and DRP in the integrated operation, in which the cost of implementing the DRP is added as an additional term to the economic dispatch program. As a result, the new objective function also determines the incentive or optimal price while minimizing the fuel cost and pollution function [6]. In [7], in addition to DR, hydrogen storage systems are employed in the operation of electricity and gas systems. The hydrogen storage system can help the electricity grid operation by storing electricity in the form of hydrogen during off-peak hours and by inversely converting hydrogen to electricity during peak hours. In addition, the stored hydrogen can be used directly in the gas system.

The importance of implementing demand response in gas and electricity systems is to maintain the supply-demand balance in both systems. For example, by implementing load response in the power grid, gas consumption of gas-fired power plants is reduced. Consequently, the need for boosting the pressure is decreased, and hence the gas consumption of the compressors is reduced. On the other hand, load response in the gas network reduces the required power for electrically driven compressors (i.e., linkage between gas and electricity systems, which is a demand in the electricity system). On the other hand, using this tool can act as a storage device in the operation of the integrated network. Nowadays, despite the high investment cost of creating electric or gas storage or hybrid, familiarity and use of this method has become very important. Therefore, this chapter has been paid to this issue. According to the subjects mentioned above, this chapter aims to answers the following questions:

- What are the types of demand responses in energy systems?
- How can the efficiency of the energy system be increased by properly implementing the demand response?
- What is the role of DR modeling in improving line capacity and reducing costs in energy systems?

The remainder of this chapter is divided into six main sections. In the second part, previous research on DR and operation of the integrated power and gas systems will be presented. Section 10.3 describes the types of DRPs, and in Sect. 10.4, types of linear and nonlinear DR modeling as well as an integrated model of the incentive-based program (emergency demand response) and price-based program using the problem of economic distribution of load and cost of pollution are developed for the simultaneous operation of power and gas systems. In Sect. 10.4, the solution method is presented, and in Sect. 10.5, the results and numerical studies will be presented to determine the applicability and reliability of the proposed model. Finally, the conclusions and summary of this chapter will be presented in Sect. 10.6.

10.1.1 Status Quo, Challenges, and Outlook

Currently, the operation for electricity and gas networks is carried out by two different entities. In addition, a limited load response program has been implemented only in the electricity network. Some of the main challenges that prevent simultaneous operation and load response in both networks are as follows: (i) organizational bureaucracy, (ii) software problems, and (iii) lack of comprehensive expansion of intelligent measurement infrastructure to implement load response.

However, due the expansion of renewable power plants due to international agreements to reduce air pollution, which imposes challenges to the energy system (i.e., high uncertainties associated with the production of renewable power plants), the share of flexible gas-fired power plants as one of the main sources to complement the lack of renewables will increase. These issues accelerate the importance of integrating the operation of electricity and gas networks. On the other hand, due to the benefits of load response, the development of smart grids, and the economic inefficiency of developing production and transmission infrastructure for limited hours of the year, demand response on both networks will be implemented soon.

10.2 Literature Survey

In this section, some articles on the simultaneous operation of power and gas systems and DR are reviewed. Several references have examined the issue of optimal operation of power and gas systems asynchronously, and many articles optimize this issue synchronously. In asynchronous operation, first, the problem of optimal operation of the power system is solved. After obtaining the production rate of gas-fired power plants, the total gas consumption is calculated. The consumption of gas-fired power plants is added to the demand for gas, and consequently the problem of optimal operation of the gas network is solved. The obtained results are checked so that there is no unloading in the gas network due to the increased demand for gas-fired power plants. At this stage, if there is load shedding, the subproblem of operation of the power system is limited until the amount of load shedding in the gas network is zero due to the demand for gas-fired generators. In simultaneous optimal operation, the objective function is the sum of the objective functions of power and gas systems, and the constraints of both networks are taken into account in the optimization problem. The following is a review of published articles on demand response. Most research on DR included a review of types of DRPs, the simultaneous study of DRPs with distributed generation and renewable energy, and discussion of the effects of DR on improving network reliability.

10.2.1 Asynchronous Operation

Reference [8] has introduced and examined the effect of the gas network on the problem of unit commitment with the asynchronous operation of power and gas systems, considering the transient characteristics of the gas flow. The results obtained from this study show that models that consider only the steady state of natural gas flow cannot model the properties of gas transmission lines well and cause the obtained results to be unusable. In another study, in addition to providing accurate modeling of power and gas systems and optimizing performance simultaneously with asynchronous strategy, the efficiency of elements such as flexible gas power plants and electrical energy storage devices in balancing production and demand was investigated [9]. The results of this study show that the use of electrical storage systems can have the most significant cost reduction in the operation of power and gas systems. Also, in this study, the British power and gas systems have been proposed as a case study, and the results show that there is a reduction in costs in the accidental case compared to the definitive case. In other studies, a stochastic model for simultaneous operation of power and gas systems has been introduced [10, 11]. In these studies, scenarios based on the line and generator outages have been investigated. The models proposed in these studies are optimized under asynchronous strategy and linearization of nonlinear constraints. The results show the effect of random programming in reducing the consequences of possible outputs and the effect of hourly DR on reducing operating costs, respectively. In another study, a robust optimization model is introduced to optimize the asynchronous operation of power and gas systems [12]. In this case, the operating costs for the worst-case scenario are minimized. In addition, a relaxation method is used to solve this complex problem, and the power system subproblem and the gas grid problem are solved asynchronously. In this study, the nonlinear problem of the gas network is approximated as a linear fragment. Finally, a case study shows the effects of the gas network on the problem of unit commitment and shows the positive effect of flexible network equipment, such as storage devices, against the random and variable nature of wind generators.

10.2.2 Simultaneous Operation

A number of articles have optimized the problem of simultaneous operation of the power system simultaneously. For example, in [13], a multi-period model for the simultaneous operation of power and gas systems is presented. In this model, the variable nature of gas flows, reservoirs, and the characteristics of power generation output changes are considered, and the costs related to gas supply, gas volume management within pipelines (linepack), storage performance, power generation, and loading are minimized. The results obtained from this study show that in case of an outage of several gas terminals, there is less loading in the simultaneous operation strategy. In this paper, due to the dependence of the Irish gas network on the British gas network, the operation of power and gas systems of the two countries is studied simultaneously [14]. Also, the effect of gas storage on improving the operation of the electrical network has been introduced as a turning point in this research. In reference [15], in addition to providing accurate modeling of power and gas systems, the advantages of using multidirectional compressors and operation under simultaneous strategy have been studied. This study defines a number of scenarios for the UK power and gas systems in 2030 to fully identify the benefits of simultaneous operating strategy over the asynchronous operation. The obtained results also show a significant improvement in the simultaneous strategy. Another article published in this field has introduced a model for optimizing the simultaneous operation of power and gas systems, and because the gas network subproblem is a complex problem of non-convex and nonlinear integers, a linearization method has been used in this study. In reference [16], a variable relaxation method is used to calculate accurate and fast responses during the simultaneous operation of power and gas systems. In another study, a new economic distribution model for power and gas systems is proposed in which the Wilmouth Equation is relaxed. The results of the relaxation of the inaccurate mixed problem are compared with the results of the complex integer programming approximated as linear fractions. The results show that performing relaxation calculates the results faster and more accurately than linearization [17]. A number of articles have addressed the uncertainty surrounding the simultaneous operation of power and gas systems. In [18], in order to optimize the operation of power and gas systems, which includes the uncertainty of DR and wind energy, it has used interval optimization. In [19], a robust optimization model is introduced for the simultaneous operation of power and gas systems. The nonlinear equations of the gas network are modeled as linear fractions, and the operation is solved under the worst-case scenario of demand and production of wind generators using the decomposition method. The results obtained in this study show the effectiveness of a robust model for situations where the load and output of renewable generators differ from the predicted value.

10.2.3 Demand Response Concepts

In [20], various aspects of demand response, including participants, challenges, and potential demand response ability to control frequency and voltage on a large scale, are investigated. In [21] a new DR algorithm is introduced. In this algorithm, the price expected by the customer as an index of pricing and the use of smart grid infrastructure to respond to these price changes are presented. This algorithm automatically programs the performance of home appliances. The algorithm also manages the home energy storage system in such a way that the extra energy is stored at low-cost hours and consumed at high-price hours. The simulation results show that the implementation of the proposed algorithm has been able to reduce power consumption during peak hours in a way that ultimately leads to a reduction in the total cost of household energy to 39%. Reference [22] examines and analyzes DRPs under smart grids and renewable resources. In this paper, with a new approach, renewable energy planning with DRP is examined. Reference [23] provides plausible models for renewable energy generation that include unforeseen errors due to wind speed and sunlight by possible distribution functions. In this model, all loads, including domestic, industrial, and commercial loads, are considered to implement the DRPs. A two-step objective function is also used to minimize costs. In [24], a stochastic operational planning model for energy resources and storage in the smart distribution system is introduced. In the introduced model, the demand is met by the generators, storage devices, and loads participating in the DRPs. The simulation results in [24] show that the introduced model reduces the costs of generation and consumption of power in the network. In [25] different types of response programs, their benefits, and main applications are stated separately. Next, the mathematical model for tariff-based response programs and incentives is presented and compared and prioritized the effects of various programs on Iran's peak curve in 2007 in terms of cost reduction and reliability using strategy success indicators. In [26] a variety of linear and nonlinear models of price-based programs such as time of use (TOU) is presented, and the effect of each model on reducing costs and increasing network reliability is examined. Then the effect of different elastic values and prices on cost and load curve is investigated.

The model in [26] is only suitable for price-based programs and does not apply to incentive-based programs. As mentioned earlier, responsive programs fall into two general categories: price-based and incentive-based. One of the important issues in incentive-based programs is determining the optimal incentive rate, and in price-based programs, determining the optimal price rate. If the incentive rate or price is set inappropriately, it may impose additional costs on the production side and create a new peak in other hours [27]. The focus of dynamic economic emission dispatch (DEED) issue is on generating power plants without considering the consumer side. On the other hand, DRPs can change consumption patterns to improve productivity and reduce price volatility. The main focus of the DRPs is consumption control. Therefore, for a more comprehensive study, the combination of these two seems very useful. However, so far, few studies have been done on the optimal combination of DRPs and the problem of economic load distribution [28]. A complete work has been done to optimally model the DRP and the problem of economic cost-dynamic load distribution in the reference [29]. The main drawback of their work is the ambiguity of the final model. In the proposed model, customers are rewarded throughout the day (even during off-peak hours) to reduce demand. This approach is not feasible from practical, accuracy, and economic perspectives. Furthermore, in this work, shiftable loads are excluded, and only reducible and disconnected loads are considered [29].

10.3 Types of Demand Response Programs

10.3.1 Price-Based Programs

10.3.1.1 Time of Use (TOU) Program

In the TOU program, the hours of the day are divided into several fixed time intervals, each of which has a different price. This price difference serves as an incentive for the consumer to reduce some of his expenses or move them to other time periods. These prices can change for hours of a day, days of a week, or different seasons of a year [30]. Today, in most countries, the price of energy has a fixed rate at all times. This has many disadvantages. With stable energy prices, it makes no difference to consumers because they have no incentive to consume energy differently (peak, medium, and low load), while they can easily reduce their load from peak load times or transfer it to low load times with medium load by modifying the consumption pattern. Also, the fixed energy rate does not differentiate between the high prices of energy produced at peak times and low and medium periods. In this case, customers with low consumption during peak times actually incur additional costs of energy consumption by customers during peak times [31].

10.3.1.2 Real-Time Pricing (RTP) Program

In this program, the selling price of energy per hour depends on the production costs at the same time; therefore, the price will be low at low demand period, medium at medium demand period, and expensive at peak demand period. The implementation of this program causes the consumers (especially the consumers who can move their power consumption) to adapt their consumption to the price, and as a result, the system load is reduced, and they are often transferred from peak hours to other hours [32]. Also, in this plan, energy consumption is recorded by consumers on an hourly basis by a power meter, and their bills are calculated based on the current price of energy in the market [33]. The time period for updating the energy price is crucial. The shorter the time for updating the energy price, the higher the efficiency and effectiveness of the response program [42]. Pricing and usage periods of TOU program are determined in advance. In other words, TOU program is not effective in emergencies when the network load needs to be reduced. In other words, suppose the network load needs to be reduced at a time other than peak load, in which case there is no incentive for customers to reduce their load. RTP, on the other hand, has a dynamic pricing process with an update time of 1 hour or less [29]. The RTP is highly dependent on the energy market model. Although this program is one of the best response methods, it has a complex execution process. The RTP method is most widely used by large consumers, as it is difficult for home consumers to deal with these real wholesale market prices in practice [46].

10.3.1.3 Critical Peak Pricing (CPP)

In this program, in some hours during the year when energy consumption is high, a very high price is set for energy, the hours that this program runs are less than 1% of the hours of the year, and consumers are notified at least 1 day in advance [29]. While RTP is an ideal pricing method, CPP is much more effective in critical times. In this method, in supercritical times, by considering prices many times the average energy price, they lead customers to reduce the load in these periods [20, 34]. This program runs only on limited days of the year and for a few hours. At CPP, in noncritical times, customers receive discounts. In order to meet the need to reduce consumption during the critical hours of the year when the hours of energy consumption are very high (e.g., due to extreme heat) or the reliability of the system is at risk (the occurrence of an accident), using critical time pricing is a good solution. Consumers are usually notified of this price increase from the day before.

10.3.2 Incentive-Based Programs

10.3.2.1 Emergency Demand Response Program (EDRP)

When the independent power system operator forecasts a short-term load for the network under its management or after a load or price jump, it is likely that price jumps will be repeated, using the emergency demand response program (EDRP) to try to reduce peak consumption and prevent price jumps. Major consumers who wish to cut part of their load participate when independent system operator (ISO) announces in the program and receive a significant amount of reward as a reward for this reduction [35]. Of course, participation in the program is voluntary, and there will always be consumers who will not reduce their consumption regardless of the amount of the prize. This uncertainty in reducing the demand will make market management difficult [43]. Nevertheless, because the amount of the incentive prize is determined in advance and the consumers will be fined if they do not participate, the program has been very popular in the energy markets [24]. In the EDRP, in case of emergency, the energy distribution company sends warning signals and invites consumers to reduce or stop consumption. Participation in this program is generally optional for consumers, and they will not be penalized if they do not reduce their consumption. Participants who reduce their consumption at the request of the energy distribution company are encouraged.

10.3.2.2 Interruptible/Curtailable (IC) Program

In this method, consumers contract with the energy distribution company or operator to reduce or cut off their energy consumption whenever the energy distribution company or operator wants. In other words, in this way, consumers provide a kind of rotating storage for the system. Customers who participate in the program receive energy credit or a discount for reducing their consumption at the time of the accident, and participants who do not reduce their consumption when necessary are penalized. This DRP can be implemented in two cases by the customer or by the energy distribution company. In the first case, the energy distribution company sends an alert signal through the AMI system to the consumers who have already expressed their consent to participate in this plan. Consumers must reduce their consumption to the amount promised in the contract after receiving the signal. Otherwise, they will be fined according to the provisions of the contract. In the second case, according to the option of remote disconnection of meters, the energy distribution company will directly cut off the energy desired by the consumers. Of course, they warn customers in advance about the time and period of the outage. The second case has higher reliability for the distribution company because it will know the exact amount of reduced load [35].

10.3.2.3 Demand Bidding (DB) Program

Auction or repurchase is a mechanism in which consumers are enabled to receive discounts from a distribution company in exchange for bids based on consumption pattern modification. In this method, which is known as the demand bidding method, major consumers offer a reduced amount of cargo along with their desired price to the network operator, and after the market operation, if the offer is accepted, the subscriber is obliged to execute the contract, and if the request is not fulfilled, the subscriber will be fined. In other words, consumers can sell the purchased energy at a higher price [36].

10.3.2.4 Capacity Market (CM) Program

The operator must always maintain some of its production capacity as a reserve so that in the event of an event in the system, it can compensate for the power shortage. System storage can also be provided from the consumption side. This means that the energy distribution company has already signed a contract with the consumer that wants to participate in these programs, and they undertake to reduce their consumption by the amount of power or energy specified in the contract in the event of an event in the network. This program is like insurance, which may not need to be reduced in practice, but the incentive amount for consumers' readiness to disconnect will always be paid. In this program, consumers will be fined if they do not reduce their consumption on time [37].

10.3.2.5 Ancillary Service (AS) Program

In the ancillary service program, consumers offer their disconnection or reduction of load to the ISO as a reserve in the market. If their offers are accepted, they will receive a market-clearing price to keep their interrupted demand ready. Also, if the load of these consumers needs to be reduced, participants will be called by ISO. In this case, they will also be paid the current market price of energy. These types of loads must have a high response speed so that they can act quickly when an event occurs. Ancillary service participants can be industrial consumers whose energy consumption goes down very quickly and without damaging the equipment. These loads in the power system can be large water pumps, electric arc furnaces, and air compressors.

10.3.2.6 Direct Load Control (DLC) Program

In the direct load control program, the remote network operator turns off or controls the customers' electrical devices. These include household appliances such as air conditioners and water heaters, as well as small commercial appliances that do not cause much damage to the quality of the system if the consumption is stopped. Incentives are paid to customers in proportion to the time period selected for load control. This program has been successfully implemented for decades in the United States and the Russian heating system [23].

10.4 Method of Determining the Incentive or Price of the Demand Response Program in the Operation of Power and Gas Systems

10.4.1 Demand Response Modeling

Modeling incentive-based DRPs to price in order to accurately predict the load curve is very important for ISO; therefore, in modeling, in addition to the linear model, a variety of nonlinear models of DRP, including power, exponential, and logarithmic models, should be considered. Since the problem of optimizing customer profits is a nonlinear problem, therefore it is necessary to develop nonlinear economic models of responsive loads in order to bring the final models closer to reality. Among the introduced models, the linear model is the simplest and most common model introduced but consequently has more errors than the real model [45]. The power model is a concave and nondecreasing model that is used in many modelings and is closer to reality [3]. There are small differences in load changes with the power system during execution, which are ignored in this section.

By using a variety of linear and nonlinear models and considering a variety of conditions such as maximum changes and predicted peak load, the power company will be able to predict the amount of load consumption with the least error and with higher reliability. By doing this, the power company will gain a deeper and better understanding of the behavior of the power system and will be able to better provide the required reservation supplies. In other words, the developed models can be used for the company in such a way that they choose the conservative model in order to predict the characteristics of the load curve with the least error (especially peak load changes and energy consumption). Simply put, any model that has the most peak load and changes is used as a suitable model for load prediction. In general, nonlinear response models have a more conservative and reliable behavior than other models [4].

DR modeling based on price elasticity matrix (PEM) is one of the most common and powerful methods in this field. Also, to achieve optimal consumption on the demand side, elasticity is defined as the sensitivity of demand to price changes according to (10.1) [38, 47]:

$$\mathrm{E}\left(t,t'\right)=\frac{\rho_0\left(t'\right)}{d_0(t)}\times\frac{\partial d(t)}{\partial\rho\left(t'\right)}\begin{cases}\mathrm{E}\left(t,t'\right)\le 0 & t=t'\\ \mathrm{E}\left(t,t'\right)\ge 0 & t\ne t'\end{cases}\tag{10.1}$$

In the above relation, E is the price elasticity matrix, $d(t)$ and $d_0(t)$ are consumer demand after and before DRP implementation, and $p(t')$ and $p_0(t')$ are power prices after and before implementation during the time t.

For 24 h a day, internal and reciprocal elasticity can be represented as a 24 × 24 matrix according to (10.2):

$$\begin{bmatrix} \frac{\Delta d(1)}{d_0(1)} \\ \frac{\Delta d(2)}{d_0(2)} \\ \vdots \\ \frac{\Delta d(24)}{d_0(24)} \end{bmatrix} = \begin{bmatrix} E(1,1) & \dots & E(1,24) \\ \vdots & \ddots & \vdots \\ E(24,1) & \cdots & E(24,24) \end{bmatrix} \times \begin{bmatrix} \frac{\Delta \rho(1)}{\rho_0(1)} \\ \frac{\Delta \rho(2)}{\rho_0(2)} \\ \vdots \\ \frac{\Delta \rho(24)}{\rho_0(24)} \end{bmatrix} \quad (10.2)$$

10.4.1.1 Linear Modeling of Demand Response

Incentive response programs provide incentives for consumers to reduce their consumption during peak hours. The total payment incentive to consumers per hour is equal to (10.3) [35, 37]:

$$\mathrm{INC}\left(\Delta d(t)\right) = \mathrm{inc}(t) \times \left[\Delta d(t)\right] \quad (10.3)$$

where inc(t) is the rate of reward paid to consumers over a period of time; in some DRPs, customers who are committed to participating in the program but do not reduce or transfer their load during the specified hours, contrary to the contract, are penalized by (10.4):

$$\mathrm{PEN}\left(\Delta \mathrm{d}\,(\mathrm{t})\right) = \mathrm{pen}\,(\mathrm{t}) \times \left\{\mathrm{IC}\,(\mathrm{t}) - \left[\Delta \mathrm{d}\,(\mathrm{t})\right]\right\} \quad (10.4)$$

where IC(t) is the amount of load that the consumer is committed to reducing or shifting; the total benefit of the consumer is obtained in relation (10.5):

$$\mathrm{NP}(t) = B\left(d(t)\right) - d(t)p(t) + \mathrm{INC}\left(\Delta d(t)\right) - \mathrm{PEN}\left(\Delta d(t)\right) \quad (10.5)$$

where B is the profit that customers make by consuming the load; to maximize profits, the derivative of (10.5) must be zero:

$$\frac{\partial \mathrm{NP}}{\partial d(t)} = \frac{\partial B\left(d(t)\right)}{\partial d(t)} - \rho(t) + \frac{\partial \mathrm{INC}}{\partial d(t)} - \frac{\partial \mathrm{PEN}}{\partial d(t)} = 0 \quad (10.6)$$

$$\frac{\partial B\left(d(t)\right)}{\partial d(t)} = \rho(t) + \mathrm{pen}(t) - \mathrm{inc}(t) \quad (10.7)$$

To determine the optimal consumption by which consumers get the maximum benefit, we have according to (10.8) the B Taylor series :

$$B\left(d(t)\right) = \mathrm{B}\left(\mathrm{d}_0(t)\right) + \frac{\partial \mathrm{B}\left(\mathrm{d}_0(t)\right)}{\partial d(t)}\left[d(t) - \mathrm{d}_0(t)\right] + \frac{1}{2}\frac{\partial^2 \mathrm{B}\left(\mathrm{d}_0(t)\right)}{\partial d^2(t)}\left[d(t) - \mathrm{d}_0(t)\right]^2 \tag{10.8}$$

By deriving from (10.8):

$$B\left(d(t)\right) = \mathrm{B}\left(\mathrm{d}_0\ (\mathrm{t})\right) + \frac{\partial \mathrm{B}\left(\mathrm{d}_0(t)\right)}{\partial d(t)}\left[d(t) - \mathrm{d}_0(t)\right] + \frac{1}{2}\frac{\partial^2 B\left(d(t)\right)}{\partial d^2(t)}\left[d(t) - \mathrm{d}_0(t)\right]^2 \tag{10.9}$$

By combining (10.7) and (10.9), the single-period load model (loads that can be reduced or interrupted, taking into account only diameter elements (PEM)) is calculated according to (10.10):

$$d(t) = \mathrm{d}_0(t) \times \left[1 + \frac{\rho(t) - \rho_0(t) + \mathrm{inc}(t) - \mathrm{pen}(t)}{\rho_0(t)}\right] \tag{10.10}$$

Also, the multi-period load model (transferable loads considering only non-diagonal elements (PEM)) is calculated according to (10.11):

$$d(t) = \mathrm{d}_0(t) \times \left[1 + \sum_{\substack{t' = 1 \\ t' \neq t}}^{24} \frac{\rho\left(t'\right) - \rho_0\left(t'\right) + \mathrm{inc}\left(t'\right) - \mathrm{pen}\left(t'\right)}{\rho_0\left(t'\right)}\right] \tag{10.11}$$

Finally, the hybrid model is calculated according to (10.12) with respect to (10.10) and (10.11), which include single-period and multi-period load models:

$$d(t) = \mathrm{d}_0(t) \times \left[1 + \sum_{t'=1}^{24} \frac{\rho\left(t'\right) - \rho_0\left(t'\right) + inc\left(t'\right) - pen\left(t'\right)}{\rho_0\left(t'\right)}\right] \tag{10.12}$$

10.4.1.2 Power Modeling of Demand Response

For power modeling, according to (10.8) [33, 35]:

$$B\left(d_{\text{pow}}(t)\right) = \mathrm{B}\left(\mathrm{d}_0(t)\right) + \frac{\rho_0(t) \times d_{\text{pow}}(t)}{1 + E^{-1}\left(t, t'\right)} \times \left[\left(\frac{d_{\text{pow}}(t)}{\mathrm{d}_0(t)}\right)^{E^{-1}\left(t,t'\right)} - 1\right] \tag{10.13}$$

By deriving from (10.13):

$$\frac{\partial B\left(d_{\text{pow}}(t)\right)}{\partial d(t)} = \frac{\rho_0(t)}{1 + E^{-1}\left(t, t'\right)} \times \left[\left(\frac{d_{\text{pow}}(t)}{\mathrm{d}_0(t)}\right)^{E^{-1}\left(t,t'\right)} - 1 + E^{-1}\left(t, t'\right) \times \left(\frac{d_{\text{pow}}(t)}{\mathrm{d}_0(t)}\right)^{E^{-1}\left(t,t'\right)}\right] \tag{10.14}$$

By combining (10.7) and (10.14):

$$\rho(t) + \mathrm{inc}(t) - \mathrm{pen}(t) = \frac{\rho_0(t)}{1 + E^{-1}\left(t, t'\right)} \times \left[\left(\frac{d_{\text{pow}}(t)}{\mathrm{d}_0(t)}\right)^{E^{-1}\left(t,t'\right)} - 1 + E^{-1}\left(t, t'\right) \times \left(\frac{d_{\text{pow}}(t)}{\mathrm{d}_0(t)}\right)^{E^{-1}\left(t,t'\right)}\right] \tag{10.15}$$

With simplification:

$$\frac{\rho(t) + \mathrm{inc}(t) - \mathrm{pen}(t)}{\rho_0(t)} = \left[\left(\frac{d_{\text{pow}}(t)}{\mathrm{d}_0(t)}\right)^{E^{-1}\left(t,t'\right)} - \frac{1}{1 + E^{-1}\left(t, t'\right)}\right] \tag{10.16}$$

The second term of (10.16) is very small and negligible. According to (10.16) and for the single-period model as (10.17):

$$d_{\text{pow}}(t) = \mathrm{d}_0(t) \times \left(\frac{\rho(t) + \mathrm{inc}(t) - \mathrm{pen}(t)}{\rho_0(t)}\right)^{E\left(t,t'\right)} \tag{10.17}$$

Also, the multi-period model is (10.18):

$$d_{pow}(t) = d_0(t) \times \prod_{\substack{t'=1 \\ t' \neq t}}^{24} \left(\frac{\rho\left(t'\right) + \text{inc}\left(t'\right) - \text{pen}\left(t'\right)}{\rho_0\left(t'\right)} \right)^{E(t,t')} \tag{10.18}$$

Finally, the hybrid model with respect to (10.17) and (10.18) is obtained as follows:

$$d_{pow}(t) = d_0(t) \times \prod_{t'=1}^{24} \left(\frac{\rho\left(t'\right) + \text{inc}\left(t'\right) - \text{pen}\left(t'\right)}{\rho_0\left(t'\right)} \right)^{E(t,t')} \tag{10.19}$$

10.4.1.3 Exponential Modeling of Demand Response

In a similar way to power modeling, exponential modeling with respect to (10.8) is as follows [33, 35]:

$$B\left(d_{exp}(t)\right) = B\left(d_0(t)\right) + \rho_0(t) \times d_{exp}(t) \times \left[1 + E^{-1}\left(t, t'\right) \times \left(Ln\left(\frac{d_{exp}(t)}{d_0(t)}\right) - 1\right)\right] \tag{10.20}$$

With calculating the derivative:

$$\frac{\partial B\left(d_{exp}(t)\right)}{\partial d_{exp}(t)} = \rho_0(t) \times \left[1 + E^{-1}\left(t, t'\right) \times \left(Ln\left(\frac{d_{exp}(t)}{d_0(t)}\right) - 1\right)\right] + \rho_0(t) \times d_{exp}(t) \times \left[E^{-1}(t,t) \times \frac{1}{d_{exp}(t)}\right] \tag{10.21}$$

By combining (10.21) and (10.22):

$$\rho(t) - \rho_0(t) + \text{inc}(t) - \text{pen}(t) = \rho_0(t) \times \left[1 + E^{-1}\left(t, t'\right) \times \left(\text{Ln}\left(\frac{d_{exp}(t)}{d_0(t)}\right) - 1\right)\right] + \rho_0(t) \times E^{-1}(t,t) \tag{10.22}$$

With simplification:

$$\frac{\rho(t) - \rho_0(t) + \text{inc}(t) - \text{pen}(t)}{\rho_0(t)} \times E\left(t, t'\right) = Ln\left(\frac{d_{exp}(t)}{d_0(t)}\right) \tag{10.23}$$

The single-period model is obtained as (10.24):

$$d_{\text{exp}}(t) = \text{d}_0(t) \times \exp\left(\left(\frac{\rho(t) - \rho_0(t) + \text{inc}(t) - \text{pen}(t)}{\rho_0(t)}\right) \times E\left(t, t'\right)\right) \tag{10.24}$$

Moreover, the multi-period model is calculated as (10.25):

$$d_{\text{exp}}(t) = \text{d}_0(t) \times \sum_{\substack{t' = 1 \\ t' \neq t}}^{24} \exp\left(\left(\frac{\rho\left(t'\right) - \rho_0\left(t'\right) + inc\left(t'\right) - pen\left(t'\right)}{\rho_0\left(t'\right)}\right) \times E\left(t, t'\right)\right) \tag{10.25}$$

Finally, the hybrid model, which is obtained as (10.26) with respect to single-period and multi-period models, is:

$$d_{\text{exp}}(t) = \text{d}_0(t) \times \sum_{t'=1}^{24} \exp\left(\left(\frac{\rho\left(t'\right) - \rho_0\left(t'\right) + \text{inc}\left(t'\right) - \text{pen}\left(t'\right)}{\rho_0\left(t'\right)}\right) \times E\left(t, t'\right)\right) \tag{10.26}$$

10.4.1.4 Logarithmic Modeling of Demand Response

Finally, the logarithmic model with respect to (10.8) is obtained as (10.27) [33, 35]:

$$B\left(d_{\log}(t)\right) = \text{B}\left(\text{d}_0\left(\text{t}\right)\right) + \rho_0\left(\text{t}\right) \times \text{d}_0\left(\text{t}\right) \times E\left(t, t\right) \times \exp\left(\frac{d_{\log}(t) - \text{d}_0\left(\text{t}\right)}{\text{d}_0\left(\text{t}\right) \times E\left(t, t\right)} - 1\right) \tag{10.27}$$

By deriving from expression (10.27):

$$\frac{\partial B\left(d_{\log}(t)\right)}{\partial d(t)} = \rho_0(t) \times \text{d}_0(t) \times E\left(t, t'\right) \times \frac{1}{\text{d}_0(t) \times E\left(t, t'\right)} \exp\left(\frac{d_{\log}(t) - \text{d}_0(t)}{\text{d}_0(t) \times E\left(t, t'\right)}\right) \tag{10.28}$$

Combining (10.7) and (10.28):

$$\frac{\rho(t) + \text{inc}(t) - \text{pen}(t)}{\rho_0(t)} = \exp\left(\frac{d_{\log}(t) - \text{d}_0(t)}{\text{d}_0(t) \times E\left(t, t'\right)}\right) \tag{10.29}$$

With simplification:

$$\frac{d_{\log}(t) - \mathrm{d}_0(t)}{\mathrm{d}_0(t) \times E(t, t')} = \mathrm{Ln}\left(\frac{\rho(t) + \mathrm{inc}(t) - \mathrm{pen}(t)}{\rho_0(t)}\right) \tag{10.30}$$

Therefore, the single-period model is obtained as (10.31):

$$d_{\log}(t) = \mathrm{d}_0(t) \times \left(1 + \mathrm{Ln}\left(\frac{\rho(t) + \mathrm{inc}(t) - \mathrm{pen}(t)}{\rho_0(t)}\right) \times E(t, t')\right) \tag{10.31}$$

The multi-period model is calculated as (10.32):

$$d_{\log}(t) = \mathrm{d}_0(t) \times \left(1 + \sum_{\substack{t' = 1 \\ t' \neq t}}^{24} \mathrm{Ln}\left(\frac{\rho(t') + \mathrm{inc}(t') - \mathrm{pen}(t')}{\rho_0(t')}\right) \times E(t, t')\right) \tag{10.32}$$

Finally, the hybrid model is obtained as (10.33):

$$d_{\log}(t) = \mathrm{d}_0(t) \times \left(1 + \sum_{t'=1}^{24} \mathrm{Ln}\left(\frac{\rho(t') + \mathrm{inc}(t') - \mathrm{pen}(t')}{\rho_0(t')}\right) \times E(t, t')\right) \tag{10.33}$$

10.4.2 The Simultaneous Operation Model of Power and Gas Systems

This section describes the indicators, parameters, and decision variables used in the model of optimal operation of power and gas systems in Table 10.1. In the following, the objective functions and constraints are described [15, 39].

In this section, the constraints on the simultaneous operation of power and gas systems are explained in detail.

10.4.2.1 Gas Flow Along a Pipe

Gas flow modeling is based on the movement of gas flow in one dimension, because the changes in gas properties along the radius are much less than the changes in the pipeline path. The following conditions are assumed when modeling gas flow:

- The tube is horizontal.
- The speed and temperature are constant along the pipe.

Table 10.1 Nomenclature

Index		s	Terminal nodes index
c	Compressors index	t	Time index
g	Generator index	p	Pipelines index
i	Bus index	q	Storage index
j	Bus index	**Parameters**	
l	Transmission lines index	a_g, b_g, c_g	Cost coefficients for generator g
n	Nodes index	B_l	Susceptance of line l (℧(
$\cos t_g^{statup/shutdown}$	Startup/shutdown cost of generating unit g ($)	$Q_c^{comp\ max\ /\ min}$	Maximum/minimum gas flow rate to compressor c (m^3/h)
d_g, e_g	Coefficients point-valve effect for generator g	$Q_p^{pipe\ min\ /\ max}$	Maximum/minimum bounds for gas flow within pipeline p (m^3/h)
$d_{0_{n,t}}^{gas}$	Initial demand of n node and t time at the gas system (mcm)	$Q_q^{output\ max}$	Maximum output of the q gas storage (m^3/h)
$d_{0_{i,t}}^{elec}$	Initial demand of i bus and t time at the power system (MW)	$Q_s^{max\ /\ min}$	Maximum/minimum capacity of gas flow rate of terminal s ($8.5*10^6$ m^3/h)
$Diameter_p$	Diameter of pipe p (mm)	R	Gas constant for natural gas (518 J/kg.K)
H_v	Gas heating value	RD_g	Ramp-down of the generator g (MW/h)
$GL_q^{min\ /\ max}$	Max/min gas level of storage q (m^3)	RP^{max}	The maximum ratio of the inlet to outlet pressure in compressors (1.5)
$Lenght_p$	Length of pipe p (m)	RU_g	Ramp-up of the generator g (MW/h)
$LP_{p,t}^0$	Initial gas stored in the p pipe at time t (mcm)	SUR_g	Startup ramp of the generator g (MW/h)
n_g	Number of POZs for the unit g	SDR_g	Shutdown ramp of the generator g (MW/h)
$P_l^{max\ /\ min}$	Thermal limit line l at time t (MW)	T^{normal}	Gas temperature under standard condition (288°K)
$P_g^{max\ /\ min}$	Maximum/minimum output power for generator g (MW)	$T_g^{on/off}$	Minimum on/off time of thermal power plant g (h)
$PW_{i,t}$	Output power of wind power plant bus i at time t (MW)	$TL_{l,i}$	Transfer limitation related to line l and bus i (MW)
$P_g^{U^R}$	Upper limits of the Rth POZ (MW)	Z	Compressibility factor for natural gas (0.95)
$P_g^{L^R}$	Lower limits of the Rth POZ (MW)	β^{comp}	Polytrophic exponent of a gas compressor
$Pr_n^{min\ /\ max}$	Maximum/minimum pressures at node n (Pa)	η_p	Pipeline efficiency (80%)

(continued)

Table 10.1 (continued)

Index		s	Terminal nodes index
$Pr_c^{comp\ max}$	Maximum pressures of compressor c (Pa)	η^{comp}	Compressor efficiency (80%)
$\rho_0(t)$	Initial power price at hour t ($)	$Pr_{p,t}^{average}$	The average pressure inside the pipe p at time t)Pa)
$\rho(t)$	Power price at hour t ($)	$Pr_{n,t}$	Pressure of node n at time t)Pa)
σ	Thermal efficiency of the gas generator	$Pr_{c,t}^{out}$	Outlet gas pressure from compressor c at time t (Pa)
χ^{normal}	Gas density under standard condition (0.713 kg/m^3)	$PL_{l,t}$	Power flow of the transmission line l at time t (MW)
Variables		$Pr_{c,t}^{in}$	Inlet gas pressure to compressor c at time t (Pa)
$Cost_t^{gas}$	Gas price at time t ($/mcm)	pen(t)	Penalty for non-execution of DR at time t ($)
$Cost_{EDRP}^{elec}$	Cost of implementing EDRP in the power system ($)	$Q_{q,t}^{in}$	Inlet gas from storage q at time t (m^3/h)
$Cos\ t_{TOU}^{elec/gas}$	Cost of implementing TOU in the power/gas system ($)	$Q_{p,t}^{pipe}$	Gas flow of the pipeline p at time t (m^3/h)
$d_{i,t}^{elec}$	Power demand at node i, time t (MW)	$Q_{q,t}^{out}$	Outlet gas from storage q at time t (m^3/h)
$d_{n,t}^{gas}$	Gas demand at node n, time t (mcm)	$Q_{c,t}^{comp}$	Gas flow through c compressor at time t (m^3/h)
$GL_{q,t}$	Gas level of storage q at time t (m^3)	$Q_{s,t}^{sup}$	Gas flow rate of terminal at node s time t (m^3/h)
inc_t^{elec}	Incentives applied to the power system at time t ($/MW)	$Q_{p,t}^{out.pipe}$	Outlet gas flow from the pipeline p at time t (m^3/h)
inc_t^{gas}	Gas incentives at time t ($/mcm)	$Q_{g,t}^{gen}$	Gas demand of thermal generator g at time t (m^3/h)
$I_{g,t}$	Commitment status of power plant g at time t (0 OR 1)	$Q_{p,t}^{in.pipe}$	Inlet gas flow to the pipeline p at time t (m^3/h)
$LP_{p,t}$	Linepack pipeline p at time t (m^3/h)	$SRPT_t$	Spinning reserve requirement at time t (MW)
$P_{g,t}$	Generated power of the g generator at the time t (MW)	TOF	Total objective function
$P_{c,t}^{comp}$	Consumption power of c compressor at time t (MW)	$V_{p,t}$	Volume of gas at pipeline p at time t (m^3)
ΔP_i	Change of power in bus i (MW)	$\theta_{l,t}$	Voltage angle line l at time t (rad)
ΔP_l	Change of power in line l (MW)		

- The diameter of the pipe is small compared to the radius of curvature.
- Changes in the pipe cross-sectional area along the gas flow path are negligible [40].

The gas flow along a pipe is calculated by Equation (10.36). Equation (10.36) is calculated by combining the equations of continuity (10.34) and the equations of motion (10.35):

$$\frac{\partial Q}{\partial x} = -\frac{A}{\rho \mathrm{ZR} T_g} \times \frac{\partial \Pr}{\partial t} \tag{10.34}$$

$$\frac{\partial \Pr}{\partial x} = -\rho \times \frac{\partial \upsilon}{\partial t} - \rho \times \frac{\left(\partial \upsilon^2\right)}{\partial x} - \frac{\left(2 \times f \times \rho \times \upsilon^2\right)}{\mathrm{D}} \tag{10.35}$$

$$Q = \upsilon \times A \tag{10.36}$$

By combining and simplifying (10.35, 10.36, and 10.37), the flow of gas in a horizontal tube with (10.38) is calculated:

$$M = \rho \times Q = \rho^{\mathrm{std}} \times Q^{std} \tag{10.37}$$

$$\frac{\partial \Pr}{\partial x} = -\frac{2 \times f \times \rho^{\mathrm{std}} \times Q^{std} \times \left|Q^{std}\right|}{\mathrm{A}^2 \times \mathrm{D}} \tag{10.38}$$

In order to represent the derivatives of (10.34) and (10.38), a finite difference approach has been applied [40]. The steady-state average pressure of a pipeline and the average gas flow in a pipeline are formulated in (10.39, 10.40, and 10.41):

$$\Pr^{\mathrm{avg}} = \frac{1}{2} \times (\Pr + \Pr_X) \tag{10.39}$$

$$\Pr_T^{\mathrm{avg}} = \frac{1}{2} \times \left(\Pr_T + \Pr_{T,X}\right) \tag{10.40}$$

$$Q_T^{\mathrm{avg}} = \frac{1}{2} \times \left(Q_T + Q_{T,X}\right) \tag{10.41}$$

By applying (10.39, 10.40, and 10.41) and the gas equation of state (10.42) [40], (10.34) and (10.38) are converted to ordinary differential equations used for calculating gas flow through a pipe (10.43) and (10.44):

$$Z \times R = \frac{p}{\rho \times T_g} = \frac{p^{std}}{\rho \times T_g^{std}} \tag{10.42}$$

$$\frac{(Q^{std}_{X.T}\text{-}Q^{std}_{T})}{\Delta x} = -\frac{A}{\rho^{std} \times Z \times R \times T_g} \times \frac{(Pr^{avg}_{T}\text{-}Pr^{avg})}{\Delta t} \tag{10.43}$$

$$\frac{(Pr_{X.T} - Pr_T)}{\Delta x} = -\frac{2 \times f \times (\rho^{std})^2 \times Z \times R \times T_g \times Q_T^{avg,std}}{A^2 \times D \times p_T^{avg}} \tag{10.44}$$

In this research, Panhandle A method is used to calculate the gas flow in high pressure networks. Therefore, friction is defined in (10.45) [40]:

$$\sqrt{\frac{1}{f}} = 6.872 \times Re^{0.073} \times \eta^{pipe} \tag{10.45}$$

$$Re = \frac{\rho \times \upsilon \times D}{\mu_p} \tag{10.46}$$

On the other hand, the gas flow can be calculated through (10.47):

$$Q = A \times \upsilon = \frac{\pi \times D^2}{4} \times \upsilon \Rightarrow \upsilon = \frac{4 \times Q \times \rho}{\pi \times D^2} \tag{10.47}$$

By substituting (10.47) into (10.46), (10.48) is obtained, and by assuming the physical parameters of gas are constant, (10.49) is achieved:

$$Re = \frac{4 \times Q \times \rho}{(\mu_p \times D \times \pi)} \tag{10.48}$$

$$Re = C \times \frac{Q}{D} \tag{10.49}$$

where C is constant. Thus, through combining (10.45, 10.46, 10.47, 10.48, and 10.49), the friction factor is simplified (10.50):

$$\sqrt{\frac{1}{f}} = \frac{6.872 \times C^{0.073} \times Q^{0.073} \times \eta^{pipe}}{D^{0.073}} \tag{10.50}$$

In a natural gas network, (10.44) can be written as follows (10.51):

$$\frac{\mathrm{Pr}_{l.t}^{\mathrm{out}} - \mathrm{Pr}_{l.t}^{\mathrm{in}}}{\mathrm{Le}_{\mathrm{l}}} = -\frac{2 \times f \times \left(\rho^{\mathrm{std}}\right)^2 \times Z \times R \times T_g \times \left(Q_{\mathrm{l.t}}^{\mathrm{avg}}\right)^2}{A^2 \times D \times \left(\frac{1}{2}\left(\mathrm{Pr}_{l.t}^{\mathrm{out}} + \mathrm{Pr}_{l.t}^{\mathrm{in}}\right)\right)} \tag{10.51}$$

Finally, by substituting (10.50) for (10.51), a relation (10.52) is obtained:

$$\left(\mathrm{Pr}_{l,t}^{\mathrm{in}}\right)^2 - \left(\mathrm{Pr}_{l,t}^{\mathrm{out}}\right)^2 = \frac{18.43 \times \mathrm{Le}_l \times \left(Q_{l,t}^{\mathrm{avg}}\right)^{1.854}}{\left(\eta_l^{\mathrm{pipe}}\right)^2 \times D_l^{4.854}} \tag{10.52}$$

10.4.2.2 Gas Network Constraints

In (10.53), the gas injection limit is introduced. Equation (10.54) shows the amount of gas consumed by each gas-fired power plant in relation to the production of electrical power. Equation (10.55) shows the equilibrium constraint for the gas network. Equation (10.56) shows the flow rate of gas along the pipelines by the pressure difference between the two nodes (Panhandle A equation [40]). To get acquainted with the method of calculating Panhandle A, it is necessary to refer to Sect. (6.4.2.1). Compressors are also used to increase the pressure of the gas network. Equation (10.57) shows the driving power of these compressors, which should be considered as load in the equilibrium of the power system. In this equation, the in and out superscripts represent the input and output flows. In fact, the performance of compressors is limited by the ratio of output flow to input pressure, maximum flow rate, and maximum power used (Eqs. (10.58 and 10.59)). In addition, Eqs. (10.60 and 10.61) show the constraints on the pressure limit at the nodes and the flow of gas passing through the lines, and the constraints on the gas storage facilities are listed in (10.62, 10.63, 10.64, 10.65, and 10.66) [15, 39]. The volume of gas within the pipelines must also be considered in the gas network. Equation (10.67) shows that the linepack of the lines is proportional to the average line pressure. Under dynamic conditions, the gas flow enters, and the output changes in proportion to changes in injection and demand. According to the law of density of materials, the change in the total volume of gas is equal to the change between the flow in and out of the pipeline; thus, (10.67) is replaced by (10.68) [40]:

$$\mathrm{Q}_s^{\mathrm{min}} \le Q_{s,t}^{\mathrm{sup}} \le \mathrm{Q}_s^{\mathrm{max}} \tag{10.53}$$

$$\mathrm{GL}_q^{\mathrm{min}} < \mathrm{GL}_{q,t} < \mathrm{GL}_q^{\mathrm{max}} \tag{10.54}$$

$$\mathrm{GL}_{q,t} = \mathrm{GL}_{q,t-1} + \left(Q_{q,t}^{\mathrm{output}} - Q_{q,t}^{\mathrm{input}}\right) \tag{10.55}$$

$$0 < Q_{q,t}^{\text{out}} < \mathrm{Q}_q^{\text{max output}} \tag{10.56}$$

$$0 < Q_{q,t}^{\text{in}} < \mathrm{Q}_q^{\text{max input}} \tag{10.57}$$

$$Q_{g,t}^{\text{gen}} = \sigma \times \mathrm{H_v} \times P_{g,t} \tag{10.58}$$

$$\sum_s Q_{s,t}^{\text{sup}} - \sum_p Q_{p,t}^{\text{pipe}} - \sum_c Q_{c,t}^{\text{comp}} + \sum_q Q_{q,t}^{\text{in}} = \sum_n d_{n,t}^{\text{gas}} + \sum_g Q_{g,t}^{\text{gen}} + \sum_q Q_{q,t}^{\text{out}} \tag{10.59}$$

$$\left(\mathrm{Pr}_{p,t}^{\text{out}}\right)^2 - \left(\mathrm{Pr}_{p,t}^{\text{in}}\right)^2 = \frac{18.43 \times \text{Lenght}_p}{\left(\eta_p\right)^2 \times \text{Diameter}_p^{4.854}} \times \left(Q_{p,t}^{\text{pipe}}\right)^{1.854} \tag{10.60}$$

$$\mathrm{Pr}_{c,t}^{\text{comp}} = \frac{\beta^{\text{comp}} \times Q_{c,t}^{\text{comp}}}{\eta^{\text{comp}}} \left[\left(\frac{\mathrm{Pr}_{c,t}^{\text{out}}}{\mathrm{Pr}_{c,t}^{\text{in}}}\right)^{\frac{1}{\beta^{\text{comp}}}} - 1\right] \tag{10.61}$$

$$1 \le \left(\frac{\mathrm{Pr}_{c,t}^{\text{out}}}{\mathrm{Pr}_{c,t}^{\text{in}}}\right) \le \mathrm{RP}^{\max} \tag{10.62}$$

$$0 \le Q_{c,t}^{\text{comp}} \le \mathrm{Q}_c^{\text{compmax}} \tag{10.63}$$

$$0 \le \mathrm{Pr}_{c,t}^{\text{comp}} \le \mathrm{Pr}_c^{\text{compmax}} \tag{10.64}$$

$$\mathrm{Pr}_n^{\min} \le \mathrm{Pr}_{n,t} \le \mathrm{Pr}_n^{\max} \tag{10.65}$$

$$\mathrm{Q}_p^{\text{pipe min}} \le Q_{p,t}^{\text{pipe}} \le \mathrm{Q}_p^{\text{pipe max}} \tag{10.66}$$

$$\mathrm{LP}_{p,t} = \frac{\mathrm{Pr}_{p,t}^{\text{average}} \times V_{p,t}}{\chi^{\text{normal}} \times \mathrm{ZTR}^{\text{normal}}} \tag{10.67}$$

$$\mathrm{LP}_{p,t} = \mathrm{LP}_{p,t}^0 + \sum_0^t \left(Q_{p,t}^{\text{pipe.in}} - Q_{p,t}^{\text{pipe.out}}\right) \tag{10.68}$$

10.4.2.3 Power System Constraints

In the power system, Eqs. (10.69) and (10.70) show the minimum on/off time limits for thermal units. The power supply-demand balance constraint is shown in (10.71). The limitation of production capacity of the thermal units is indicated in (10.72). Also, the rate limitation of increase/decrease generation of the generator is indicated in constraints (10.73 and 10.74), respectively. The power passing through the lines is shown in (10.75). The capacity limit passing through each line is indicated (10.76). Also, the reservations required per hour during the planning period is presented in (10.77). The amount of rotating storage of the power system is shown (10.78). The prohibited operating zones (POZs) constraint of power plant g is expressed in (10.79). It is worth mentioning that the incentive range of each network is considered from 0.1 to 10 times the initial price in the same network [21, 26]:

$$\sum_t I_{g,t} \geq \mathrm{T}_g^{\mathrm{on}} \left(I_{g,t} - I_{g,t-1} \right) \tag{10.69}$$

$$\sum_t \left(1 - I_{g,t} \right) \geq \mathrm{T}_g^{\mathrm{off}} \left(I_{g,t-1} - I_{g,t} \right) \tag{10.70}$$

$$\sum_g P_{g,t} + \sum_i Pw_{i,t} = \sum_i d_{i,t}^{\mathrm{elec}} + \sum_c P_{c,t}^{\mathrm{comp}} \tag{10.71}$$

$$I_{g,t} \times \mathrm{P}_{\mathrm{g}}^{\mathrm{min}} \leq P_{g,t} \leq I_{g,t} \times \mathrm{P}_{\mathrm{g}}^{\mathrm{max}} \tag{10.72}$$

$$P_{g,t} - P_{g,t-1} \leq \mathrm{RU_g} I_{g,t-1} + \mathrm{SUR_g} \left(I_{g,t} - I_{g,t-1} \right) \tag{10.73}$$

$$P_{g,t-1} - P_{g,t} \leq \mathrm{RD_g} I_{g,t} + \mathrm{SDR_g} \left(I_{g,t-1} - I_{g,t} \right) \tag{10.74}$$

$$\mathrm{PL}_{L,t} = \mathrm{B_L} \left(\theta_{L,t}^{\mathrm{in}} - \theta_{L,t}^{\mathrm{out}} \right) \tag{10.75}$$

$$\mathrm{P}_L^{\mathrm{min}} \leq P_{L,t} \leq \mathrm{P}_L^{\mathrm{max}} \tag{10.76}$$

$$\sum_g I_{g,t} \times P_{t,g}^{\mathrm{max}} + \sum_i Pw_{i,t} \geq \mathrm{SRPT}_t + \sum_i d_{i,t}^{\mathrm{elec}} + \sum_c P_{c,t}^{\mathrm{comp}} \tag{10.77}$$

$$\mathrm{SRPT}_t = 0.1 \times \sum_i d_{i,t}^{\mathrm{elec}} \tag{10.78}$$

$$\begin{cases} P_g^{\min} \le P_{g,t} \le P_g^{L^1} & g = 1, 2, \ldots, \mathrm{Ng} \\ P_g^{U^{R-1}} \le P_{g,t} \le P_g^{L^R} & t = 1, 2, \ldots, 24 \\ P_g^{U^{n_g}} \le P_{g,t} \le P_g^{\max} & R = 2, 3, \ldots, n_g \end{cases} \tag{10.79}$$

10.4.2.4 Simultaneous Implementation of EDRP and DEED Models in Power and Gas Systems

In this section, the formulation of the optimal integrated model of DEED and EDRP is presented, in which DR is implemented in the power system and gas network. Implementing multiple responses simultaneously reduces production and pollution costs and determines the average optimal reward rate simultaneously [3, 5]. The fuel cost of power generation units is usually approximated by a quadratic function. By applying the point-valve effect to the problem because of the steam power plants, a sine function is added to the fuel cost of the steam power plant. The general expression for the total fuel cost over the entire 24-hour distribution period is shown in Eq. (10.80):

$$\begin{aligned} &\mathrm{TOF} \\ &= \sum_{t=1}^{24} \begin{cases} \omega_F \times \sum_{g=1}^{N_g} \left[\left(a_g + b_g P_{g,t} + c_g (P_{g,t})^2 + \left| d_g \sin \left(e_g \left(P_{\min} - P_{g,t} \right) \right) \right| \right) \times I_{g,t} + \cos t_g^{\mathrm{startup}} \left(I_{g,t} - I_{g,t-1} \right) + \cos t_g^{\mathrm{shutdown}} \left(I_{g,t-1} - I_{g,t} \right) \right] \\ \omega_E \times \sum_{g=1}^{N_g} \mathrm{ppf}(g) \times \left[\alpha_g + \beta_g P_{g,t} + \gamma_g (P_{g,t})^2 + \eta_g \exp \left(\delta_g P_{g,t} \right) \right] + \omega_F \times \mathrm{Cost}_{\mathrm{EDRP}}^{\mathrm{elec}} + \cos t_t^{\mathrm{gas}} + \omega_{FF} \times \mathrm{Cost}_{\mathrm{EDRP}}^{\mathrm{gas}} \end{cases} \end{aligned} \tag{10.80}$$

All the terms and conditions introduced in the previous section must also be met in this section.

10.4.2.5 Implementation of Integrated TOU and DEED Models in Both Power and Gas Systems

In combining TOU and DEED, instead of the optimal reward, the optimal price is obtained in different periods. For this purpose, the parameter is used to change the price in different periods according to (10.81). In other words, one of the main goals of the optimization algorithm is to determine the optimal value of the parameter. The optimal pricing of TOU program can be defined according to (10.81). By definition, the designing method of load program is to change the selling price of power in different periods. Therefore, we define the price matrix of peak and valley periods by variables as (10.81):

$$\begin{bmatrix} \rho_{peak} \\ \rho_{off-peak} \\ \rho_{valley} \end{bmatrix} = \begin{bmatrix} \rho^{0}_{\text{peak}} + \sigma \\ \rho^{0}_{\text{off-peak}} \\ \rho^{0}_{\text{valley}} - \sigma \end{bmatrix} \tag{10.81}$$

where ρ^{0}_{peak}, $\rho^{0}_{\text{off-peak}}$, and ρ^{0}_{valley} are the price of power before the implementation of TOU program in the peak-, medium-, and low-load periods of the power system, respectively. In the gas network, the price of the peak, medium, and low load hours after the implementation of TOU program will be similar to the power system, with the difference that the price variable is σ, instead of γ in the gas network. Price-based DRPs provide incentives for consumers to reduce their consumption during peak hours and shift their consumption to nonpeak hours. As there is no direct payment as a reward or as a penalty in TOU program, there is no executive cost. However, to examine the economic viability of the program, we must apply the cost through (10.82) in the form of reducing the revenue of the power company in the ultimate goal function. In other words, the cost of implementing TOU program, in this case, is according to (10.82). This cost is the cost of an opportunity to shift consumption for the power utility company per hour [34, 35]:

$$\text{Cos}\ t^{\text{elec}}_{\text{TOU}} = \sum_{t} \rho_0\,(\text{t}) \times \text{d}_0\,(\text{t}) - \rho(t) \times d(t) \tag{10.82}$$

All the terms and conditions introduced in the previous section in this chapter must also be met. In (10.83), the parameter limits σ are defined. The parameter can be increased as much as the price in the period of low load. Because if it is more than this value, according to (10.81), the price in the low-load period is negative, which is not a logical approach and is contrary to the definition of TOU program:

$$0 \leq \sigma \leq \rho^{0}_{valley} \tag{10.83}$$

10.5 Method of Solving the Integrated Operation Model of Power and Gas Systems

The model of simultaneous operation of power and gas systems to solve the combined problem of DRPs and DEED is using mixed integer and nonlinear programming. Figure 10.1 shows the flowchart of how to integrate the DRP and DEED. The method of determining the optimal incentive or tariff for DR in the simultaneous operation of power and gas systems is based on power and gas supply costs, which is shown in the flowchart of Fig. 10.1. First, in network management, input data are defined, including the production cost of each generator, the DR model and gas supply cost of each terminal, and gas and power transmission network constraints. Integrated network management then performs simultaneous optimization of the objective function by observing the constraints. After optimization, the share of

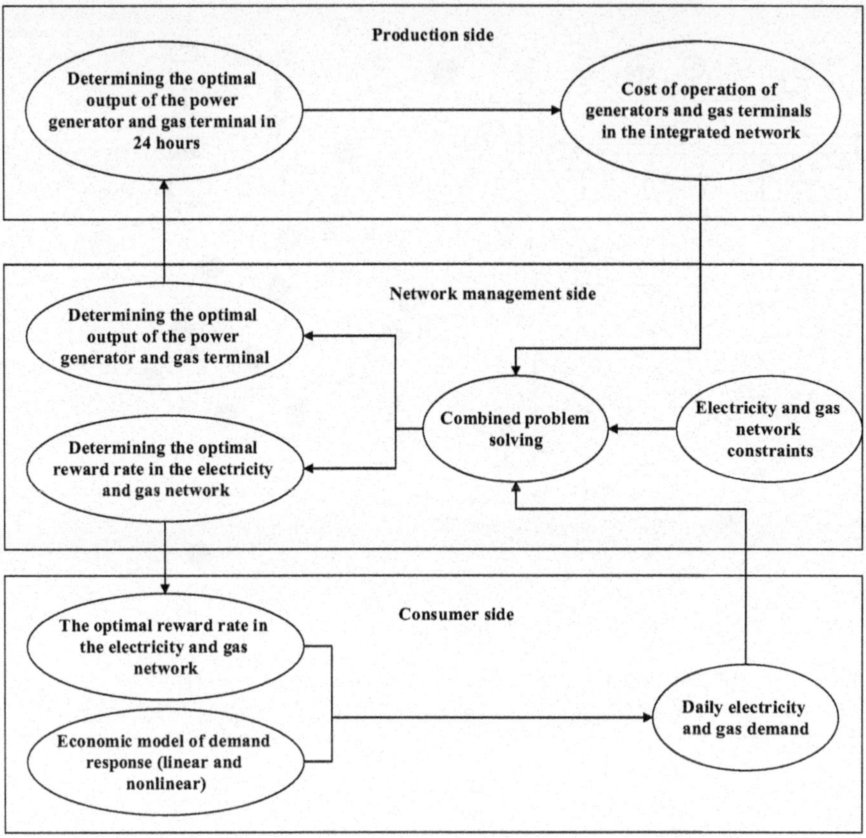

Fig. 10.1 How production and consumption are related in the optimal incentive/price determination algorithm

each generator and gas terminal in meeting the demand of the integrated network and the amount of incentives or tariffs for DR is determined. It should be noted that due to asynchronous optimization, if the demand for the gas network is not met, the consumption of generators will be reduced first. Therefore, as the production of gas-fired generators decreases, so does the production of other non-gas-fired generators. In simultaneous operation, the non-power supply-demand of the gas network is added as a penalty coefficient to the objective function.

10.6 Discussion

In this section, a comparison is made between the effects of execution and non-execution of DR in power and gas systems during the simultaneous operation of the integrated power and gas systems. In this section, the linear DR model is

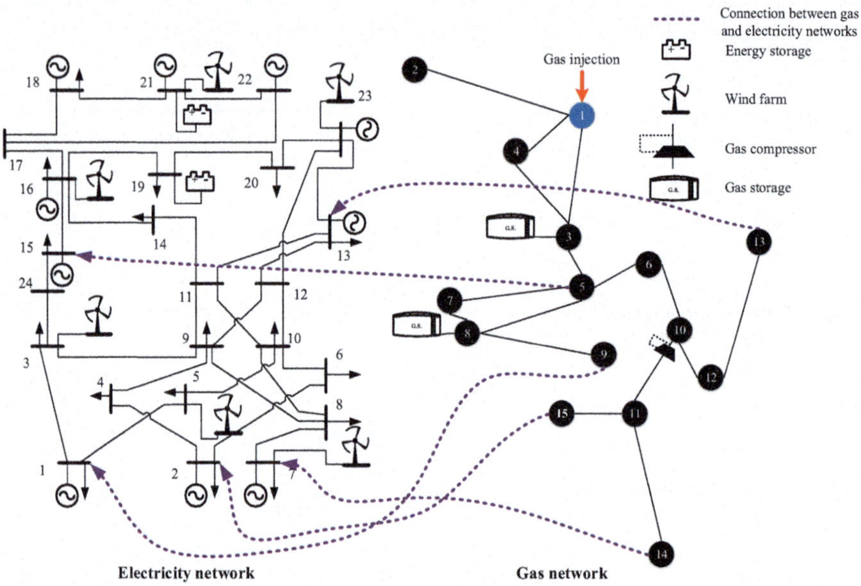

Fig. 10.2 The studied integrated electricity and gas systems [41]

Table 10.2 PEM values and state of hours in electricity and gas systems [41]

	Valley	Off-peak	Peak
Electricity system	1–8	(8–11) and (16–20)	(12–15) and (21–24)
Gas system	1–6	(10–16) and (22–24)	(7–9) and (17–21)
Valley	−0.1	0.01	0.012
Off-peak	0.01	−0.1	0.016
Peak	0.012	0.016	−0.1

implemented. The studied system is shown in Fig. 10.2, and the load response PEM coefficients are in Table 10.2 (more details are provided in [41]).

Figures 10.3 and 10.4 show the gas and power consumption profiles of the network when performing EDRP in both networks, respectively. As shown in Figs. 10.3 and 10.4, linear modeling is the best model in terms of peak reduction. Figure 10.5 shows the status of total linepack changes over 24 h in different scenarios. According to Figs. 10.3, 10.4, 10.5, DR implementation reduces the demand peak in power and gas systems. This reduces the distance between the peak and the valley and smoothes the gas network consumption curve, and as a result, the linepack changes in the gas network pipelines are reduced.

According to Table 10.3, it can be seen that the implementation of the EDRP in the simultaneous operation of power and gas systems reduces the total operational cost of these networks. This reduction is not only due to the reduction of gas costs due to the implementation of DR in the gas network but also the cost of the entire power system. The further reduction of the total cost of the power system is due to the fact that reducing the peak consumption of the gas network removes the gas

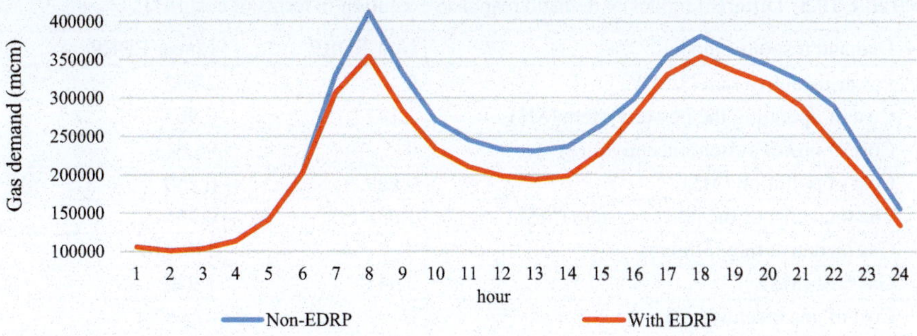

Fig. 10.3 Gas consumption curve due to emergency demand response in both networks [41]

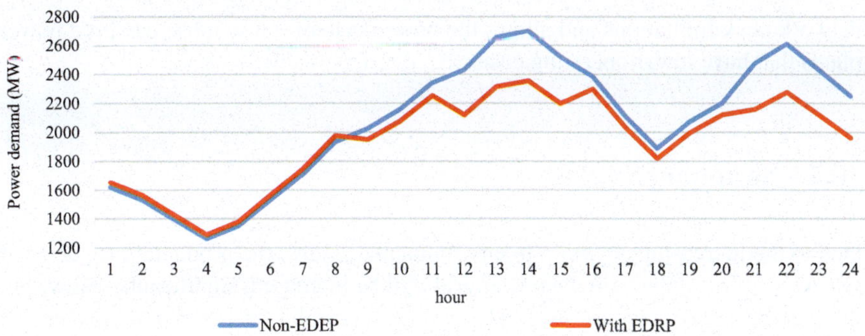

Fig. 10.4 Power consumption curve due to emergency demand response in both networks [41]

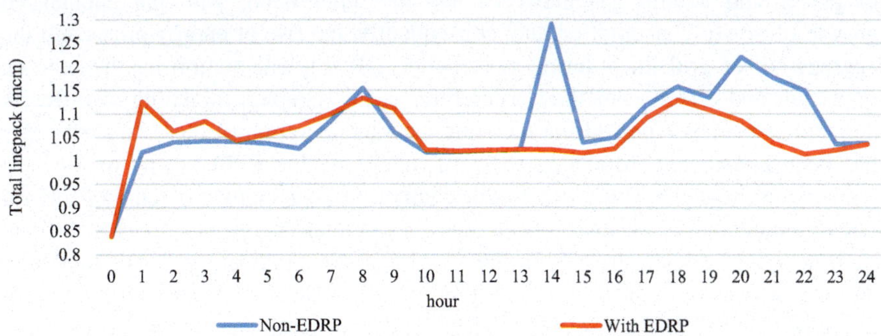

Fig. 10.5 Linepack changes of gas pipelines due to emergency demand response in both networks [41]

Table 10.3 Different modes of demand response execution in terms of cost [41]

Demand response model	Non-EDRP	Linear EDRP
Power incentive rate ($/MW)	0	24.92
Cost of operating the power system (M$)	1.147	0.965
Cost of power system incentives ($)	0	98,643
Cost of pollution (M$)	0.969	0.750
The total cost of the power system (M$)	1.058	0.956
Gas incentive rate ($/mcm)	0	278,410
Gas cost (M$)	2.941	1.976
Cost of gas incentives ($)	0	47,987
The cost of the entire gas network (M$)	2.941	2.024
Total cost (M$)	3.998	2.980

network peak limitations and allows the power system to use more gas-fired power plants that have lower operating costs.

10.7 Summary

Due to the increasing attention to environmental issues, the production capacity of renewable power plants in the power grid will be increased, significantly. However, despite the uncertainty in the production of renewable power plants, the use of flexible power plants such as gas power plants to compensate for the lack of potential production is inevitable. In other words, flexible gas power plants are intermediaries between power and natural gas networks and necessitate the simultaneous operation of power and natural gas networks. On the other hand, with the unbalanced growth of power consumption and, consequently, the rise of energy prices and the aggregation of grid lines during peak hours, DR implementation has become an important and effective method. Therefore, in this chapter, modeling of different demand responses types during the simultaneous operation of these two networks has been considered. Modeling the DRP based on the PEM is one of the best methods in this field. Therefore, in this chapter, various models based on PEM for DR implementation are investigated, and the optimal energy incentive or tariff in each network is determined from the simultaneous combination of DR modeling and integrated economic dispatch. The proposed model also deals with the role of linepack as storage in the gas network to deal with rapid changes in gas demand, which can be used at specific times to ensure the security of supply in the gas network. Based on the simulation results, the implementation of DR in power and gas systems has improved the consumption profiles in both networks. Finally, the improved consumption profile has reduced the total cost and congestion of pipes and lines. However, ISO should consider the characteristics of each model when selecting a demand response model. Because if the estimator is chosen incorrectly according to the consumer behavior, the integrated operation will suffer from a lack of production and an increase in the final cost, which is contrary to the purpose of demand response. Furthermore, due to importance of reducing the amount of unsupplied energy for ISO, it is necessary to use a conservative demand estimation model (lower estimate in reducing peak demand) such as the power model.

References

1. Zantye, M. S., Arora, A., & Hasan, M. F. (2019). Operational power plant scheduling with flexible carbon capture: A multistage stochastic optimization approach. *Computers & Chemical Engineering, 130*, 106544.
2. Ameli, H., Ameli, M. T., & Hosseinian, S. H. (2017a). Multi-stage frequency control of a microgrid in the presence of renewable energy units. *Electric Power Components and Systems, 45*(2), 159–170.
3. Moshari, A., Yousefi, G. R., Ebrahimi, A., & Haghbin, S. (2010). *Demand-side behavior in grid environment*. In IEEE innovative smart grid technologies, conference on 2010.
4. Smith, V., & Kiesling, L. (2015, August). *A market-based model for ISO-sponsored demand response programs*. A white paper prepared for the multi-client study.
5. Siano, P. (2014). Demand response and smart grids-A survey. *Renewable and Sustainable Energy Reviews, 30*, 461–478.
6. Sun, J., Palade, V., Wu, X. J., Fang, W., & Wang, Z. (2014, February). Solving the power economic dispatch problem with generator constraints by random drift particle swarm optimization. *IEEE Transactions on Industrial Informatics, 10*(1), 222–231.
7. Mansour-Saatloo, A., Agabalaye-Rahvar, M., Mirzaei, M. A., Mohammadi-Ivatloo, B., Abapour, M., & Zare, K. (2020). Robust scheduling of hydrogen based smart micro energy hub with integrated demand response. *Journal of Cleaner Production, 267*, 122041.
8. Qadrdan, M., et al. (2017). Efficacy of options to address balancing challenges: Integrated gas and electricity perspectives. *Applied Energy, 190*, 181–190.
9. Liu, C., Shahidehpour, M., & Wang, J. (2011). Coordinated scheduling of electricity and natural gas infrastructures with a transient model for natural gas flow. *Chaos: An Interdisciplinary Journal of Nonlinear Science, 21*(2), 025102.
10. Alabdulwahab, A., et al. (2015). Stochastic security-constrained scheduling of coordinated electricity and natural gas infrastructures. *IEEE Systems Journal, 11*(3), 1674–1683.
11. Zhang, X., et al. (2016). Hourly electricity demand response in the stochastic day-ahead scheduling of coordinated electricity and natural gas networks. *IEEE Transactions on Power Systems, 31*(1), 592–601.
12. He, C., et al. (2017). Robust co-optimization scheduling of electricity and natural gas systems via ADMM. *IEEE Transactions on Sustainable Energy, 8*(2), 658–670.
13. Ahmadi, A., Nezhad, A. E., & Hredzak, B. (2019). Security-constrained unit commitment in presence of lithium-ion battery storage units using information-gap decision theory. *IEEE Transactions on Industrial Informatics, 15*(1), 148–157.
14. Ringkjøb, H. K., Haugan, P. M., & Solbrekke, I. M. (2018). A review of modelling tools for energy and electricity systems with large shares of variable renewables. *Renewable and Sustainable Energy Reviews, 96*, 440–459.
15. Ameli, H., Qadrdan, M., & Strbac, G. (2017). Value of gas network infrastructure flexibility in supporting cost effective operation of power systems. *Applied Energy, 202*, 571–580.
16. Joung, M., & Kim, J. (2016). Assessing demand response and smart metering impacts on long-term electricity market prices and system reliability. *Applied Energy, 101*, 441–448.

17. Ameli, H., Qadrdan, M., & Strbac, G. (2019). Coordinated operation strategies for natural gas and power systems in presence of gas-related flexibilities. *Energy Systems Integration, 1*(1), 313.
18. Bai, L., et al. (2016). Interval optimization based operating strategy for gas-electricity integrated energy systems considering demand response and wind uncertainty. *Applied Energy, 167*, 270–279.
19. Chuan, H., et al. (2017). Robust coordination of interdependent electricity and natural gas systems in day-ahead scheduling for facilitating volatile renewable generations via power to-gas technology. *Journal of Modern Power Systems and Clean Energy, 5*(3), 375–388.
20. Hansen, J., Knudsen, J., & Annaswamy, A. M. (2014). *Demand response in smart grids: Participants, challenges, and a taxonomy*. Massachusetts Institute of Technology. Available at: http://web.mit.edu/aaclab/pdfs/ij-cdc2014.pdf.
21. Li, X. H., & Ho Hong, S. (2014). User-expected price-based demand response algorithm for a home to-grid system. *Energy, 64*, 437–449.
22. Aghaei, J., & Alizadeh, M. I. (2013). Demand response in smart electricity grids equipped with renewable energy sources: A review. *Renewable and Sustainable Energy Reviews*, 64–72.
23. Mazidi, M., Zakariazadeh, A., Jadid, S., & Siano, P. (2017). Integrated scheduling of renewable generation and demand response programs in a microgrid. *Energy Conversion and Management, 86*, 1118–1127.
24. Zakariazadeh, A., Jadid, S., & Siano, P. (2017). Stochastic operational scheduling of smart distribution system considering wind generation and demand response programs. *Electrical Power and Energy Systems*, 218–225.
25. Parsa Moghaddam, M., Abdollahi, A., & Rashidinejad, M. (2012). Flexible demand response programs modeling in competitive electricity markets. *Applied Energy, 88*, 3257–3269.
26. Aalami, H. A., Parsa Moghaddam, M., & Yousefi, G. R. (2015). Evaluation of nonlinear models for time-based rates demand response programs. *Electrical Power and Energy Systems, 65*, 282290.
27. Elaiwa, A. M., Xia, X., & Shehata, A. M. (2012). *Solving dynamic economic emission dispatch problem with valve-point effects using hybrid DE-SQP*. In Power engineering society conference and exposition in Africa (Power Africa), IEEE Johannesburg, 9–13 July 2012.
28. Alipour, M., Zare, K., & Mohammadi-Ivatloo, B. (2014). Short-term scheduling of combined heat and power generation units in the presence of demand response programs. *Energy*, 1–13.
29. Nwulu, N. I., & Xia, X. (2015). Multi-objective dynamic economic emission dispatch of electric power generation integrated with game theory based demand response programs. *Energy Conversion and Management, 89*, 963–974.
30. Falsafi, H., Zakariazadeh, A., & Jadid, S. (2013). "The role of demand response in single and multi-objective wind-thermal generation scheduling": A stochastic programing. *Energy, 64*, 853867.
31. Mandala, K. K., Mandalb, S., Bhattacharyac, B., & Chakraborty, N. (2015). Non-convex emission constrained economic dispatch using a new self-adaptive particle swarm optimization technique. *Applied Soft Computing, 28*, 188–195.
32. Behrangrad, M. (2015). A review of demand side management business models in the electricity market. *Renewable and Sustainable Energy Reviews, 47*, 270–283.
33. Ashfaq, A., Yingyun, S., & Zia Khan, A. (2014). *Optimization of economic dispatch problem integrated with stochastic demand side response*. In IEEE international conference on Intelligent Energy and Power Systems (IEPS).
34. Nikzad, M., & Mozafari, B. (2014). Reliability assessment of incentive- and priced-based demand response programs in restructured power systems. *Electrical Power and Energy Systems, 56*, 83–96.
35. Devlin, J., et al. (2017). A multi vector energy analysis for interconnected power and gas systems. *Applied Energy, 192*, 315–328.
36. Benasla, L., Belmadani, A., & Rahli, M. (2016). Spiral optimization algorithm for solving combined economic and emission dispatch. *International Journal of Electrical Power & Energy Systems, 62*, 163–174.

37. Jiang, S., Ji, Z., & Shen, Y. (2017). A novel hybrid particle swarm optimization and gravitational search algorithm for solving economic emission load dispatch Problems with various practical constraints. *International Journal of Electrical Power & Energy Systems, 55*, 623–644.
38. Federal Energy Regulatory Commission. (2008, December). *Assessment of demand response and advanced metering staff report*. [Online]. Available: http://www.ferc.gov/legal/staff-reports/12-08-demand-response.pdf
39. Ameli, H., Qadrdan, M., & Strbac, G. (2017). Value of gas network infrastructure flexibility in supporting cost effective operation of power systems. *Journal of Applied Energy, 202*, 571–580.
40. Ameli, H. (2018). *Value of flexibility options in transition to lower carbon natural gas and power systems*. Thesis for doctor of philosophy, Imperial College London.
41. Davary, M. M. (2021, February). *Optimal Modeling of demand response in the integrated network of electricity and gas for minimizing fuel costs and emission of greenhouse units*. Master's thesis, Department of Electrical Engineering, Shahid Beheshti University, Tehran, Iran.
42. Cui, H., Xia, W., Yang, S., & Wang, X. (2020). Real-time emergency demand response strategy for optimal load dispatch of heat and power micro-grids. *International Journal of Electrical Power & Energy Systems, 121*, 106127.
43. Wang, F., Ge, X., Yang, P., Li, K., Mi, Z., Siano, P., & Duić, N. (2020, December). Day-ahead optimal bidding and scheduling strategies for DER aggregator considering responsive uncertainty under real-time pricing. *Energy, 213*, 118765.
44. Zhong, W., Xie, K., Liu, Y., Yang, C., Xie, S., & Zhang, Y. (2021, January). Distributed demand response for multienergy residential communities with incomplete information. *IEEE Transactions on Industrial Informatics, 17*(1), 547–557.
45. Wang, F., Xiang, B., Li, K., Ge, X., Lu, H., Lai, J., & Dehghanian, P. (2020, March). Smart households' aggregated capacity forecasting for load aggregators under incentive-based demand response programs. *IEEE Transactions on Industrial Applications, 56*(2), 1086–1097.
46. Real-time emergency demand response strategy for optimal load dispatch of heat and power micro-grids. (2020). IJEPES.
47. The impact of customers' participation level and various incentive values on implementing emergency demand response program in micro grid operation (2020). IJEPES.

Chapter 11
Two-Stage Stochastic Market Clearing of Energy and Reserve in the Presence of Coupled Fuel Cell-Based Hydrogen Storage System with Renewable Resources

Masoud Agabalaye-Rahvar, Amin Mansour-Saatloo, Mohammad Amin Mirazaei, Behnam Mohammadi-Ivatloo, Kazem Zare, and Amjad Anvari-Moghaddam

11.1 Introduction

Environmental issues compelled managers to generate the required electricity with a minimal level of emission production. Related to the International Renewable Energy Agency (IRENA) reports, renewable's contribution would set nearly 86% of the whole power production by 2050 [1]. Among the renewable energy sources (RESs), wind energy sources (WESs) are the prominent technology that supplies more than one-third of the whole electricity demand by 2050 to decarbonize energy systems in the subsequent three decades. However, deploying WESs at a high penetration level leads to some challenges in power system scheduling. Variability and uncertainty nature are the crucial challenges of WESs to system operators, which require considerable flexibility [2, 3]. As Fig. 11.1 highlighted the importance of integrating WES by coupling hydrogen storage system (HSS) technology as much

M. Agabalaye-Rahvar (✉) · A. Mansour-Saatloo · M. A. Mirazaei · K. Zare
Faculty of Electrical and Computer Engineering, University of Tabriz, Tabriz, Iran
e-mail: m.agabalaye97@ms.tabrizu.ac.ir; amin_mnsr97@ms.tabrizu.ac.ir; kazem.zare@tabrizu.ac.ir

B. Mohammadi-Ivatloo
Department of Electrical and Electronics Engineering, Muğla Sıtkı Koçman University, Muğla, Turkey

Faculty of Electrical and Computer Engineering, University of Tabriz, Tabriz, Iran
e-mail: bmohammadi@tabrizu.ac.ir

A. Anvari-Moghaddam
Department of Energy Technology, Aalborg University, Aalborg, Denmark
e-mail: aam@energy.aau.dk

© The Author(s), under exclusive license to Springer Nature Switzerland AG 2022
V. Vahidinasab, B. Mohammadi-Ivatloo (eds.), *Whole Energy Systems*, Power Systems, https://doi.org/10.1007/978-3-030-87653-1_11

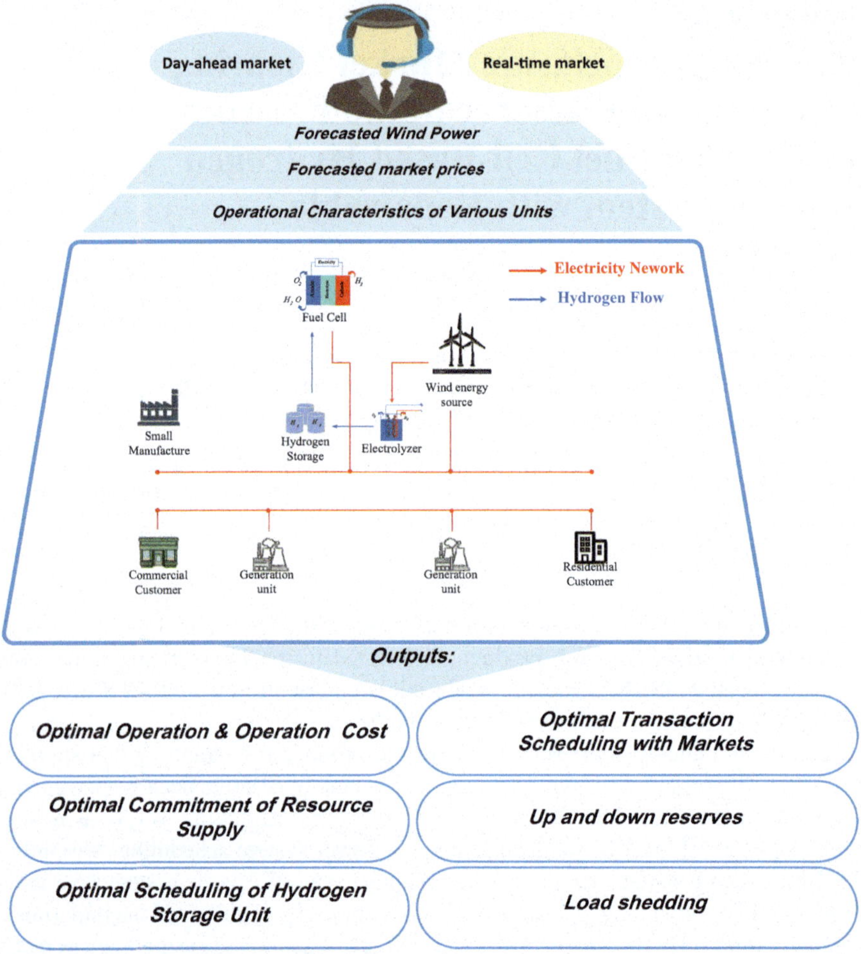

Fig. 11.1 The graphical representation for the whole integration of the hydrogen storage system into the electric power system

as possible into the electric power systems, achieving optimal scheduling in various energy markets is being indispensable.

The target of operational flexibility is supporting the mentioned challenges in a secure way with the least expenditures by deploying sufficient flexible resources. Generic solutions could be categorized into two main groups as novel market mechanism design and incorporation of flexible facilities. The first group emphasizes designing modern market mechanisms to provide flexibility procurement in system operation and management [4–6]. The second one is related to the integration of flexible facilities like environmental-friendly energy storages (EFESs), demand response programs (DRPs), and plug-in electric vehicles (PEVs) into the generation structure [7–10]. It is worth noting that the interest in applying these flexible technologies worldwide is more considerable in recent decades. In [9], authors have proposed optimal energy management for a renewable-based microgrid (RMG) by considering PEVs as a flexible load to take part in DRPs to achieve maximum benefit in both energy and reserve markets. So, in [10], the integration of HSS technology as a clean energy source (CES) with WES by the interaction of DRP in the day-ahead energy market has been investigated. The novel HSS technology has become widely utilized due to its capability to support several critical energy challenges and also provides tremendous participation to RESs.

In recent decades, various researchers have studied the structure and implementation of HSS technology in power system optimization problems. In the power system integrated with HSS and WES introduced by [11], a risk-averse strategy accompanied by proposed downside risk constraint is analyzed. A modern power to hydrogen technology has been proposed in [12] for the optimal operation of multi-energy microgrids (MGs) in multi-energy markets. A robust scheduling methodology has been presented in [13] for a micro-energy hub (mEH), which is integrated with HSS and integrated demand response program (IDR) technologies. A bi-objective scheduling framework specified applying HSS in both economic and environmental aspects [14]. Also, in real operating projects, the interest of deploying HSS in a coordinated path has grown rapidly, similar to the research viewpoint. For instance, German transmission and gas net operators have represented an investment-ready scheme for a 100 MW electrolyzer and hydrogen pipeline, which could get online by 2023 [15]. In South Australia, a 30 MW electrolyzer incorporated with ammonia storage is organized in 2018 [16].

A lot of researches have centralized on the day-ahead security-constrained scheduling of electricity networks. All reviewed papers are separated into two principal groups. The first group is denoted the schedule of a day-ahead energy market problem. A bi-level programming approach is proposed in [17] to minimize consumer payment in the pool-based electricity market. In this approach, the generation's scheduling with a decreasing of the total consumer payment is stated in the upper level; however, the lower level included the determination of local marginal prices (LMPs) with a multi-period optimal power flow (MOPF). The hourly scheduling of centralized and decentralized energy storage systems (ESSs) for day-ahead electricity markets has been analyzed through [18]. Ref [19] has indicated the robust optimal strategy of the introduced energy system integrated with power-to-gas (P2G), gas-fired units, and WESs in energy markets. In [20], the authors have presented the optimal hourly network-constrained and dispatch of producing units in the day-ahead optimization problem by considering the charging/discharging schedule of EV batteries and user driving essentials. Two-stage network-constrained unit-commitment programming for the day-ahead energy market scheduling with DRPs is presented in [21]. In this paper, the first stage determines a network-constrained unit-commitment problem with DRP, whereas the second stage is relevant to the specification of the demand shifting/curtailment during the time horizon based on netload changes. Ref [20] has outlined an effective solution methodology called Benders' decomposition to solve the stochastic security-constrained AC unit-commitment problem. The proposed problem in [20] is modeled as a two-stage stochastic programming problem, in which the first stage denotes the day-ahead energy market and the second stage demonstrates real-time operation. In [22], an optimal hourly day-ahead scheduling of combined heat and power (CHP) units and electrical/thermal energy storages (EESs/TESs), taking into account security constraints, have been addressed. Two-stage stochastic mixed-integer linear programming (SMILP) is deployed to solve the problem described in [22]. The coordinated multi-carrier energy system (MCES) with a robust/information gap decision theory (IGDT) methodology has been developed in [23] by considering the TESs facility. The second group of papers is relevant to the scheduling problem in day-ahead energy and reserve markets. Various time-based rate DRPs (TBRDRPs) are investigated in [24] to identify these effects on the day-ahead energy and reserve scheduling in the islanded residential MGs equipped with RESs and EVs. A novel predicted interval (PI) called adjustable intervals (AIs) is introduced in [25] to handle the wind power uncertainties and proposed an AI optimization method for energy and reserve market clearing accompanied by WES. Co-optimization of daily security-constrained energy and reserve markets under WES has been addressed in [26], which is formulated as a two-stage risk-averse optimization technique. To consideration of gas network besides the electricity network, a two-stage stochastic security-constrained model in energy and reserve markets with the integration of ESS and WES is represented by [27]. Most of the analyzed papers in the literature are not considered the integration of EFES and WES in the scheduling problem. Therefore, the importance of integrating HSS and WES in the energy and reserve market-clearing problem is much more indispensable and should be taken into account.

As the investigation and analysis of related works in the literature, a research gap is the coordination operation of integrated HSS and WES in the energy and reserve market-clearing process as the two-stage stochastic network-constrained energy and reserve scheduling. Thus, the present chapter concentrated on the proposed problem, along with consideration of fluctuations dependent on WES and electrical demand. Also, the effects of coupled HSS and WES on the load shedding and wind power curtailment are investigated to obtain minimal curtailments of the system. A suitable scenario decrement methodology is deployed to prevent the presentation of all produced scenarios and decrease the computational time. Finally, a comprehension illustrative of the proposed framework is demonstrated in Fig. 11.2. Thus, the principal contributions of the current chapter could be stated in the following:

1. A set of flexibility suppliers, including HSS technology and reserve services, is considered to increment the system's operational flexibility.
2. The integrated fuel cell-based HSS and WES are taken into account in the presented system to reduce the load shedding and wind power curtailment.
3. A two-stage stochastic network-constrained energy and reserve scheduling approach is proposed that the energy expenditure, scheduled and deployed reserves, load shedding, and wind power curtailment are considered.
4. The Monte Carlo simulation (MCS) algorithm is applied to satisfy the variability characteristic of electrical demand and WES to determine the system's hourly required reserve.

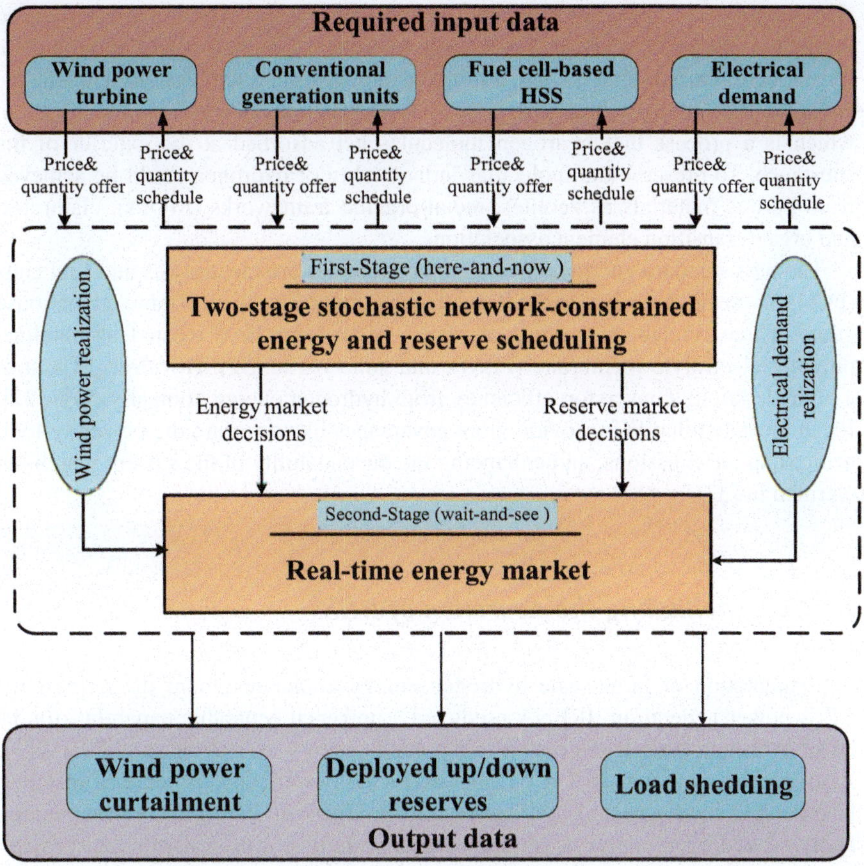

Fig. 11.2 The comprehension illustrative of the presented framework

The remainder of this chapter is categorized as the following: Sect. 11.2 presents the mechanism and impacts of HSS technology on system flexibility. The mathematical problem formulation is introduced in Sect. 11.3. Various case studies and simulation results are presented in Sect. 11.4. Eventually, in Sect. 11.5 is performed the conclusion such that Sect. 11.6 denotes the status quo, challenges, and outlooks.

11.2 Hydrogen Storage System Technology

11.2.1 Mechanism of HSS

Although more previous papers mainly focused on metal hydrides and carbon materials [28, 29], the other storing hydrogen substances like chemical storage and physisorption are introduced in subsequent years [30, 31]. Hydrogen is produced through a chemical reaction by storing hydrogen materials such as ammonia, formic acid, carbohydrates, and liquid organic hydrogen carriers (LOHC) in the chemical-type storages. The other procedure to generate hydrogen is physisorption, which is a process that hydrogen molecules get adsorbed at the exterior of the substance. To preserve the molecular individuality of hydrogen could be achieved from porous materials as zeolites, metal-organic frameworks (MOFs), clathrates, and organotransition element associations.

The best standard energy conversion technologies are electrolysis and fuel cells (FCs). A clean procedure called electrolysis produces no utilizable greenhouse gases. Three fundamental types of water electrolyzer have existed as alkaline, polymer electrolyte membrane (PEM), and solid oxide [32]. However, FCs have a high electricity conversion efficiency from hydrogen energy, to nearly 50% [32]. It can be stated that FCs provide more advantages like covering the power system, decreasing air emissions, and enhancing the dependability of the system which are clarified in [33].

11.2.2 Facilitating the Accretion of RESs

Implementation of large-scale hydrogen energy could increment the request for RESs power generation. IRENA predicted a universal economic capability for 19 EJ of hydrogen from RESs power in the whole energy utilization up to 2050 [15]. Also, approximately 4–16 TW of solar and wind power capacity must be installed to procreate hydrogen and hydrogen-based products in 2050 [15]. Consequently, applying hydrogen energy at an appropriate scale of power system could provide remarkable outcomes for the power section and allocate further opportunities for RESs development.

11.2.3 Impacts of HSS on Increasing System Flexibility

Hydrogen electrolyzers could construct extra flexibility to a network-constrained power system. Modernized electrolyzers have the ability to ramp up and down in a few minutes or even seconds. PEM electrolyzers, in contrast to alkaline, are capable of responding faster, which is the one reason why they are prominent in the forthcoming studies. The problems related to power grid congestion and RESs electricity curtailment are solved by locating strategically electrolyzers. For instance, considering such strategic electrolyzers could be settled in the North sea area by developing offshore wind energy [15]. Thus, the countries have a selection to transfer RESs electricity through a copper wire or embedded in the form of hydrogen.

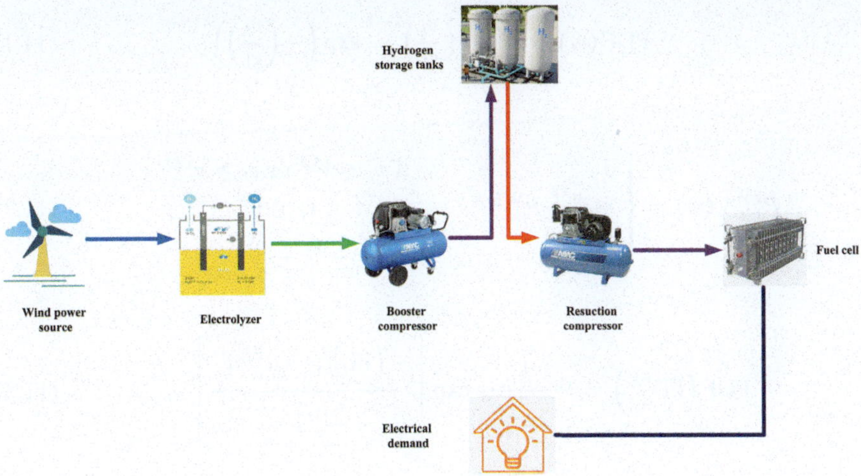

Fig. 11.3 The schematic composition of HSS technology

By these explanations, hydrogen energy could be generated from extra electricity production of other energy reservoirs, especially WES power production during low electricity demand or price via utilizing electrolyzers. The stored hydrogen in tanks could be reconverted to power through fuel cell-based HSS to satisfy peak demand. Therefore, a schematic structure of HSS technology has been illustrated in Fig. 11.3.

11.3 Problem Formulation

11.3.1 Modeling of Uncertain Nature

To tackle the fluctuations of WES and electrical demand, various methodologies have been studied and investigated by researchers [34, 35]. Wind power production has an undetermined characteristic due to variable wind speed data. The most practical approach to wind speed scenario generation is the Weibull probability density function (PDF) presented in Eq. (11.1). Finally, the output wind power production is calculated through Eq. (11.2). Besides, normal PDF is deployed to model the unpredictability of electrical demand, and this relation is defined as Eq. (11.3).

By these formulations, a vast quantity of scenarios is generated through the executing MCS algorithm. Thus, the fast backward forward selection approach is utilized to decrement the generated scenarios.

$$\mathrm{PDF}\left(v_t\right)=\left(\frac{\hat{k}}{\mathrm{sf}}\right)\left(\frac{v_t}{\mathrm{sf}}\right)^{\hat{k}-1}\exp\left(-\left(\frac{v_t}{\mathrm{sf}}\right)\right)^{\hat{k}} \tag{11.1}$$

$$P_{w,t}^{\mathrm{FWT}}\left(v_t\right)=\begin{cases}0 & \text{if } v_t \leq v^{\mathrm{CI}} \text{ or } v_t \geq v^{\mathrm{CO}} \\ \frac{v_t-v^{\mathrm{CI}}}{v^R-v^{\mathrm{CI}}} P_w^{\mathrm{WT},R} & \text{if } v^{\mathrm{CI}} \leq v_t \leq v^R \\ P_w^{\mathrm{WT},R} & \text{if } v^R \leq v_t \leq v^{\mathrm{CO}}\end{cases} \tag{11.2}$$

$$\mathrm{PDF}\left(\mathrm{PD}_{d,t}^{\mathrm{EL}}\right)=\frac{1}{\sqrt{2\pi\sigma_d^{\mathrm{EL}}}}\exp\left[-\frac{\left(\mathrm{PD}_{d,t}^{\mathrm{EL}}-\mu_d^{\mathrm{EL}}\right)^2}{2\left(\sigma_d^{\mathrm{EL}}\right)^2}\right] \tag{11.3}$$

where $\hat{k}$, sf are the shape and scale factors of the Weibull PDF; v^{CI},v^{CO}, and v^R are the cut-in, cut-out, and rated speed required for wind turbine w; $P_{w,t}^{\mathrm{FWT}}$, $P_w^{\mathrm{WT},R}$ are the forecasted and rated power of wind power turbine w; $\mathrm{PD}_{d,t}^{\mathrm{EL}}$ is the forecasted electrical demand d at time t; and μ_d^{EL} and σ_d^{EL} are the mean value and standard deviation of the electrical demand d.

11.3.2 Objective Function

The proposed energy and reserve scheduling problem includes the objective function, conventional generating units, and integrated fuel cell-based HSS with WES in both the first and second stages. The objective function in (11.4) minimizes the system's entire operation expenditure in the attendance of uncertainties by modeling the day-ahead and real-time market clearings. The power generation of conventional units contains energy costs, start-up costs, and scheduled up and down capacity reserve costs, which are indicated in the first line (11.4). The energy costs related to conventional units are considered the piecewise linear fuel costs function, located in the first term of the first line. The second line of (11.4) shows the costs relevant to discharging power and up and down capacity reserve via fuel cell-based HSS. The third line of (11.4) models the real-time market costs, which are denoted by the restorative conductions embedded in scenarios. Some principal actions that can be used to overcome the uncertainty of the electrical demand and WES are deployed up and down reserves by the traditional units and fuel cell-based HSS. The costs related to the load shedding and wind power curtailment in any scenario are presented in the third line's last two terms.

$$\begin{aligned}
&\text{Minimize} \\
&\sum_{t=1}^{NT}\sum_{g=1}^{NCU}\left[P_{g,t}^{CU}C_{g,t}^{CUEg}+SUC_{g,t}^{CU}+R_{g,t}^{CUSUP}C_{g,t}^{CUSUP}+R_{g,t}^{CUSDN}C_{g,t}^{CUSDN}\right] \\
&+\sum_{t=1}^{NT}\sum_{h=1}^{NHS}\left[P_{h,t}^{dis}C_{h,t}^{HSEg}+R_{h,t}^{HSSUP}C_{h,t}^{HSSUP}+R_{h,t}^{HSSDN}C_{h,t}^{HSSDN}\right] \\
&+\sum_{t=1}^{NT}\sum_{k=1}^{NK}p_k\left[\begin{array}{l}\sum\limits_{g=1}^{NCU}\left(r_{g,t,k}^{CUDUP}C_{g,t}^{CUDUP}-r_{g,t,k}^{CUDDN}C_{g,t}^{CUDDN}\right) \\ +\sum\limits_{h=1}^{NHS}\left(r_{h,t,k}^{HSDUP}C_{h,t}^{HSDUP}-r_{h,t,k}^{HSDDN}C_{h,t}^{HSDDN}\right) \\ +\sum\limits_{w=1}^{NW}P_{w,t,k}^{WPCurt}C_{w,t}^{WPCurt}+\sum\limits_{d=1}^{ND}Voll_{d,t}^{EL}LS_{d,t,k}^{EL}\end{array}\right]
\end{aligned} \tag{11.4}$$

where t, g, h, k, w, and d are indices of the time periods, conventional units, hydrogen storage systems, scenarios, wind power turbines, and electrical demands, respectively; $C_{g,t}^{CUEg}$,$C_{g,t}^{CUSUP}$, and $C_{g,t}^{CUSDN}$ denote the offer costs of energy, scheduled upward reserve, and scheduled downward reserve for the conventional unit g at time t, respectively; $SUC_{g,t}^{CU}$ is the start-up cost for the conventional unit g at time t; $P_{g,t}^{CU}$, $R_{g,t}^{CUSUP}$, and $R_{g,t}^{CUSDN}$ are the power production, scheduled upward reserve, and scheduled downward reserve for the conventional units g at time t, respectively; $P_{h,t}^{dis}$, $R_{h,t}^{HSSUP}$, and $R_{h,t}^{HSSDN}$ are the discharge power, scheduled upward reserve, and scheduled downward reserve for the HSS unit h at time t, respectively; $C_{h,t}^{HSEg}$, $C_{h,t}^{HSSUP}$, and $C_{h,t}^{HSSDN}$ are the offer costs of energy, scheduled upward reserve, and scheduled downward reserve for the HSS unit h at time t, respectively; $r_{g,t,k}^{CUDUP}$ and $r_{g,t,k}^{CUDDN}$ represent the deployed upward and downward reserves for the conventional unit g at time t in scenario k, while $r_{h,t,k}^{HSDUP}$ and $r_{h,t,k}^{HSDDN}$ represent for the HSS unit h at time t in scenario k; $C_{g,t}^{CUDUP}$ and $C_{g,t}^{CUDDN}$ are the offer costs of deployed upward and downward reserves for the conventional unit g at time t, while $C_{h,t}^{HSDUP}$ and $C_{h,t}^{HSDDN}$ are for the HSS unit h at time t; $C_{w,t}^{WPCurt}$ and $Voll_{d,t}^{EL}$ are the costs of the wind power curtailment in wind turbine w at time t and load shedding of electrical demand d at time t; and $P_{w,t,k}^{WPCurt}$ and $LS_{d,t,k}^{EL}$ are the amount of wind power curtailment w at time t in scenario k and load shedding of electrical demand d at time t in scenario k.

11.3.3 First Stage

In the first stage of the proposed problem, the electric system constraints include conventional units, fuel cell-based HSS, WES, and transmission network constraints that are represented in (11.5, 11.6, 11.7, 11.8, 11.9, 11.10, 11.11, 11.12, 11.13, 11.14, 11.15, 11.16, 11.17, 11.18, 11.19, 11.20, 11.21, 11.22, 11.23, 11.24, 11.25, 11.26, and 11.27). The offered up and down reserve capacity through the conventional units in the day-ahead reserve market is influenced by the generation units' maximum reserve capacity, expressed by (11.5 and 11.6). The best feature of adding up and down reserve to the dispatched power units is to offer maximum and minimum power capacity into the energy and reserve markets, as stated in (11.7 and 11.8). The conventional units' up and down ramp rate constraints between two sequential hours are introduced in (11.9, 11.10, 11.11, and 11.12). In addition, these units must be turned on or off for specific hours before switching the status to off or on, respectively, which are presented by (11.13 and 11.14).

$$0 \leq R_{g,t}^{\mathrm{CUSUP}} \leq R_g^{\mathrm{Max\ UP}} \tag{11.5}$$

$$0 \leq R_{g,t}^{\mathrm{CUSDN}} \leq R_g^{\mathrm{Max\ DN}} \tag{11.6}$$

$$P_{g,t}^{\mathrm{CU}} + R_{g,t}^{\mathrm{CUSUP}} \leq P_g^{\mathrm{Max}} I_{g,t} \tag{11.7}$$

$$P_{g,t}^{\mathrm{CU}} - R_{g,t}^{\mathrm{CUSDN}} \geq P_g^{\mathrm{Min}} I_{g,t} \tag{11.8}$$

$$P_{g,t}^{\mathrm{CU}} - P_{g,t-1}^{\mathrm{CU}} \leq \left(1 - X_{g,t}\right) R_g^{\mathrm{RU}} + X_{g,t} P_g^{\mathrm{Min}} \tag{11.9}$$

$$P_{g,t-1}^{\mathrm{CU}} - P_{g,t}^{\mathrm{CU}} \leq \left(1 - Y_{g,t}\right) R_g^{\mathrm{RD}} + Y_{g,t} P_g^{\mathrm{Min}} \tag{11.10}$$

$$X_{g,t} - Y_{g,t} \leq I_{g,t} - I_{g,t-1} \tag{11.11}$$

$$X_{g,t} + Y_{g,t} \leq 1 \tag{11.12}$$

$$\left(Z_{g,t-1}^{\mathrm{UPT}} - T_g^{\mathrm{UPT}}\right)\left(I_{g,t-1} - I_{g,t}\right) \geq 0 \tag{11.13}$$

$$\left(Z_{g,t-1}^{\mathrm{DNT}} - T_g^{\mathrm{DNT}}\right)\left(I_{g,t} - I_{g,t-1}\right) \geq 0 \tag{11.14}$$

where $R_g^{\text{Max UP}}$, $R_g^{\text{Max DN}}$ are the maximum upward and downward reserve capacities of the conventional unit g; P_g^{Max}, P_g^{Min} are the maximum and minimum power capacity of the conventional unit g; R_g^{RU}, R_g^{RD} are the ramp-up and ramp-down rates of the conventional unit g; $X_{g,\,t}$, $Y_{g,\,t}$ are the start-up and shutdown binary decision variables of the conventional unit g at time t, while $Z_{g,t-1}^{\text{UPT}}$, $Z_{g,t-1}^{\text{DNT}}$ are the up-time and down-time of conventional unit g before at time t; T_g^{UPT}, T_g^{DNT} represent the minimum up-time and down-time of the conventional unit g; and $I_{g,\,t}$ is the binary decision variable for the commitment status of the conventional unit g at time t.

HSS, like other energy storages, has two operating modes. The hourly scheduled up and down reserves offered by the fuel cell-based HSS in the day-ahead markets are indicated by (11.15 and 11.16). These reserves depend on the maximum reserve capability of these units. Constraints (11.17 and 11.18) are limited to the discharge and charge power of the HSS facility to the corresponding maximum values to take part in both energy and reserve markets. Constraint (11.19) is applied to avoid operating both modes of HSS simultaneously. The amount of hydrogen cumulative in the hydrogen tanks in every hour is calculated through (11.20), and the minimum and maximum capacity of HSS tanks restricted the hydrogen stored as expressed in (11.21). Besides, (11.22) determines the definite initial value of stored hydrogen in containers, and also, the initial and final stored hydrogen must be equal in each day, which is denoted by Eq. (11.23). Constraint (11.24) is deployed to consider the WES coupled with HSS technology.

$$0 \le R_{h,t}^{\text{HSSUP}} \le R_h^{\text{Max UP}} \tag{11.15}$$

$$0 \le R_{h,t}^{\text{HSSDN}} \le R_h^{\text{Max DN}} \tag{11.16}$$

$$0 \le P_{h,t}^{\text{dis}} + R_{h,t}^{\text{HSSUP}} \le P_h^{\text{Max,dis}} u_{h,t}^{\text{dis}} \tag{11.17}$$

$$0 \le P_{h,t}^{\text{ch}} + R_{h,t}^{\text{HSSDN}} \le P_h^{\text{Max,ch}} u_{h,t}^{\text{ch}} \tag{11.18}$$

$$u_{h,t}^{\text{dis}} + u_{h,t}^{\text{ch}} \le 1 \tag{11.19}$$

$$E_{h,t} = E_{h,t-1} + P_{h,t}^{\text{ch}} \eta_h^{\text{ch}} - \frac{P_{h,t}^{\text{dis}}}{\eta_h^{\text{dis}}} \tag{11.20}$$

$$E_h^{\text{Min}} \le E_{h,t} \le E_h^{\text{Max}} \tag{11.21}$$

$$E_{h,0} = E_h^{\text{Ini}} \tag{11.22}$$

$$E_{h,0} = E_{h,\text{NT}} \tag{11.23}$$

$$0 \leq P_{w,t}^{\text{WT}} \leq P_{w,t}^{\text{FWT}} \tag{11.24}$$

where $R_h^{\text{Max UP}}, R_h^{\text{Max DN}}$ are the maximum upward and downward reserve capacities of the HSS unit h; $P_h^{\text{Max,dis}}$, $P_h^{\text{Max,ch}}$ are the maximum discharging and charging power capacities of the HSS unit h; η_h^{dis}, η_h^{ch} are the efficiencies of the discharging and charging power of the HSS unit h; $u_{h,t}^{\text{dis}}$, $u_{h,t}^{\text{ch}}$ are discharging and charging binary decision variables of the HSS unit h at time t; E_h^{Max}, E_h^{Min}, and E_h^{Ini} are the maximum, minimum, and initial energy of the HSS unit h, respectively; $E_{h,0}$, $E_{h,\text{NT}}$ are the existing energy in the beginning and ending time of the HSS unit h; $P_{w,t}^{\text{WT}}$ is the dispatched power of wind turbine w at time t; and $P_{h,t}^{\text{ch}}$ is the charge power for the HSS unit h at time t.

In the following equations, the system constraints are taken into account. For each bus of the system, the power balance is provided with Eq. (11.25), DC power flow in each line is determined via Eq. (11.26), and also, constraint (11.27) limited the line power transmission to the line capacity limitations.

$$\sum_{g=1}^{\text{NCU}_b} P_{g,t}^{\text{CU}} + \sum_{w=1}^{\text{NW}_b} P_{w,t}^{\text{WT}} + \sum_{h=1}^{\text{NHS}_b} P_{h,t}^{\text{dis}} - \sum_{h=1}^{\text{NHS}_b} P_{h,t}^{\text{ch}} - \sum_{d=1}^{\text{ND}_b} \text{PD}_{d,t}^{\text{EL}} = \sum_{l=1}^{\text{NL}_b} \text{PT}_{l,t} \tag{11.25}$$

$$\text{PT}_{l,t} = \frac{\delta_{b,t} - \delta_{\tilde{b},t}}{X_l} \tag{11.26}$$

$$-\text{PT}_l^{\text{Max}} \leq \text{PT}_{l,t} \leq \text{PT}_l^{\text{Max}} \tag{11.27}$$

where $PT_{l,t}$, $\delta_{b,t}$ are the power transmitted of the line l at time t and the voltage angle of the bus b at time t and PT_l^{Max},X_l are the maximum capacity and the reactance of transmission line l.

11.3.4 Second Stage

In this stage of the proposed problem, the uncertainty and forecasted errors related to the electrical demand and WES are considered by deploying the decreased scenarios. Meanwhile, the scheduled up and down hourly reserves in the first stage are changing to meet the variations of electrical demand and WES fluctuations. In constraints (11.28, 11.29, 11.30, and 11.31), the interconnection relationships among the deployed and scheduled hourly reserves from the first and second stages are indicated. In each considered scenario, the deployed up and down reserves cannot transgress the scheduled reserves from the conventional units and also from fuel cell-based HSS technology. Consequently, Eqs. (11.32, 11.33, and 11.34) are expressing the power produced by conventional units and HSS facility in a per scenario, which is obtained through the summation of scheduled power from the first stage and deployed reserves from the second stage. The up and down ramp rate limitations of conventional units between sequential hours for each scenario are defined as (11.35 and 11.36). Constraint (11.37) is described as considering the hydrogen stored in HSS in each hour and scenario in which the amount of hydrogen stored is restricted by (11.38). In addition, the initial and final values should be equal with each other in each scenario and per day, as denoted in (11.39). When the deployed reserves in each scenario could not maintain the power balance, the system's indispensable load shedding must be applied. So, the wind power curtailment and load shedding in each scenario are limited by (11.40 and 11.41). Finally, the real-time power balance constraint is shown in (11.42), and each line power transferred in any scenario is specified as (11.43 and 11.44).

$$0 \leq r_{g,t,k}^{\mathrm{CUDUP}} \leq R_{g,t}^{\mathrm{CUSUP}} \tag{11.28}$$

$$0 \leq r_{g,t,k}^{\mathrm{CUDDN}} \leq R_{g,t}^{\mathrm{CUSDN}} \tag{11.29}$$

$$0 \leq r_{h,t,k}^{\mathrm{HSDUP}} \leq R_{h,t}^{\mathrm{HSSUP}} \tag{11.30}$$

$$0 \leq r_{h,t,k}^{\mathrm{HSDDN}} \leq R_{h,t}^{\mathrm{HSSDN}} \tag{11.31}$$

$$P_{g,t,k}^{\mathrm{CU}} = P_{g,t}^{\mathrm{CU}} + r_{g,t,k}^{\mathrm{CUDUP}} - r_{g,t,k}^{\mathrm{CUDDN}} \tag{11.32}$$

$$P_{h,t,k}^{\mathrm{dis}} = P_{h,t}^{\mathrm{dis}} + r_{h,t,k}^{\mathrm{HSDUP}} \tag{11.33}$$

$$P_{h,t,k}^{\mathrm{ch}} = P_{h,t}^{\mathrm{ch}} + r_{h,t,k}^{\mathrm{HSDDN}} \tag{11.34}$$

$$P_{g,t,k}^{\mathrm{CU}} - P_{g,t-1,k}^{\mathrm{CU}} \leq \left(1 - X_{g,t}\right) R_g^{\mathrm{RU}} + X_{g,t} P_g^{\mathrm{Min}} \tag{11.35}$$

$$P^{\mathrm{CU}}_{g,t-1,k} - P^{\mathrm{CU}}_{g,t,k} \le \left(1 - Y_{g,t}\right) R^{\mathrm{RD}}_{g} + Y_{g,t} P^{\mathrm{Min}}_{g} \tag{11.36}$$

$$E_{h,t,k} = E_{h,t-1,k} + P^{\mathrm{ch}}_{h,t,k} \eta^{\mathrm{ch}}_{h} - \frac{P^{\mathrm{dis}}_{h,t,k}}{\eta^{\mathrm{dis}}_{h}} \tag{11.37}$$

$$E^{\mathrm{Min}}_{h} \le E_{h,t,k} \le E^{\mathrm{Max}}_{h} \tag{11.38}$$

$$E_{h,0,k} = E_{h,\mathrm{NT},k} = E_{h,0} \tag{11.39}$$

$$0 \le P^{\mathrm{WPCurt}}_{w,t,k} \le P^{\mathrm{WT}}_{w,t,k} \tag{11.40}$$

$$0 \le LS^{\mathrm{EL}}_{d,t,k} \le \mathrm{PD}^{\mathrm{EL}}_{d,t,k} \tag{11.41}$$

$$\begin{gathered} \sum_{g=1}^{\mathrm{NCU}_b} P^{\mathrm{CU}}_{g,t,k} + \sum_{w=1}^{\mathrm{NW}_b} \left(P^{\mathrm{WT}}_{w,t,k} - P^{\mathrm{WPCurt}}_{w,t,k}\right) + \sum_{h=1}^{\mathrm{NHS}_b} P^{\mathrm{dis}}_{h,t,k} - \sum_{h=1}^{\mathrm{NHS}_b} P^{\mathrm{ch}}_{h,t,k} \\ - \sum_{d=1}^{\mathrm{ND}_b} \left(\mathrm{PD}^{\mathrm{EL}}_{d,t,k} - \mathrm{LS}^{\mathrm{EL}}_{d,t,k}\right) = \sum_{l=1}^{\mathrm{NL}_b} \mathrm{PT}_{l,t,k} \end{gathered} \tag{11.42}$$

$$\mathrm{PT}_{l,t,k} = \frac{\delta_{b,t,k} - \delta_{\tilde{b},t,k}}{X_l} \tag{11.43}$$

$$-\mathrm{PT}^{\mathrm{Max}}_{l} \le \mathrm{PT}_{l,t,k} \le \mathrm{PT}^{\mathrm{Max}}_{l} \tag{11.44}$$

where $P^{\mathrm{CU}}_{g,t,k}$ is the power production of the conventional unit g at time t in scenario k; $P^{\mathrm{dis}}_{h,t,k}, P^{\mathrm{ch}}_{h,t,k}$ are the discharging and charging power of the HSS unit h at time t in scenario k; $E_{h,t,k}$ is the stored hydrogen energy of the HSS unit h at time t in scenario k; $P^{\mathrm{WT}}_{w,t,k}$ is the dispatched wind power turbine w at time t in scenario k; $\mathrm{PD}^{\mathrm{EL}}_{d,t,k}$ is the electrical power demand d at time t in scenario k; and $\mathrm{PT}_{l,t,k}$, $\delta_{b,t,k}$ are the power transmitted of the line l at time t in scenario k and the voltage angle of the bus b at time t in scenario k.

11.4 Case Studies

11.4.1 Test System and Data

The modified six-bus test system, as illustrated in Fig. 11.4, is composed of three gas-fueled generation units at buses 1, 2, and 6 and three electrical demands at buses 3, 4, and 5. Buses are linked to each other via seven transportation lines. All techno-economic data of the system is taken from [36]. The system is modified by using the hydrogen storage system and wind turbine with a power capacity of 30MW on bus 5. HSS technical data are reported in Table 11.1. Also, wind power curtailment and load shedding costs are 50$/MWh and 400$/MWh, respectively. Expenditures of up and down scheduled reserves for G1, G2, and G3 units are equal to 8, 10, and 11 $/MW, respectively. The start-up expenditure of these conventional units is equal to 500$/MW. Besides, the providing energy cost by the HSS unit is equal to 2$/MWh, and the scheduled reserve capacity costs by this unit are equivalent to 50% of the energy proposed cost. To deal with system uncertainties, 1000 scenarios were generated for both wind power and electrical demand based on the MCS approach using Weibull and normal PDFs, respectively. Thus, the number of scenarios was reduced to 10, utilizing the SCENRED tool in General Algebraic Modeling System (GAMS) software. The forecasted electrical demand and wind power production profiles are indicated in Fig. 11.5. According to the final reduced scenarios introduced in Table 11.2, the realization of the forecasted parameters can be accomplished.

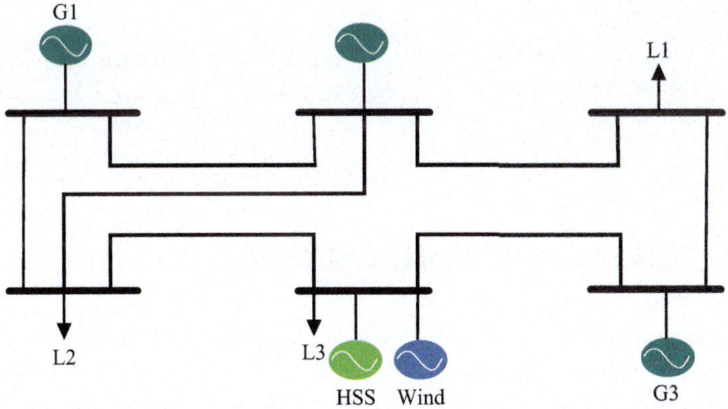

Fig. 11.4 The proposed six-bus electrical test system with WES and HSS units

Table 11.1 HSS technical data

Parameter	E_h^{Min} (MW)	E_h^{Max} (MW)	E_h^{Ini} (MW)	$P_h^{Max,dis}$ (MW)	$P_h^{Max,ch}$ (MW)	η_h^{dis} (%)	η_h^{ch} (%)
Value	40	180	70	30	30	80	80

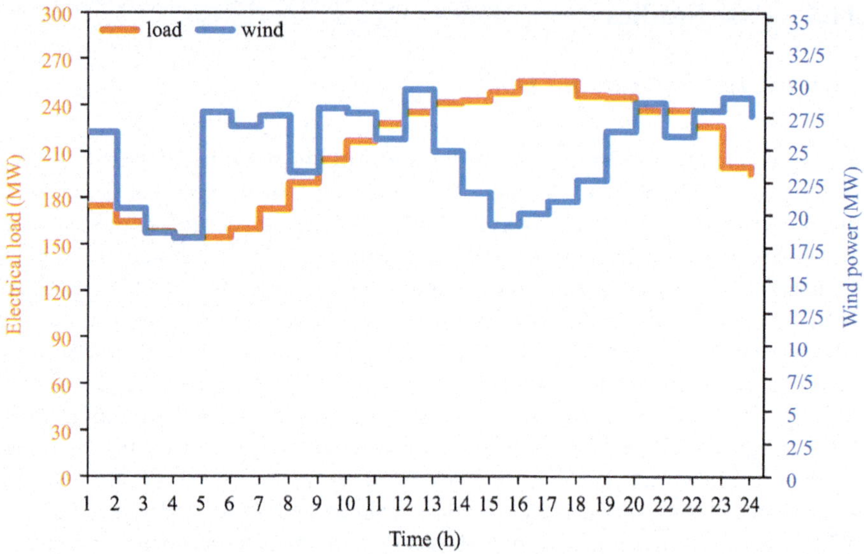

Fig. 11.5 The forecasted electrical load and wind power production profiles

Table 11.2 The probability and variations of each selected scenario

Scenario no.	Probability	Variations
1	0.26	−0.0176
2	0.12	+0.0442
3	0.08	+0.0507
4	0.07	+0.0087
5	0.14	−0.01198
6	0.09	−0.0305
7	0.02	+0.0532
8	0.07	−0.0021
9	0.07	−0.0182
10	0.08	+0.0532

11.4.2 *Simulation and Analysis of Results*

In this chapter, according to the above described two subsections, an electric six-bus test system along with a WES and fuel cell-based HSS is taken into account for investigating the application of the introduced model. The effects of the proposed coupled fuel cell-based HSS technology with WES on the costs of energy, reserve, wind power curtailment, and load shedding have been analyzed. To model the system uncertainties, i.e., the error of forecasting electrical demand and WES production, the MCS algorithm has been applied to generate the acceptable scenario sets. The presented problem formulation is modeled as a mixed-integer linear programming (MILP) model to implement in GAMS software, which is

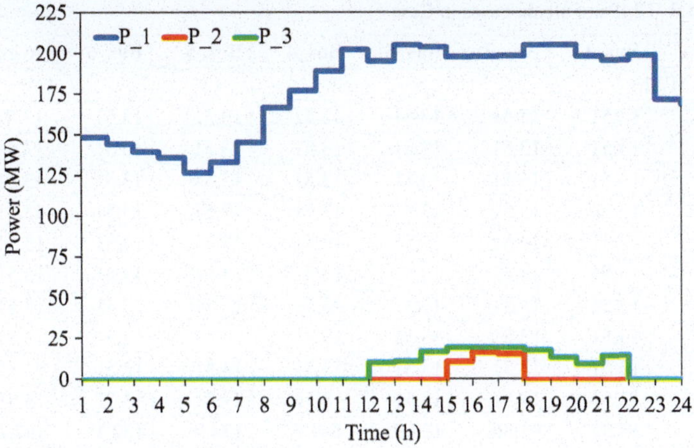

Fig. 11.6 Hourly power generation in Case 1

solved using CPLEX 12.9.0 solver. CPLEX solver especially utilizes branch and cut methodology and assigns the optimal solutions [37].

To study and analyze the obtained results of the presented problem, two case studies are taken into account as the following:

Case 1: Synchronous clearing of energy and reserve markets without consideration of fuel cell-based HSS technology

Case 2: Synchronous clearing of energy and reserve markets with consideration of fuel cell-based HSS technology

Case 1 In this case, all transmission line capacity restrictions are applied except the effects of fuel cell-based HSS on energy and reserve markets. As depicted in Fig. 11.6, the low-cost unit G1 is committed during the whole scheduling time horizon, whereas the high-cost unit G2 is committed only in peak hours, i.e., hours 15–17. The medium-cost unit G3 dispatches between hours 11 and 22 to meet the rest of the electrical demand in these hours. Table 11.3 reports all local marginal prices (LMPs) each hour to specify LMPs in the proposed modified test system with the mentioned techno-economic data. The provided LMPs are obtained via the joint energy and reserve market-clearing methodology. As reported in Table 11.3, the LMPs of all buses except bus 1 during hours 11, 14, and 18, which these rows highlighted in black, reached the maximum values, i.e., about 107 $/MWh. In reality, due to the transmission line restrictions, congestion of lines, and the obtained optimal power generations of conventional units, nearly all of the load shedding has occurred in hours 11, 14, and 18. As it can be seen in these three hours, the LMP of bus 4 has achieved the highest quantity of price compared to the rest buses. In the last column of Table 11.3, the average LMPs (Avg.LMPs) are increased by the growth of demand except in hours 11, 14, and 18.

Table 11.3 LMPs of all six buses in Case 1

Time (h)	Bus_1	Bus_2	Bus_3	Bus_4	Bus_5	Bus_6	Avg.LMPs
LMP($/MWh)							
1	13.51	13.51	13.51	13.51	13.51	13.51	13.51
2	13.51	13.51	13.51	13.51	13.51	13.51	13.51
3	13.51	13.51	13.51	13.51	13.51	13.51	13.51
4	13.51	13.51	13.51	13.51	13.51	13.51	13.51
5	13.51	13.51	13.51	13.51	13.51	13.51	13.51
6	13.51	13.51	13.51	13.51	13.51	13.51	13.51
7	13.51	13.51	13.51	13.51	13.51	13.51	13.51
8	13.51	13.51	13.51	13.51	13.51	13.51	13.51
9	13.51	13.51	13.51	13.51	13.51	13.51	13.51
10	13.51	13.51	13.51	13.51	13.51	13.51	13.51
11	**13.51**	**95.36**	**126.3**	**140.7**	**134.9**	**132.1**	**107.2**
12	13.51	15.09	15.61	15.90	15.88	15.80	15.32
13	13.51	26.53	31.46	33.75	32.83	32.83	28.41
14	**13.51**	**95.86**	**127.0**	**141.5**	**135.7**	**132.8**	**107.7**
15	13.51	23.92	27.86	29.70	28.96	28.60	25.43
16	13.51	32.63	39.86	43.23	41.88	41.22	35.39
17	13.51	32.63	39.86	43.23	41.88	41.22	35.39
18	**13.51**	**95.86**	**127.0**	**141.5**	**135.7**	**132.8**	**107.7**
19	13.51	44.23	55.86	61.28	59.10	58.04	48.67
20	13.51	15.09	15.69	15.97	15.85	15.80	15.32
21	13.51	15.69	16.52	16.91	16.75	16.68	16.02
22	13.51	25.76	30.40	32.56	31.69	31.27	27.53
23	13.51	13.51	13.51	13.51	13.51	13.51	13.51
24	13.51	13.51	13.51	13.51	13.51	13.51	13.51

Besides the conventional units providing the required energy, these units are supported up and down reserve capacities. Therefore, Figs. 11.7 and 11.8 indicated the scheduled up and down reserves provided by three generation units, respectively. During peak hours, the high-cost generation unit G2 nearly is provided all of up and down reserve capacity. However, the two other generation units are rendered up and down reserve capability during off-peak hours, which are clarified in these figures.

As stated before, the almost load shedding has taken place in hours 11, 14, and 18, which has caused to curtail demands instead of starting and contributing more expensive generation units in the reserve market. In reality, due to much-low scenario occurrence probability, an independent system operator (ISO) prefers to lessen demands to decrease energy and reserve expenditures. So, the expected load shedding and wind power curtailment are 3.838 MWh and 6.668 MWh, respectively. The related expenditures of providing energy and scheduled reserve capacity are 67833.987$ and 2679.2$, respectively, in which summation of them could report

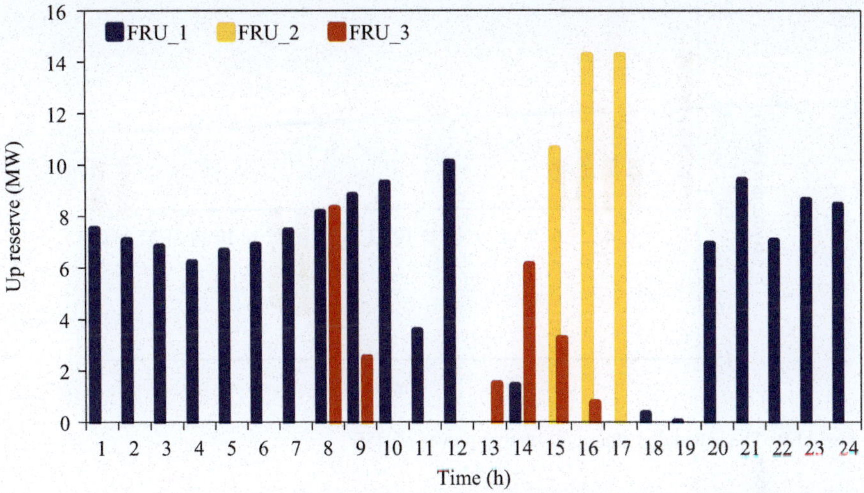

Fig. 11.7 Up scheduled reserve presented by generation units in Case 1

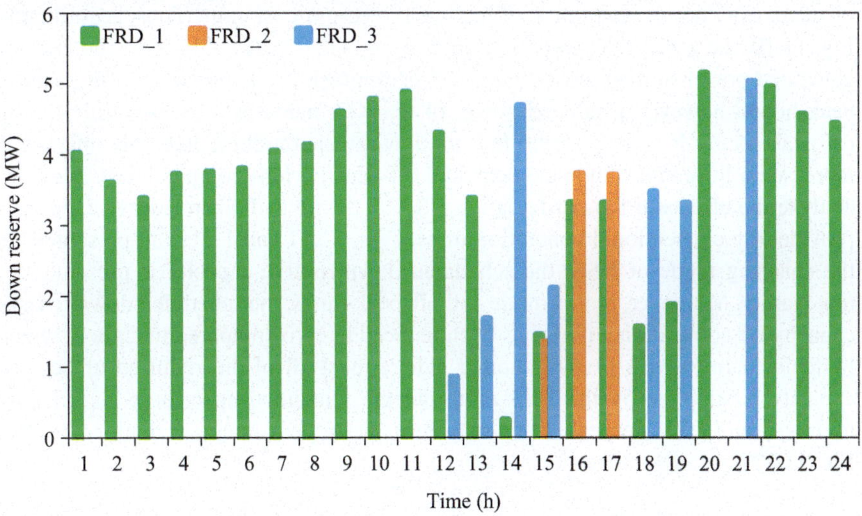

Fig. 11.8 Down scheduled reserve presented by generation units in Case 1

the entire operating cost. Besides, load shedding and wind power curtailment costs are precisely equivalent to 1533.870$ and 333.380$, respectively.

Case 2 In this case, HSS technology is considered and added to the discussed units in Case 1. The scheduled charge and discharge power of HSS is demonstrated in Fig. 11.9. As shown in this figure, during peak periods (i.e., hours 15–17), the discharging mode has happened due to reducing the costs, while the charging mode

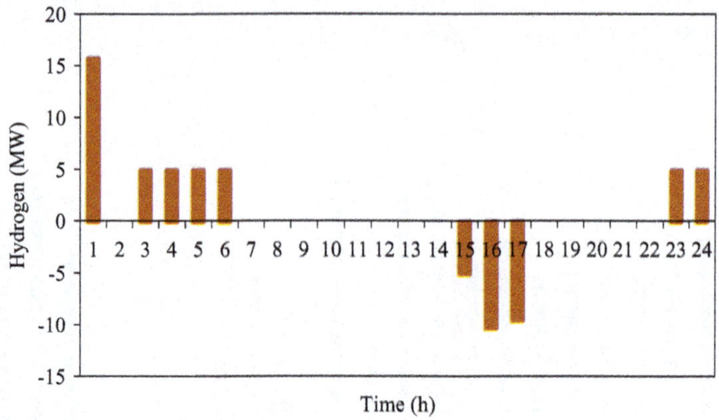

Fig. 11.9 The scheduled charge/discharge power of HSS

takes place during off-peak times (i.e., hours 1–6 and 23–24). Furthermore, the presence of HSS in the power dispatch of conventional units has a significant effect on decreasing power generation of high- and medium-cost units. Thus, according to Fig. 11.10, the generated power of high-cost unit G2 is reduced to zero, and also the generated power of medium-cost unit G3 is diminished compared to Case 1 during peak hours; however, produced power of unit G3 increases a relative little during off-peak hours. It is noteworthy that the power generation of low-cost unit G1 is increased a little minor in some peak and off-peak periods compared to Case 1. To analyze the effects of the participation HSS on the up and down reserve capacities provided by conventional generation units, Figs. 11.11 and 11.12 are presented. In the charging mode of HSS, the scheduled down reserve capacity is presented by this facility. Moreover, in this mode and off-peak times, the scheduled down reserve capacity by conventional units has been reduced to zero compared to Case 1, which is depicted in Fig. 11.11. On another side, almost all of the required up reserve capacity is supplied by the HSS unit, and the scheduled up reserve capacity by high-cost unit G2 is decreased to nearly zero during peak hours, as demonstrated in Fig. 11.12.

As well as committing the HSS facility to investigate the noticeable effects on the scheduling energy and reserve of units and their related costs, LMPs in each hour have been affected by this added unit. Therefore, all LMPs in the presence of the HSS unit are listed in Table 11.4. Similar to Case 1, the maximum value of LMPs has happened in hours 11, 14, and 18 that are highlighted in black, which are directly dependent on the congestion of transmission lines. And also, between these three congested hours, the LMP of bus 4 is the highest one compared to the other buses. In addition, in this case, LMPs of some buses are decreased compared to Case 1, in which Fig. 11.13 shows the Avg.LMPs and the reduction level in comparison with Avg.LMPs of Case 1. As it is evident in Fig. 11.13, Avg.LMPs of Case 2 during peak times, i.e., hours 15–17, approximately are reduced by 40% than Case 1.

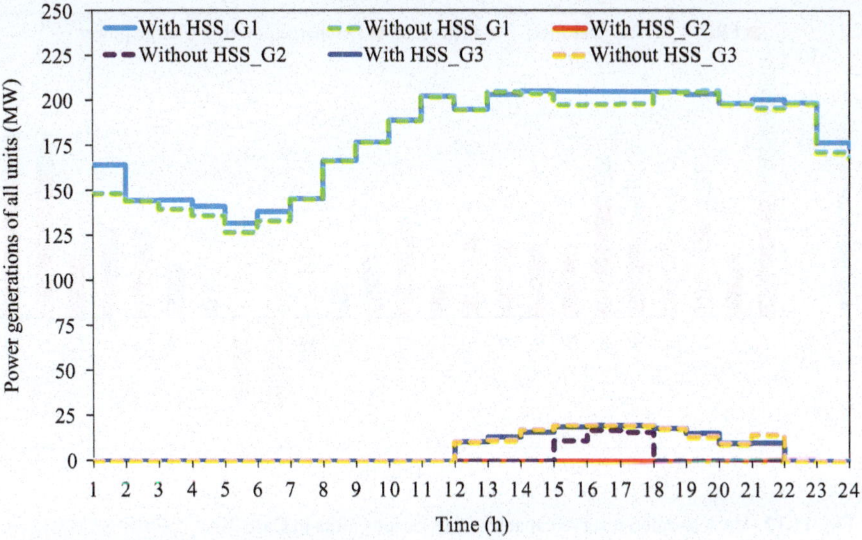

Fig. 11.10 Comparison of hourly power generation in Case 1 and Case 2 for all units

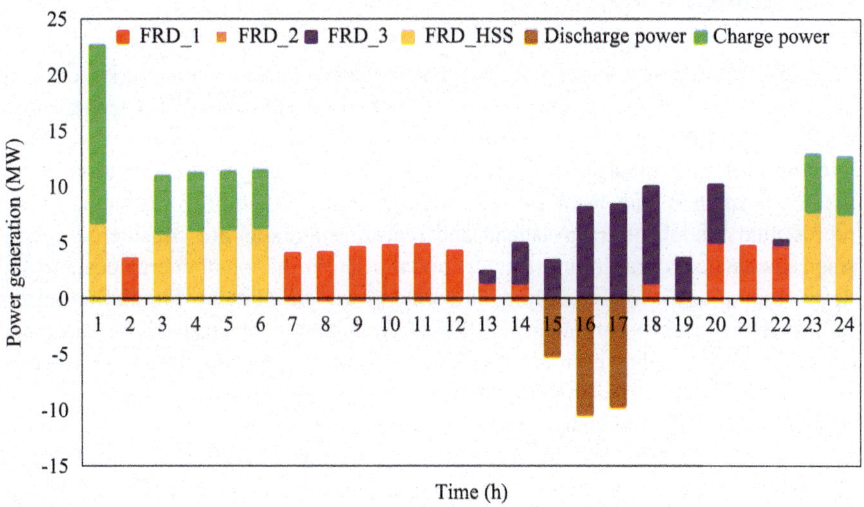

Fig. 11.11 Down scheduled reserve capacity from various units in Case 2

Finally, Table 11.5 is reported and compared the optimal obtained results with Case 1 to investigate the effects of adding HSS unit into the proposed system in terms of total operating cost, including energy and reserve costs, load shedding cost, and wind power curtailment cost. In this case, wind power curtailment has been reduced to 4.357 MWh due to the HSS unit's obligation in providing charge and discharge reserves. Also, the load shedding in this case is equivalent to 3.735 MWh.

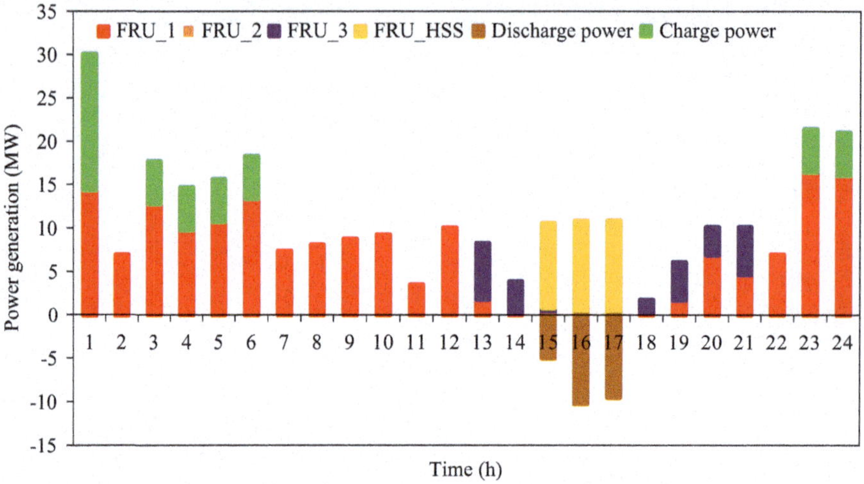

Fig. 11.12 Up scheduled reserve capacity from various units in Case 2

11.5 Conclusions

This chapter proposed a two-stage stochastic network-constrained market-clearing model by considering the integrated fuel cell-based HSS and WES technologies to supply optimal energy and reserve services. In addition, to cover the system uncertainties, i.e., forecasting electrical demand and WES power production errors, the MCS method has been utilized to realize the scheduling problem. In the represented model, both day-ahead and real-time markets are considered a two-stage stochastic programming structure and the determination of energy cost, up and down reserve costs, load shedding, and wind power curtailment. Simulation results indicated that incorporating the HSS facility in the considered six-bus modified electric test system has noticeable effects on the decrement of energy and reserve costs and also in the load shedding and wind power curtailment. So, the percentage reduced levels of total daily scheduling cost, energy cost, and reserve cost are about 3%, 1.4%, and 5.9%, respectively, in the presence of the HSS unit. Besides, electricity prices in peak hours are decreased by 40% by contributing HSS unit. Another significant effect is that it leads to a 35% and 3% reduction in wind power curtailment and electrical load shedding, respectively.

11.6 Status Quo, Challenges, and Outlooks

The status quo of the HSS facility integration into the electric power systems in a coordinated-based approach is about those finite countries such as Canada,

Table 11.4 LMPs of all six buses in Case 2

Time (h)	Bus_1	Bus_2	Bus_3	Bus_4	Bus_5	Bus_6	Avg.LMPs
LMP($/MWh)							
1	13.51	13.51	13.51	13.51	13.51	13.51	13.51
2	13.51	13.51	13.51	13.51	13.51	13.51	13.51
3	13.51	13.51	13.51	13.51	13.51	13.51	13.51
4	13.51	13.51	13.51	13.51	13.51	13.51	13.51
5	13.51	13.51	13.51	13.51	13.51	13.51	13.51
6	13.51	13.51	13.51	13.51	13.51	13.51	13.51
7	13.51	13.51	13.51	13.51	13.51	13.51	13.51
8	13.51	13.51	13.51	13.51	13.51	13.51	13.51
9	13.51	13.51	13.51	13.51	13.51	13.51	13.51
10	13.51	13.51	13.51	13.51	13.51	13.51	13.51
11	**13.51**	**95.34**	**126.3**	**140.8**	**134.2**	**132.5**	**107.2**
12	13.51	15.02	15.61	15.90	15.88	15.83	15.32
13	13.51	26.53	31.40	33.79	32.85	32.85	28.42
14	**13.51**	**95.83**	**127.3**	**141.5**	**135.2**	**132.8**	**107.7**
15	13.51	19.21	21.36	22.46	22.00	21.82	20.06
16	13.51	19.97	22.42	23.59	23.19	22.86	20.92
17	13.51	19.99	22.47	23.59	23.19	22.86	20.92
18	**13.51**	**95.86**	**127.3**	**141.5**	**135.7**	**132.8**	**107.7**
19	13.51	44.23	55.83	61.26	59.15	58.04	48.67
20	13.51	15.02	15.61	15.90	15.88	15.83	15.32
21	13.51	15.69	16.58	16.94	16.79	16.63	16.01
22	13.51	25.75	30.41	32.54	31.64	31.21	27.53
23	13.51	13.51	13.51	13.51	13.51	13.51	13.51
24	13.51	13.51	13.51	13.51	13.51	13.51	13.51

China, France, Germany, Japan, Norway, South Korea, and the USA that are taking part in developing and implementing. With regard to the International Energy Agency's Hydrogen Projects Database, approximately 320 green hydrogen production demonstration projects have been declared globally which have a whole 200 MW electrolyzer capacity. However, some technical challenges have been seen and should be eliminated or reduced significantly to improve as much as possible the full and large-scale integration of HSS technology in various power grids. These challenges and obstacles are efficiency, durability, and investment costs.

The hydrogen's long-term potential and a development roadmap predicted that 18% of global final energy demand could be supplied by hydrogen energy sources until 2050. Decarbonizing transportation is the first realization of HSS that indicates the fuel cell PEVs, high-utilization road vehicles (trucks, buses), ferry boats, and forklifts. The other applications contribute to decarbonizing rail transport, shipping, and aviation in the longer term. In the short term, the injection of hydrogen energy into the natural gas network provides a potential revenue to improve the power-to-

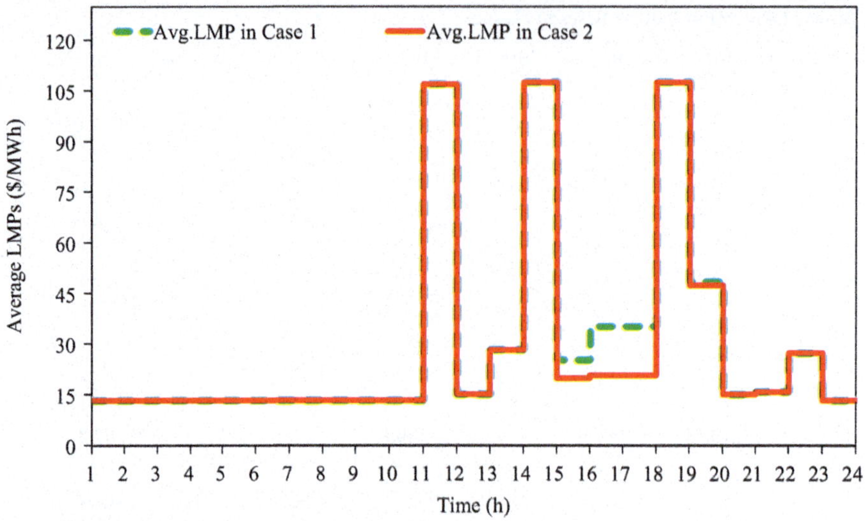

Fig. 11.13 Comparison of Avg.LMPs in Case 1 and Case 2

Table 11.5 Various costs of the proposed system in Case 1 and Case 2

Cases	Total scheduling cost	Energy cost	Reserve cost	Load shedding cost	Wind curtailment cost
Costs($)					
Case 1	73184.201	67833.98	2678.17	1533.870	333.380
Case 2	71463.84	66900.41	2521.09	1503.530	217.859

hydrogen's (P2H) economics however that keeps the promise of storing a major amount of RESs in the long term which decarbonizes demand for natural gas.

Acknowledgements This publication was partially supported by award NPRP12S-0125-190013 from the QNRF-Qatar National Research Fund, a member of The Qatar Foundation. The information and views set out in this publication are those of the authors and do not necessarily reflect the official opinion of the QNRF.

References

1. IRENA (2019) and Abu Dhabi, Future of wind: Deployment, investment, technology, grid integration and socio – Economic aspects (A Global Energy Transformation paper) International Renewable Energy Agency.
2. Heydarian-Forushani, E., Golshan, M. E. H., & Shafie-khah, M. (2016). Flexible interaction of plug-in electric vehicle parking lots for efficient wind integration. *Applied Energy, 179*, 338–349.

3. Wang, S., Yang, H., Pham, Q. B., Khoi, D. N., & Nhi, P. T. T. (2020). An ensemble framework to investigate wind energy sustainability considering climate change impacts. *Sustainability, 12*(3), 876.
4. Ela, E., et al. (2016). Wholesale electricity market design with increasing levels of renewable generation: Incentivizing flexibility in system operations. *The Electricity Journal, 29*(4), 51–60.
5. Alipour, M., Mohammadi-Ivatloo, B., Moradi-Dalvand, M., & Zare, K. (2017). Stochastic scheduling of aggregators of plug-in electric vehicles for participation in energy and ancillary service markets. *Energy, 118*, 1168–1179.
6. Daneshvar, M., Mohammadi-Ivatloo, B., Zare, K., & Asadi, S. (2020). Two-stage stochastic programming model for optimal scheduling of the wind-thermal-hydropower-pumped storage system considering the flexibility assessment. *Energy, 193*.
7. García-Garre, A., Gabaldón, A., Álvarez-Bel, C., Ruiz-Abellón, M. D. C., & Guillamón, A. (2018). Integration of demand response and photovoltaic resources in residential segments. *Sustainability, 10*(9), 3030.
8. Xu, Q., Ding, Y., & Zheng, A. (2017). An optimal dispatch model of wind-integrated power system considering demand response and reliability. *Sustainability, 9*(5), 758.
9. Aliasghari, P., Mohammadi-Ivatloo, B., Alipour, M., Abapour, M., & Zare, K. (2018). Optimal scheduling of plug-in electric vehicles and renewable micro-grid in energy and reserve markets considering demand response program. *Journal of Cleaner Production, 186*, 293–303.
10. Mirzaei, M. A., Sadeghi Yazdankhah, A., & Mohammadi-Ivatloo, B. (2019). Stochastic security-constrained operation of wind and hydrogen energy storage systems integrated with price-based demand response. *International Journal of Hydrogen Energy, 44*(27), 14217–14227.
11. Yu, D., Wang, J., Li, D., Jermsittiparsert, K., & Nojavan, S. (2019). Risk-averse stochastic operation of a power system integrated with hydrogen storage system and wind generation in the presence of demand response program. *International Journal of Hydrogen Energy, 44*(59), 31204–31215.
12. Mansour-Saatloo, A., Mirzaei, M. A., Mohammadi-Ivatloo, B., & Zare, K. (2020). A risk-averse hybrid approach for optimal participation of power-to-hydrogen technology-based multi-energy Microgrid in multi-energy markets. *Sustainable Cities and Society, 63*.
13. Mansour-Saatloo, A., Agabalaye-Rahvar, M., Mirzaei, M. A., Mohammadi-Ivatloo, B., Abapour, M., & Zare, K. (2020). Robust scheduling of hydrogen based smart micro energy hub with integrated demand response. *Journal of Cleaner Production, 267*.
14. Seyyedeh-Barhagh, S., Majidi, M., Nojavan, S., & Zare, K. (2019). Optimal scheduling of hydrogen storage under economic and environmental priorities in the presence of renewable units and demand response. *Sustainable Cities and Society, 46*.
15. Dhabi, A. (2019). *Hydrogen: A renewable energy perspective*. IRENA, International Renewable Energy Agency.
16. *ThyssenKrupp's "green hydrogen and renewable ammonia value chain*. Ammon. Ind. ammoniaindustry.com/thyssenkrupps-green-hydrogen-and-renewable-ammonia-value-chain/."
17. Fernandez-Blanco, R., Arroyo, J. M., & Alguacil, N. (2014). Network-constrained day-ahead auction for consumer payment minimization. *IEEE Transactions on Power Systems, 29*(2), 526–536.
18. Parvania, M., Fotuhi-Firuzabad, M., & Shahidehpour, M. (2014). Comparative hourly scheduling of centralized and distributed storage in day-ahead markets. *IEEE Transactions on Sustainable Energy, 5*(3), 729–737.
19. Agabalaye-Rahvar, M., Mansour-Saatloo, A., Mirzaei, M. A., Mohammadi-Ivatloo, B., Zare, K., & Anvari-Moghaddam, A. (2020). Robust optimal operation strategy for a hybrid energy system based on gas-fired unit, power-to-gas facility and wind power in energy markets. *Energies, 13*(22).
20. Nasri, A., Kazempour, S. J., Conejo, A. J., & Ghandhari, M. (2016). Network-constrained AC unit commitment under uncertainty: A benders' decomposition approach. *IEEE Transactions on Power Systems, 31*(1), 412–422.

21. Tumuluru, V. K., & Tsang, D. H. K. (2018). A two-stage approach for network constrained unit commitment problem with demand response. *IEEE Transactions on Smart Grid, 9*(2), 1175–1183.
22. Kia, M., Nazar, M. S., Sepasian, M. S., Heidari, A., & Siano, P. (2017). Optimal day ahead scheduling of combined heat and power units with electrical and thermal storage considering security constraint of power system. *Energy, 120*, 241–252.
23. Nazari-Heris, M., Mohammadi-Ivatloo, B., & Asadi, S. (2020). Optimal operation of multi-carrier energy networks considering uncertain parameters and thermal energy storage. *Sustainability, 12*(12), 5158.
24. Vahedipour-Dahraie, M., Najafi, H., Anvari-Moghaddam, A., & Guerrero, J. (2017). Study of the effect of time-based rate demand response programs on stochastic day-ahead energy and reserve scheduling in islanded residential Microgrids. *Applied Sciences, 7*(4).
25. Doostizadeh, M., Aminifar, F., Ghasemi, H., & Lesani, H. (2016). Energy and reserve scheduling under wind power uncertainty: An adjustable interval approach. *IEEE Transactions on Smart Grid, 7*(6), 2943–2952.
26. Cobos, N. G., Arroyo, J. M., Alguacil, N., & Wang, J. (2018). Robust energy and reserve scheduling considering bulk energy storage units and wind uncertainty. *IEEE Transactions on Power Systems, 33*(5), 5206–5216.
27. Mirzaei, M. A., Yazdankhah, A. S., Mohammadi-Ivatloo, B., Marzband, M., Shafie-khah, M., & Catalão, J. P. S. (2019). Stochastic network-constrained co-optimization of energy and reserve products in renewable energy integrated power and gas networks with energy storage system. *Journal of Cleaner Production, 223*, 747–758.
28. Jain, I. P., Jain, P., & Jain, A. (2010). Novel hydrogen storage materials: A review of lightweight complex hydrides. *Journal of Alloys and Compounds, 503*(2), 303–339.
29. Meregalli, V., & Parrinello, M. (2001). Review of theoretical calculations of hydrogen storage in carbon-based materials. *Applied Physics A Materials Science & Processing, 72*(2), 143–146.
30. Demirci, U. B. (2017). Ammonia borane, a material with exceptional properties for chemical hydrogen storage. *International Journal of Hydrogen Energy, 42*(15), 9978–10013.
31. Heo, Y.-J., Yeon, S.-H., & Park, S.-J. (2019). Defining contribution of micropore size to hydrogen physisorption behaviors: A new approach based on DFT pore volumes. *Carbon, 143*, 288–293.
32. Maniyali, Y., Almansoori, A., Fowler, M., & Elkamel, A. (2013). Energy hub based on nuclear energy and hydrogen energy storage. *Industrial & Engineering Chemistry Research, 52*(22), 7470–7481.
33. Mansour-Satloo, A., Agabalaye-Rahvar, M., Mirazaei, M. A., Mohammadi-Ivatloo, B., Zare, K., & Anvari-Moghaddam, A. (2021). A hybrid robust-stochastic approach for optimal scheduling of interconnected hydrogen-based energy hubs. *IET Smart Grid, 4*(2), 241–254.
34. Soroudi, A., Aien, M., & Ehsan, M. (2012). A probabilistic modeling of photo voltaic modules and wind power generation impact on distribution networks. *IEEE Systems Journal, 6*(2), 254–259.
35. Mansour-Saatloo, A., Agabalaye-Rahvar, M., Mirzaei, M. A., Mohammadi-Ivatloo, B., & Zare, K. (2021). Chapter 9 – Economic analysis of energy storage systems in multicarrier microgrids. In B. Mohammadi-Ivatloo, A. Mohammadpour Shotorbani, & A. Anvari-Moghaddam (Eds.), *Energy storage in energy markets* (pp. 173–190). Academic Press.
36. Alabdulwahab, A., Abusorrah, A., Zhang, X., & Shahidehpour, M. J. I. S. J. (2015). Stochastic security-constrained scheduling of coordinated electricity and natural gas infrastructures. *IEEE Systems Journal, 11*(3), 1674–1683.
37. Sun, W. L. L., Xu, B., & Chai, T. (2010). *The scheduling of steel-making and continuous casting process using branch and cut method via CPLEX optimization*. In 5th International conference on computer sciences and convergence information technology, pp. 716–721.

Chapter 12
Polygeneration Systems in Fossil Fuel Power Plants: The Role of Power-to-X in CO_2 Mitigation

Amirhossein Khalili-Garakani, Leila Samiee, and Kazem Kashefi

12.1 Introduction

The industries development and the rapid growth of the population due to the increase in consumables and the consequent increase in industrial waste are among the issues that have recently created huge crises in human societies. The severity of environmental pollution caused by gaseous wastes in industrial assembly centers is such that it has attracted the attention of scientific and executive resources of the world to the principled recycling of these materials. Over the past few decades, society has faced one of the most important consequences of industrial development, the recycling of gaseous waste, which accounts for a large share of total environmental pollution and global warming, and its destructive effects in creating environmental crises are quite evident.

One of the most important pollution in the energy sector based on climate, type of activity and natural resources in the region, etc. is air pollution due to the release and leakage of pollutants due to the combustion of fossil fuels. Sulfur oxides (SOx), nitrogen oxides (NOx), carbon monoxide (CO), particulate matter (PM), hydrocarbons (CH), and carbon dioxide (CO_2) are among the pollutants and greenhouse gases due to the energy activities of the energy sector which are released in the atmosphere [1]. Among them, greenhouse gases such as CO_2 cause climate change and global warming and are important.

A. Khalili-Garakani (✉)
Chemistry & Process Engineering Department, Niroo Research Institute (NRI), Tehran, Iran
e-mail: akhalili@nri.ac.ir

L. Samiee · K. Kashefi
Development & Optimization of Energy Technologies Research Division, Research Institute of Petroleum Industries (RIPI), Tehran, Iran
e-mail: Samieel@ripi.ir; Kashefik@ripi.ir

© The Author(s), under exclusive license to Springer Nature Switzerland AG 2022
V. Vahidinasab, B. Mohammadi-Ivatloo (eds.), *Whole Energy Systems*, Power Systems, https://doi.org/10.1007/978-3-030-87653-1_12

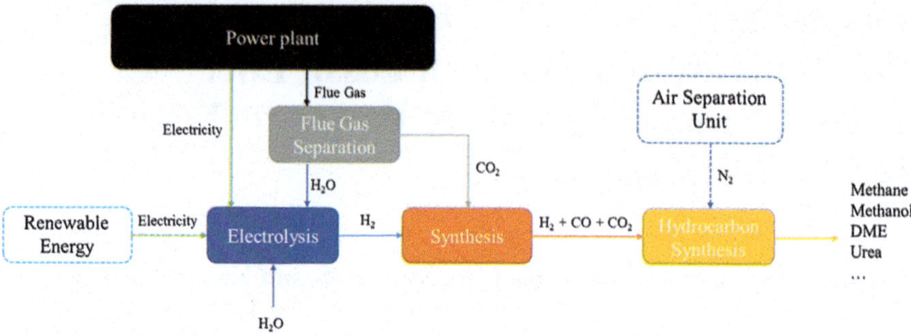

Fig. 12.1 Power-to-X process overview in a power plant

Many countries have tried to use appropriate technologies and scientific methods to utilize CO_2 as one of the main reasons of global warming. However, CO_2 utilization technologies are still developing. Our country is no exception to this rule and it is necessary to control pollution in line with the development of industries.

The overview of the power-to-X approach with using renewable energy as the source of electricity generation and CO_2 component of the flue gas has been illustrated in Fig. 12.1. This approach is now being used for hydrocarbon synthesis such as methane, methanol, DME, urea, etc.

12.1.1 Power Generation, Fuel Consumption, and CO_2 Generation in Iran

The amount of gross power generation in Iranian power plants in the period 2009–2019 is shown in Fig. 12.2. As can be seen, thermal power plants that produce power with fossil fuels are responsible for a major part of power generation in the country. There are four types of fossil fuel power plants in Iran: gas turbine cycle, steam cycle, combined cycle, and diesel power plants. However, due to the small share of diesel power plants in this study, they have been omitted. In Iran power generation industry, natural gas, diesel, and fuel oil are commonly used as fuels. According to the detailed statistics published by the Ministry of Energy, the amount of fuel consumption separately in the years 2009–2019 is according to Fig. 12.3 [1]. As can be seen, most of the fuel consumed in the country in recent years has been supplied from natural gas. Figure 12.4 shows the share of each fuel in the specific production of thermal power plants in 2019.

The table below shows the amount of CO_2 emission of each of these fuels based on the type of power plant [2]. The emission factor is presented in terms of the mass of pollutants produced from combustion of fuels in terms of the electricity generated or the thermal energy used.

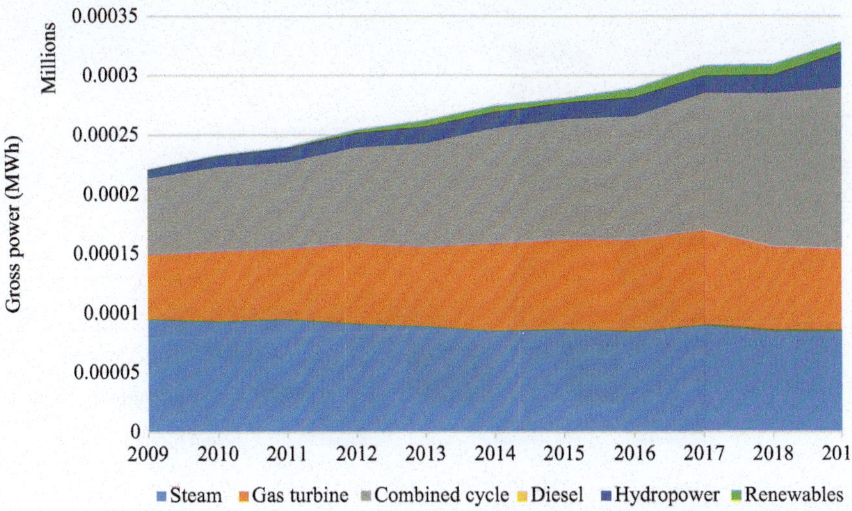

Fig. 12.2 The amount of special production in Iranian power plants during 2009–2019

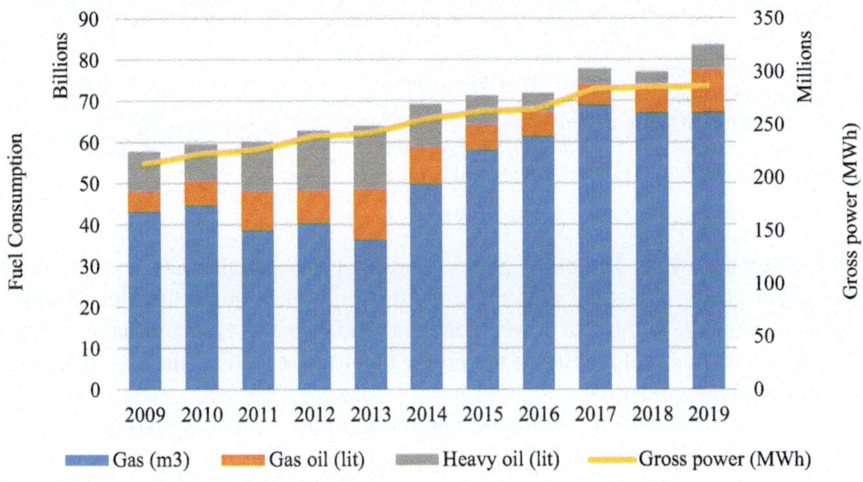

Fig. 12.3 Amount of fuel consumption and special production in thermal power plants in Iran during 2009–2019

Using the fuel consumption information and production capacity of each power plant and the emission factors reported in Table 12.1, the amount of CO_2 emissions can be calculated for each power plant. Figure 12.5 indicates the total amount of emission in three categories of steam turbine, gas turbine, and combined cycle power plants. The volume and the amount of CO_2 production play a key role in selecting the recovery and utilization process for thermal power plants. The country's power plants vary in production and CO_2 emissions from 11.5 million tons per year to

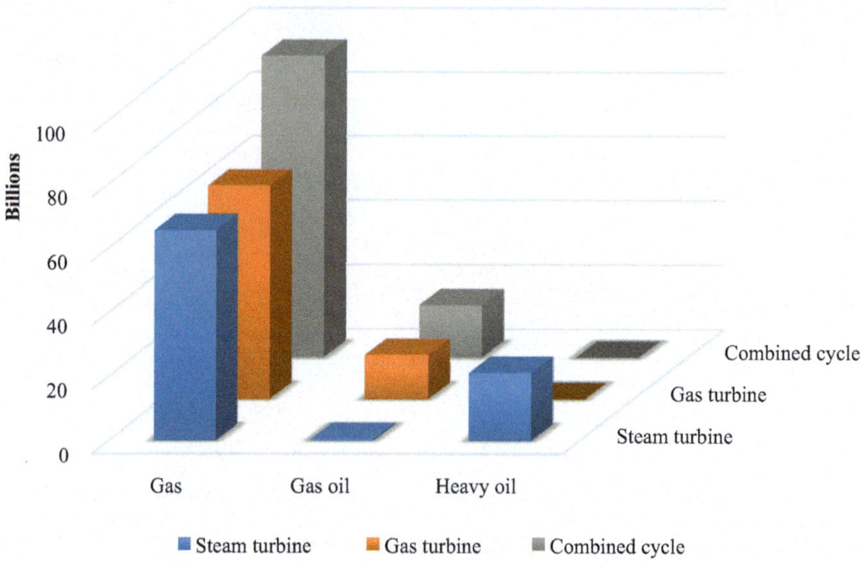

Fig. 12.4 The share of each fuel in the special production of Iranian thermal power plants

Table 12.1 The average CO_2 emission factor of combustion gases in steam, gas, and combined cycle power plants (kg/kWh) [2]

Fuel type	Steam turbine	Gas turbine	Combined cycle
Gas	0.633	0.782	0.450
Gas oil	0.860	1.048	0.622
Heavy oil	1.025	–	–

100,000 tons per year. Therefore, power plants that have a higher CO_2 emission rate than other power plants are preferred, and this difference in the amount of power plant smoke in the number of power plant units and their fuel type, etc. has occurred. The less smoke is removed from the power plant for recycling, the less efficiency the power plant will suffer.

The nominal power of the biggest combustion power plant is presented in Fig. 12.6. Besides, the rate of change in CO_2 emissions from the three index power plants in all three categories of steam turbine, gas turbine, and combined cycle during 2009–2019 is shown in Fig. 12.7. As can be seen, due to the consumption of liquid fuels in heating power plants, the amount of emissions is higher than others. However, as illustrated in Fig. 12.5, due to the lower efficiency of gas turbine power plants and the greater number of these power plants, the total emissions of these power plants account for the largest share of CO_2 production.

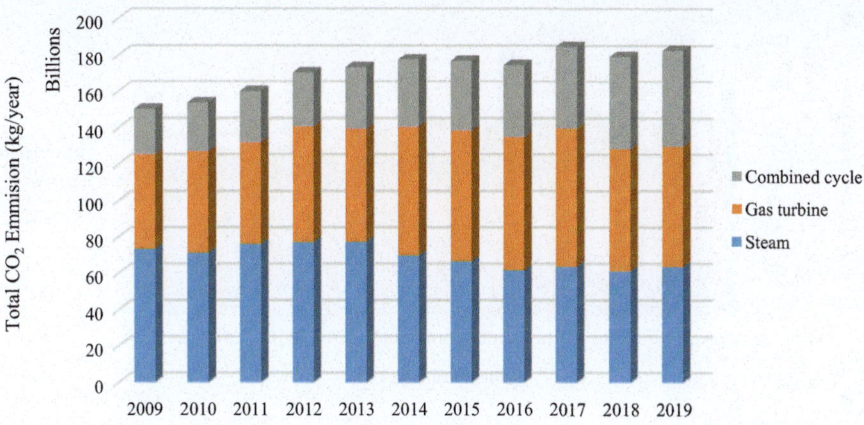

Fig. 12.5 Emission rate of carbon oxides during 2009–2019

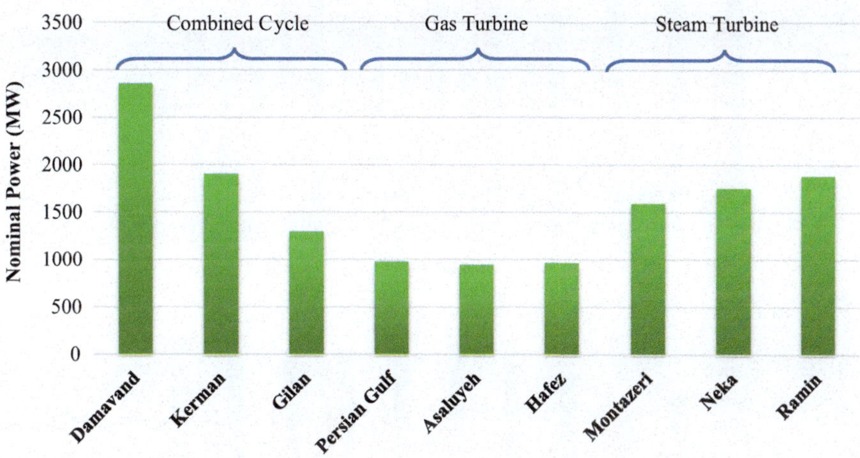

Fig. 12.6 The nominal power of the biggest combustion power plant in different categories

12.1.2 CO_2 Mitigation Techniques

In this section, the CO_2 mitigation options have been evaluated. There are three different mitigation options: (1) energy intensity reduction, (2) carbon intensity reduction, and (3) carbon sequestration.

Energy intensity reduction means increasing energy efficiency by applying energy-saving devices or using methodologies which reduce the energy consumption. A decrease in the carbon intensity of the power generation results in a reduction in the CO_2 emission per unit power generation. For example, fuel cells as a novel power generation process which has a high efficiency could be included in this category.

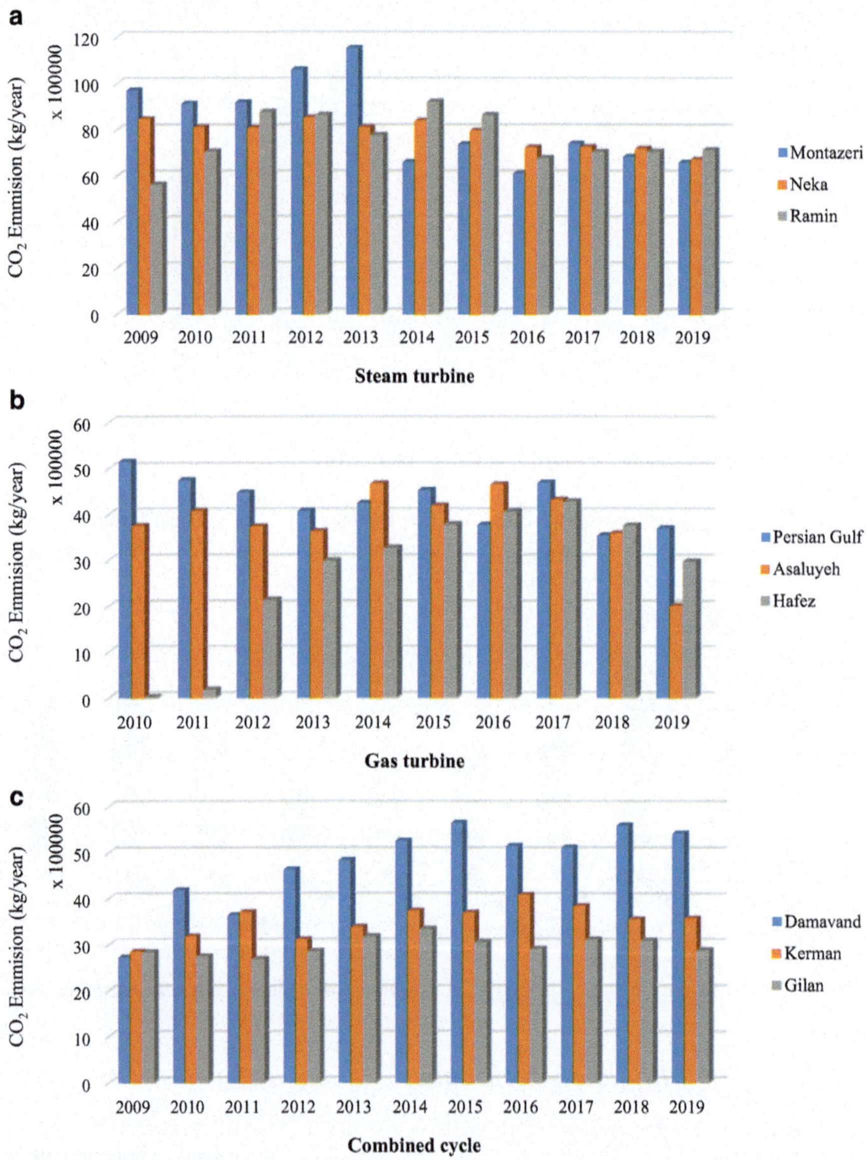

Fig. 12.7 The amount of CO_2 emitted by the biggest Iranian combustion power plant during 2009–2019

Using renewable energy resources instead of fossil fuels could reduce the carbon intensity of the process. Also, the nuclear energy resources could be included in this classification. Moreover, in the fossil fuels division, shifting fuels with high carbon contents, like coal and heavy fuel oil, to ones with low carbon contents, like natural gas, could further reduce the carbon intensity. These two classifications are very promising options as by which not only the global warming is alleviated, but also a more sustainable society can be established.

The third classification is the sequestration of carbon. Two types of options are included in this classification: the natural sinking process improvement and the direct discharging of CO_2 into the ocean or underground.

Unfortunately, there is a big difference between the expected reduction potential of CO_2 emissions in the Kyoto Protocol and what is really achieved. According to oil crisis, there is a fast progress in the early stages, the technological developments in energy saving have somewhat matured, but due to the high cost and the lack of infrastructures of the renewable energies, its use was limited. So, probably, the consumption of fossil fuels will continue, due to which the CO_2 emissions will increase for about at least the next hundred years. Therefore, applying the sequestration of CO_2 should be considered in the future, while according to the IPCC (intergovernmental panel on climate change) report, only a limited amount is estimated in the present level.

The active research works on carbon sequestration issue really began in the 1990s. There are a lot of documents and presentations for different conferences and workshops in the field of carbon sequestration, besides a large number of papers and reviews which have been published till today.

Regarding to carbon, the earth is a closed system: the total amount of carbon is constant, but circulates among different levels in the biosphere, atmosphere, lithosphere, and the ocean (Fig. 12.8). The amount of CO_2 released into the atmosphere by humans (anthropogenic) is fixed and circulated. Instead, the CO_2 concentration in the atmosphere is increasing. The fundamental of the carbon sequestration comes from increasing the sinking flux of CO_2 generated by human activities to avoid accumulating the amount of CO_2 in the atmosphere. The CO_2 sequestration alternative is separated into two ways:

1. Enhancement of natural sinking processes of CO_2 in order to fix CO_2.
2. Artificial processes for direct CO_2 sequestration.

The fossil fuels formation through photosynthesis and rock weathering natural sinking processes of CO_2. The ocean fertilization, mineral carbonation, and afforestation are among the main routes which are being concerned. Here, the CO_2 in the atmosphere would be fixed by conversion to more stable forms such as organic matter or carbonates. In the direct sequestration, the CO_2 is transported to an appropriate location and then injected in the form of CO_2 either in the ocean or underground. According to the global carbon cycle, by passing a certain period of time, the sequestered CO_2 could be emitted back into the atmosphere. This time period based on the location, method of sequestration, and form of CO_2 could be changed. The feasible sequestration process must have these specifications:

- Large potential capacity for CO_2 sequestration.
- Net reduction of CO_2 emission.
- Long-term CO_2 isolation (at least several hundred years).

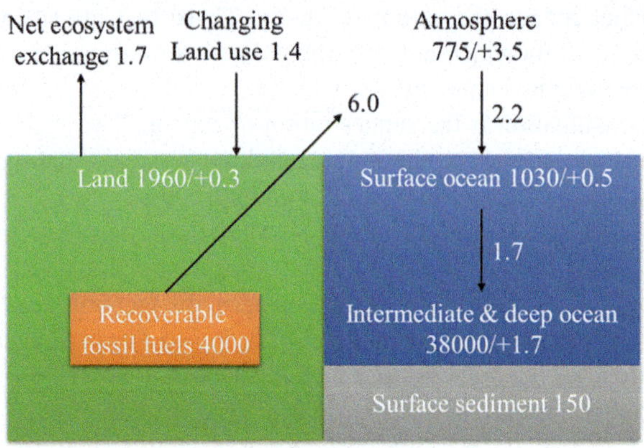

Fig. 12.8 Global carbon cycle, units: Gt-C and Gt-C/year

- Modest cost and energy consumption.
- Minimal environmental impact.

From the process design aspect in order to precisely estimate the amount of carbon sequestrated and the energy requirement, a life cycle assessment would be necessary [3].

Except the above discussed CO_2 mitigation options, there is another important approach which is carbon capture and utilization (CCU). The closed-loop carbon cycle idea is back to the 1970s and starts with electrolytic hydrogen generation utilizing nuclear power and carbon capture [4]. More recently, a "methanol economy" has been presented by Nobel laureate [5, 6] which proposes a new cycle containing direct carbon capture from the air and hydrogen generation utilizing renewable energy resources with the purpose of methanol (MeOH) production to use in mobility. There is an optimistic perspective about the future of CCU among scientific literature reviews [7, 8]. Some expressions like "teraton challenge" or "a carbon dioxide-neutral world" are stated and related to CO_2 conversion utilizing renewable energy like catalytic photoreactors. Different research programs are supported by public and private R&D about CCU concepts at different scales from pilot to industrial scale [9, 10]. Besides, there is another route called thermochemical energy storage process which tries to present a chemical closed loop of carbon (as CO_2 or fuels) in caverns or reservoirs.

Generating carbon-free energy is the main motivation of the CO_2 conversion to fuels. The replacement fuel can be utilized in different sector, power, heat, or mobility and decrease the fossil carbon extraction. Aresta et al. summarize these routes in a recent review and claimed that each atom of carbon recycled is the replacement of the fossil atom of carbon, which remains in the underground for the future and does not emit to the atmosphere [10].

The CO_2 utilization process could have different products through thermochemical, biochemical, and electrochemical conversions [11]. The technologies have different advantages, according to the amount of carbon mitigated, economic and market size of the product, and the fixation time of carbon. Carbon capture and storage (CCS) has the potential to store large volume of carbon in underground reservoirs for more than hundred years. But, it is not an economical process unless it is utilized as enhanced oil recovery (EOR). Another option is to produce cement and minerals which not only have any value but also can fix the large volume of carbon. Instead chemicals have more financial values, yet have differently mitigated capabilities and different fixation time periods, like plastics which is stable for many decades or fuels that fix the CO_2 emission till they are burned. Nevertheless, their production is a kind of energy storage for renewable energy resources and can guarantee the energy security in the countries which import foreign sources of fuel [12].

Polygeneration process is another CCU solution which focuses on the efficient generation of different products like electricity, chemicals, fuels, etc. from the specific amount of feedstock which is discussed in more detail in the next parts [13].

12.2 Polygeneration in Industry

In recent years, polygeneration processes convert feedstocks like coal, natural gas, etc. to a variety of products such as power, liquid fuels, chemicals, etc. [14]. These systems not only enhance the achieved efficiency by integrating different processes together to use benefits of synergies among them (as an instance the exothermic heat of one process can be used to provide the required heat in another endothermic reaction) but also have the flexibility to alter the amount of feed and products according to the fluctuation in the market [15]. Different studies investigate the design, simulation, thermal studies, and economic analysis [16]. The polygeneration systems have higher efficiency due to the multi-in/multi-out configuration in comparison to the process with one feedstock [17, 18].

Combined heat and power (CHP) process is an instance of polygeneration systems. In this process not only electricity but also the required heat of nearby consumers is generated. Changes in heat requirement during the year make it possible to utilize the excess heat in other processes like integrating fluidized bed combustor of the CHP system with a fluidized bed gasifier in a biomass process [19]. Hyene et al. proposed another sample of CHP-biomass integrated plant and concluded that the efficiency improvement of the new process is dependent on the capacity of not only the boiler of the CHP system but also the gasification facility in the biomass process [20]. Gustavsson et al. presented an enhanced and efficient CHP-biomass integrated process for generating biofuels like methanol and Fischer-Tropsch fuel [21].

The municipal solid waste, commercial and industrial organic waste, and agricultural waste called the refuse-derived fuel (RDF) are a source which, according to the waste management directives 2006/12/EC in Europe, must be utilized for energy generation. Only municipal solid waste reach 338 million tons by 2020 in Europe [21]. Initially, RDF waste combustion is suggested as a simple solution which could be applied to larger scales and generate heat and electrical power. Besides, it is proposed that RDF waste can be converted to biofuels through biological or thermochemical processes. But, in biological processes like digestion or fermentation, not only the separation of biodegradable waste from RDF but also long process time is required. On the other hand, gasification is suggested as the most appropriate way of RDF conversion to biofuels [21]. In most of the researches, biomethane or hydrogen is considered, but recently other liquid biofuels such as DME are presented as potential products of the CHP-biomass integrated plant [22].

The water-gas shift (WGS) reaction is applied currently in most of the polygeneration plants to intensify the H_2 content of syngas product. But, it is confirmed that WGS reaction is conducted to shortcoming in the plants: first, low heat generation due to the exothermic WGS reaction which is lost in the plants and, second, additional CO_2 emission which followed by the increasing in the capacity of the downstream carbon capture unit. Additionally, the carbon capture unit required itself has consequential energy losses. This unit is needed for separating CO_2 from the generated syngas and flue gas of the plant. So, by the side of effect on efficiency, the shortcoming of negative environmental impact must be considered [18]. The flexible nature of polygeneration process can help to overcome these consequences. According to the literature, different integrated flowsheets of polygeneration-carbon capture processes are proposed. Besides, these integrated designs permit the alteration of the amount of products according to the fluctuation in the markets [18].

As an instance, Li et al. proposed an integrated plant which can generate power and synthetic natural gas [23]. The new design has not only a lower life cycle energy use but also lower GHG emission compared to an ultra-supercritical coal power plant. Another cost-effective example is a polygeneration plant of urea and power which is combined with a CCS unit [23]. Also, Jana et al. through life cycle assessment analysis proposed a sustainable power, ethanol, heating, and cooling polygeneration plant which used rice straw as feedstock [24]. Besides, Huang et al. presented an efficient coal-based polygeneration plant for producing methanol and power [25]. A new chemical looping combustion plant is proposed by Khojasteh Salkuyeh which can generate methanol, DME, and power simultaneously from coal and natural gas [18].

Most of the proposed polygeneration systems are based on coal as feedstock [13], but recently, the oxyfuel combustion process which uses natural gas and biomethane as fuels is considered as the low emission technologies (LETs) [26]. In addition, some novel ways for improving the utilization of natural gas in high efficient combined cycle power plants (as high as 57%) are proposed [27].

Chemical looping due to CO_2/H_2O splitting using methane to generate syngas (CO/H_2) is one of the novel approaches [26]. This method is a fascinating option to substitute the solar thermochemical redox cycle, which is applied to generate syngas through water splitting by oxygen careers (OC) [28]. In the solar thermochemical redox cycle, OC is a metal oxide, which required a very concentrated heat to pass endothermic thermal reduction (TR) stage. This concentrated heat is provided by solar energy under a very high vacuum pressure [29, 30]. In the next stage, the reduced OC goes through an oxidation process by the incoming CO_2/H_2O and generating syngas and fulfills the continuous cycle. This step is exothermic and has atmospheric and a lower temperature operating condition compared to reduction stage. So, temperature and pressure swing among the two stages is possible. The process of reduction of fuel with methane proceeds not only near atmospheric operating conditions which prevent pressure swing but also lower the temperature of reduction stage, which avoids a large temperature swing among the two stages [26]. Another advantage of applying the methane reduction (or biomethane) stage in chemical looping is the ability to operate day and night unlike the technology utilized solar energy resources.

Recently, dimethyl ether (DME) is mostly considered as future synthetic fuel through resemblance with diesel. In spite of the fact that DME has a lower latent heat of vaporization (LHV) compared to diesel and pressurization which is required to store in liquid state at ambient conditions, some futures like its physical properties and chemical structure provide an interesting fuel. The other advantage of DME is low emissions (such as NO_x, hydrocarbon (HC), SO_x, and particulate) during combustion. However, there are some drawbacks like the redesign of the injector and fuel pump systems and analysis of the compatibility of sealant materials which make the fuel switching from diesel to DME fairly challenging task. As a short-term solution to extend the applicability and economic perspective of DME usage, the blending of DME with conventional mobility fuel is proposed [31]. The similar physicochemical properties of DME to liquid petroleum gas (LPG) make it possible to be used in the LPG-based mobility sector. Besides being used as fuel, DME can be utilized as feedstock in the synthesis of aromatics, gasoline, olefins, and other chemicals or applied as a refrigerant [32]. However, the cost of the feed (commonly natural gas) has the main role in constraining the price of DME for its usage as synthetic fuel. The future outlook illustrates the equal rise of the diesel and natural gas prices. So, DME relative economic competitiveness to diesel fuel in the future would be tainted as its production cost would be increased. Considering this issue, investigation for the new source of methane from biomass or other feedstocks like associated gas could possibly enhance not only economical but also environmental perspective of DME synthesis process [33, 34].

Hankin and Shah proposed a process of DME and methanol synthesis from CO_2 and H_2O [35]. In their process, syngas is generated by water electrolysis and solid oxide electrolysis for CO. Also, Khojasteh Salkuyeh and Adam II presented a new chemical looping configuration which integrates the coal gasification and natural gas reforming to produce power, methanol, and DME [18]. Moreover, producing DME utilizing syngas generated from CO_2/H_2O splitting can be a fascinating route of clean fuel production [36].

12.2.1 Power-to-X and CO_2 Mitigation

Power-to-X process can help large-scale industries and technologies such as mobility, manufacturing, and chemicals to get rid of fossil fuels as feedstock or energy delivery. This applies to energy carriers like hydrogen and hydrocarbons, besides the available raw materials, like methanol, etc.

As an instance the hydrogen use in production of synthetic fertilizers, reforming process, and chemical industry surpass 70 million metric tons annually. It is anticipated that according to extended applicability of hydrogen as an energy carrier, its demand grows faster and stronger in the near future.

Besides, the opportunity to grow considerable attention to the application of CO_2 as a carbonaceous material increases the demand for hydrogen. Nowadays, hydrogen is mainly produced through steam methane reforming, which utilizes methane and water to generate syngas and CO_2. However, electrolysis process provides hydrogen production without CO_2 emission. As a matter of fact, extension of renewable and other carbon-free energy resources makes available the carbon-free hydrogen production. Furthermore, it is a route to reserve excess electricity generated through renewable energy resources.

Likewise, renewable electricity can be applied to generate routine hydrocarbons which commonly derived from coal or petroleum. These carbon-free generated hydrocarbons may be utilized as fuels or feedstock to produce an extensive variety of everyday products. For instance, carbon monoxide generated from carbon dioxide can be utilized as feedstock of Fischer-Tropsch process and substitute the routine syngas produced from fossil resources.

For example, the configuration of a power-to-X concept in a combustion power plants (CPP) is presented in Fig. 12.9. It contains an amine-based scrubber, CO_2 purification, water electrolysis and hydrogen storage and hydrocarbon synthesis (such as methanol, etc.), and at last product upgrading and storage. Several companies presented different demo plants in the last 30 years. The direct CO_2-based methanol synthesis of the Lurgi or RITE and indirect CAMERE process present a separate reverse water-gas shift reactor in the 1990s. Besides, Mitsui and CRI after 2000 utilize renewable energy resources for higher capacities [37]. For example, the CO_2-based methanol plant in Iceland has the capacity of 4000 t/a [23].

Another sample of power-to-X plant is illustrated in Fig. 12.10. It contains fluidized-bed steam drying of lignite, gasification, syngas scrubber, acid gas removal unit and hydrocarbon synthesis (such as methanol, etc.) and supplementary plants called air separation unit (ASU), and optional water electrolysis unit. Gasification is one of the favored technologies which is fully investigated by many researchers. Some issues like investment and operating cost reduction and flexibility enhancement must be considered for improvement of the applicability of this technology. The high cost of the technology is due to some supplementary units such as gas purification, solids handling, and effluent streams treatment.

The growing interest in polygeneration power plants is focused on product flexibility and resource utilization for higher overall efficiency and improving the

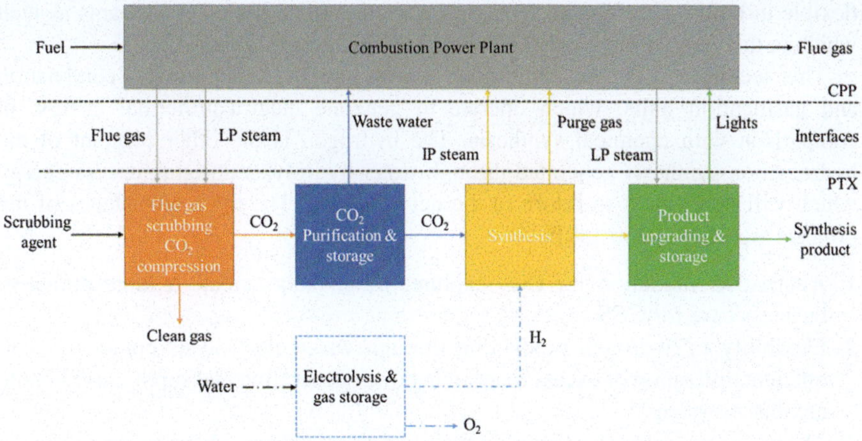

Fig. 12.9 Process flow sheet "power-to-X" concept

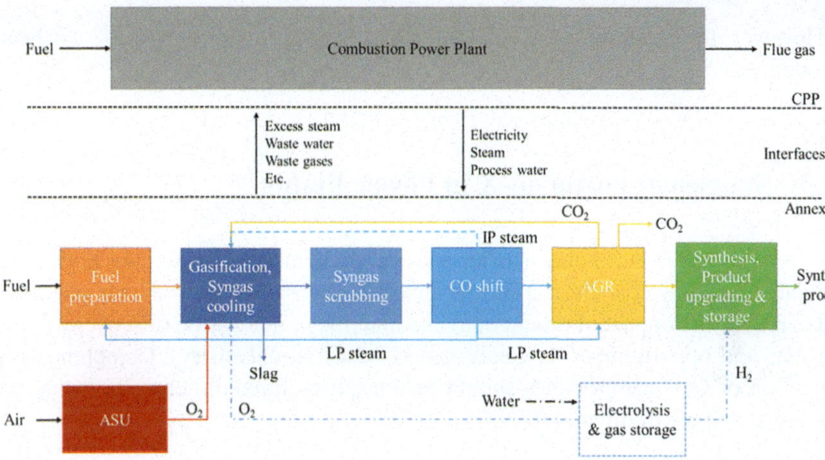

Fig. 12.10 Process flow sheet; AGR, acid gas removal; ASU, air separation unit

overall power plant economy. Polygeneration plants which combine gasification and synthesis can utilize excess electricity in supplement units like water electrolysis. Clausen et al. proposed a power-to-X plant which utilizes biomass gasification, water electrolysis, and autothermal reforming of natural gas or biogas for methanol synthesis [33]. The other plant is presented by Herdem et al. which combined gasification and alkaline water electrolysis in order to generate hydrogen with low CO_2 emissions [38].

Besides, there is an investigation which has the support and fund of the European Union's Horizon 2020. The project has 9 participants with different specializations and an 11 million Euro budget. The purpose of the plant is the methanol synthesis with low carbon dioxide/carbon monoxide ratios as the feedstock. The unit must be flexible in both operation and feedstock and can utilize high CO_2 streams as feed which in this project is provided from a fossil fuel power plant [39].

This technology is also deliberated for use in existing biomass combustion and gasification units, which operate to generate electrical/thermal energy, in comparison with chemical synthesis. The hydrogen is the other reactant of the hydrocarbon synthesis generated through water hydrolysis utilizing excess energy which will be difficult to return to the network grid. The main advantages of the projects are listed below [39]:

1. Alternative validation for CO_2 capture, which helps create a more attractive business case for CCS.
2. Flexibility in hydrogen production through water electrolysis comes up with efficient utilization of excess renewable power generation and supplying valuable auxiliary services.
3. The use of renewable methanol in fuels can play a role in achieving the European Commission's ambitious goals of using renewable energy in 2020 and 2030 and improve energy security and governance by reducing fossil fuel imports.

However, the accurate technical data and details of this process have not been reported yet.

12.3 Suggested Power-to-X in Power Plants

CO_2 as a raw material for chemical production offers many opportunities both commercially (as a low-cost alternative raw material, to produce new valuable materials and to reduce the impact of production) and in terms of innovation (discovering new concepts and opportunities for catalysts and industrial chemistry). Comprehensive evaluation of CO_2 technologies should be done from three dimensions: technical, economic, and environment [40].

- **Technical Evaluation.**

Technology maturity, geographical constraints, and the degree of independence from fossil fuels are important criteria for the technical evaluation of CO_2 technologies. In the technical review, two important questions must be answered: (1) Will the technology be operational within the next 10 years? (2) What is the level of readiness of the technology and has it been operational at least on a pilot scale?

To assess the level of readiness (TRL) or maturity of a technology, the whole system must be considered comprehensively. For example, in the methanol production process, the CO_2 uptake system, water electrolysis, and CO_2 catalytic conversion are involved. Therefore, to evaluate the TRL of CO_2 to methanol conversion technology, the TRL of all three components must be identified. Finally, the TRL of the whole system is equal to the lowest TRL detected.

Geographical constraints are another important factor in the technical study of a technology. This index is separate from the "technological maturity" index. TRL, for example, has a CO_2 methanol production technology of around 8–9, especially since a company in Iceland is producing renewable methanol. In this particular case, most of the conditions (cheap and low carbon energy (H_2 emissions), CO_2 source) were provided for the deployment of technology, which may not currently be the case in other parts of the world despite the maturity of technology.

Independence from fossil fuels is another important factor in the technical review of a CO_2 technology. This index refers to reactants involved in technology such as H_2, etc. (can be considered for ease of study of auxiliary and catalyst factors) and determines whether independence from fossil sources is possible. It is acceptable. Independence from fossil fuels can be direct or indirect. Indirect independence from fossil fuels relates to the cases where (i) the reactants necessary for the synthesis of a particular substance can be produced from a CO_2-based pathway, for example, dimethyl carbonate can be produced from methanol from the CO_2 reaction. Produced with H_2 (from renewable energy), it is obtained that (ii) H_2 is the only reactive agent (except CO_2) and can be produced from renewable energy [40].

– **Economic Evaluation.**

"Market size" is an important indicator of economic goals and provides information to the decision-maker on how large the potential market for targeted products is. This is one of the first questions for artisans to explore the potential for CO_2 use. Market size is often the driving force behind (economic/industrial), for example, larger markets create greater incentives for emerging technologies and make more profit when selling. "Market size" is an index that has been referred to several times in CCU sources as the "economic index" [40].

"Relative added value" is an indicator of the economic availability of a process. This index is obtained from the difference between the product price and the variable operating cost (materials, electricity, and heat required) and is expressed as a percentage by dividing the variable operating cost. Relative value added between 50 and 100% indicates the economic feasibility of a process and close to its profitability. Relative value added between 100 and 200% indicates the commercial and economic availability of the product.

"Competition with other technologies" is an indicator to answer the question of whether CO_2 technology is competitively priced with other alternative processes, products, or technologies and has similar output, low technologies or products. Carbon, such as hybrid and electric vehicles and other green construction products, are among these alternatives. This indicator complements "relative value added" because a process may be economically viable, but not necessarily competitive with other options. In addition, competitiveness is a conceptual economic value, while "relative value added" is an economic indicator that shows the value of CO_2 output [40].

Both economic sustainability and competitiveness are required for the long-term sustainability of CO_2-using technologies. Depending on how the "Competition with Other Technologies" index is evaluated, it gives more indications to the decision-maker in the market through the use of the SWOT matrix [40].

– **Environmental, Health, and Safety Evaluation.**

"CO_2 uptake potential" is an indicator to determine the maximum use potential of CO_2 (based on stoichiometry) for a particular product. This index is measured by the amount of CO_2 required to produce one unit of product, multiplied by the production volume. "CO_2 uptake potential" is an indicator that has been mentioned several times in CCU sources under the heading "Environmental Indicator." It is worth noting that this is not a CO_2 reduction potential because CO_2 reduction depends on the whole process and not just how many moles of CO_2 can be modified by the reaction [40].

The "Environmental Potential" index adds more information to the previous index because it takes into account global warming and the effects of fossil fuel depletion. Since the goal of the CCU is to reduce CO_2 and the independence of fossil resources, this indicator must be considered in CCU environmental evaluation. The reduction in fossil fuels also indicates that a CO_2-based alternative consumes more or less fossil fuels than the conventional method. The reduction in the use of fossil fuel equivalents is calculated as equivalent to oil [40].

There are certain risks associated with CCU processes. The main risk is the incidental distribution of volatile or hazardous components during operations or conversion reactions. In addition, safety in the workplace is becoming increasingly important for employees. Safety and health are intertwined with emerging technologies. Thus, in addition to the environmental contribution of the processes, the "safety and health" issues that may arise with the deployment of such technologies are explored, even if they are largely described in the proposed framework [40].

In this report, chemical, mineralization, and biological conversion technologies of CO_2 into several current products have been evaluated based on the mentioned index and available information and documents.

12.3.1 Urea

Nowadays global food support completely depends on synthetic fertilizers. They required to enhance soil restoration and increase the productivity in modern agriculture [41]. The role of nitrogen synthetic fertilizers is more crucial as nitrogen is commonly the restrictive nutrition which prohibits developing highly concentrated farms [42]. According to the forecasted population expansion, food scarcity is inevitable without such synthetic fertilizers [43, 44]. Due to these concerns, incremental consumption of synthetic fertilizers within agriculture is inescapable in the future.

– **Method of Production.**

In addition to the main challenges noted above, fertilizer production from non-fossil feedstock is another concern. The solution is the integration of sustainable inputs and utilization of renewable energy. For this purpose and in order to remove related CO_2 emissions, the electrolysis process which utilized renewable energy resources could be applied, instead of H_2 generation by reforming process (Eq. 12.1). Then, this H_2 reacted with externally sourced N_2 (Eq. 12.2) and CO_2 (Eqs. 12.3 and 12.4) to produced urea. As a result the urea production reduces CO_2 or carbon footprint and can be called carbon-neutral.

$$H_2O \overset{\Delta H_r=+286kJmol^{-1}}{\longleftrightarrow} H_2 + \frac{1}{2}O_2 \tag{12.1}$$

$$3H_2 + N_2 \overset{\Delta H_r=-45.7kJmol^{-1}}{\longleftrightarrow} 2NH_3 \tag{12.2}$$

$$2NH_3 + CO_2 \overset{\Delta H_r=-117kJmol^{-1}}{\longleftrightarrow} NH_2COONH_4 \tag{12.3}$$

$$NH_2COONH_4 \overset{\Delta H_r=+16kJmol^{-1}}{\longleftrightarrow} CO(NH_2)_2 + H_2O \tag{12.4}$$

But the formal reaction conditions of urea generation are challenging (170–220 °C, 150 bar) [45]. Barzagli et al. proposed a new route with relatively mild condition which not only utilized renewable resources but also reduced operating, investment, and environmental cost which they called it blue urea process (referred to electrolytic origins of the H_2) [46].

– **Environmental Evaluation.**

According to the investigation, the blue urea production process as utilized renewable energy and carbon capture systems has the capability to reduce CO_2 emissions around 21% (or 17% when equipped with direct air capture) when compared to the formal process. So, the blue urea process not only produces synthetic fertilizers but also reduces GHG emissions [47].

12.3.1.1 Case Study of Petrochemical Plant for Urea Production

Assaluyeh region, with its vast natural gas resources, has become one of the hubs of the country's oil, gas, and petrochemical industry.

The existence of numerous gas processing units and petrochemical production units has led to the growth of greenhouse gas emissions in this region. Currently, in this region, due to the existence of the mentioned industries, about 14 million tons of CO_2 are emitted to the atmosphere annually.

In some of these industries, such as ethane extraction units, carbon dioxide with a very high purity of more than 92% is produced and then released into the atmosphere, while the carbon dioxide produced can be fed to some downstream production units such as urea units. Production of urea chemical fertilizer or injection into oil wells can be used to increase the efficiency of extraction from the reservoir, which not only has economical profits but also will prevent the entry of CO_2 into the atmosphere and its destructive environmental effects. Emissions of greenhouse gases cause several problems, the most important consequence of which is global warming and climate change. Studies show that if the current trend of greenhouse gas emissions continues, world temperatures in 2100 will increase by about 8.5 °C.

In this regard, with the help of the National Petrochemical Company, which has made efforts to reduce carbon dioxide in other petrochemical units, the plan to produce urea chemical fertilizer from carbon dioxide sources in Assaluyeh region while reducing greenhouse gases in the region also creates added value. In petrochemical plant, the direct production of urea is done by absorbing the excess ammonia of the existing units in the country.

- **Environmental Evaluation.**

The results show that with the use of the surplus ammonia, the need of ammonia production units in the country, and carbon dioxide emissions from other units, urea production units can be designed and built, and thus the release of CO_2 into the atmosphere can be significantly reduced.

With the construction of a urea unit with a capacity of 2330 tons per day, 570,000 tons of CO_2 annually are absorbed from other sources of production such as ethane extraction units and converted to urea. This capacity can be increased to a daily production of 3250 tons of urea and the removal of 830,000 tons of carbon dioxide by applying design changes [48].

12.3.2 Dimethyl Ether

Dimethyl ether (DME: CH_3OCH_3) is the simplest aliphatic ether. It is a colorless, odorless, and highly flammable gas under environmental conditions. DME can be used as a propellant in aerosols; a precursor in the synthesis of other organic compounds such as dimethyl sulfate, methyl acetate, and light olefins; and clean fuel for combustion engines and turbines. It used a gas source of hydrogen for use in fuel cells as well as a hydrogen carrier.

Due to the similarity of the physical properties of DME with LPG, this material can be a good alternative to LPG fuel. On the other hand, DME can be used as diesel fuel due to its cetane number and high oxygen content (35% by weight). It is worth noting that this material, due to the lack of C-C bond in its structure, emits very little soot in the exhaust gas of the diesel engine. As a result, DME can be used as fuel in various sectors of the heat and power generation, mobility, and industry. Also, due to the nontoxicity of dimethyl ether, this substance can be a good alternative to methanol. DME can be converted to light and aromatic olefins in the presence of zeolite catalysts. The comparison of DME and other fuels is illustrated in Table 12.2.

Table 12.2 Comparison of DME and other fuels [1]

Properties	DME	Propane	Methane	Methanol	Diesel
Chemical formula	CH_3OCH_3	C_3H_8	CH_4	CH_3OH	C_{10}-C_{18}
Boiling point (°C)	−25.1	−42.0	−161.5	64.5	170–180
Liquid Vis. (kg/ms @25 °C)	0.12–0.15	0.2	–	–	2–4
Sp. gravity	1.59	1.52	0.55	–	–
Vapor press. (atm. @25 °C)	6.1	9.3	1@-161.5 °C	1@64.7 °C	0.035@21 °C
Explosion limit	3.4–17	2.1–9.4	5–15	–	0.6–6.5
Auto ignition temperature (°C)	350	504	632	470	–
Cetane number	55–60	5	0	5	40–50
Net calorific value (kcal/kg)	6900	11,100	12,000	5024	10,000
Net calorific value (kcal/m^3)	14,200	21,600	6600	–	–
Energy density* (MJ/lit)	18.97	23.5	7.88	15.69	35.08

Table 12.3 DME fuel-tested turbines in General Electric [49]

Gas turbine	Combustion technology	Nox emission, ppmvd @15% o2
6B, 6FA, 7EA, 7FA, 9E, 9FA	Diffusion	37
6B, 7EA, 9E	Dry low nox	25

General Electric has tested DME fuel in its products as illustrated in Table 12.3 [49]. According to the results, the company offers its E and F series to generate electricity from DME. For example, the efficiency test and comparison with natural gas performed by the BP Company in a combined cycle power plant with DME fuel (8% methanol and 3% water) is given in Table 12.4. The combined cycle tested consisted of three General Electric Frame 9A gas turbines and a heating turbine with a total capacity of 1000 MW [49].

– **Method of Production.**

There are two methods for catalytic synthesis of DME in the gas phase: direct and indirect. In the direct process, the synthesis gas (mixture of CO and H_2) is converted to DME by dual-acting catalysts. In the indirect process, first the synthesis gas is converted to methanol by $CuO/ZnO/Al_2O_3$ catalysts, and then in the next step, DME is obtained from the methanol dehydration reaction by solid-acid catalysts. An illustration of these two methods is shown in Fig. 12.11.

The indirect method, which is the common method of producing DME, takes place in two catalytic steps. In the first reactor, the synthesized gas is converted to methanol, and in the second reactor, methanol is converted to dimethyl ether by dehydration. Nowadays, most companies like Haldor Topsoe, Lurgi, Mitsubishi Gas

Table 12.4 BP test results in a 1000 MW combined cycle power plant [49]. (performed by Flour Daniel (USA) and funded by BP using GE 9FA gas turbine)

Fuel	Natural gas	Fuel grade DME
Fuel feed to the plant gate	Gas at 60 °F and 500 psia	Liquid at 60 °F and 77 psia
Fuel flow, per gas turbine, lb./hr	103,401	190,010
Air flow per gas turbine, lb./hr	4806,227	4,806,227
Total gas turbine power output, MW	683.7	708.7
Total steam turbine power, MW	384.0	384.5
Total aux. Power consumption, MW	23.9	24.6
Net plant power output, MW	1043.8	1068.6
Net heat rate, Btu/kWh (LHV)	6278	6106
Power generation efficiency, LHV %	54.4	55.9
Flue gas stack temperature	205	189
CO_2 generated, lb./MWh	819.5	959.9
LHV at 77 °F, kcal/kg	Natural gas, 11,800	Liquid DME, 6420

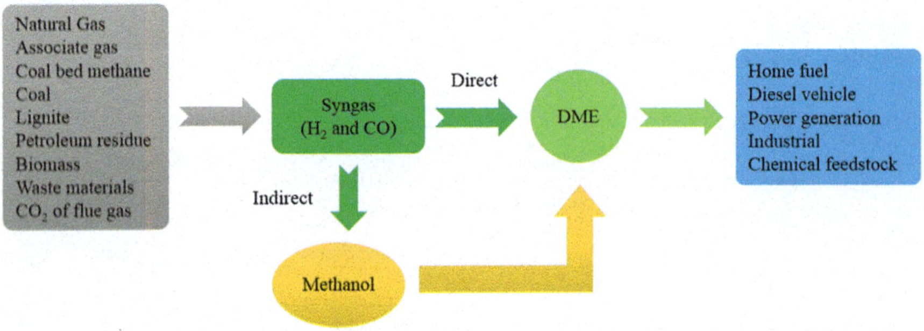

Fig. 12.11 Methods for synthesis of dimethyl ether

Chemical Company, China Southwestern Research Institute of Chemical Industry, Toyo Engineering Corporation, and China Energy applied indirect processes.

The reaction is commonly performed with catalysts based on alumina in a fixed-bed reactor. The temperature of this reactor is between 250 and 400 degree Celsius, and the pressure can change between 10 and 25 bar. Under these conditions, the conversion efficiency of methanol to DME is between 70 and 85% [50]. The selectivity is also usually high and only a small amount of formaldehyde is produced. In the indirect method, the cost of production is closely related to the price of methanol. Two-acting catalysts for single-stage synthesis are generally a combination of catalysts for methanol production reactions from synthetic gas ($CuO/ZnO/Al_2O_3$) and synthesis of DME from methanol dewatering (solid-acid catalysts) [51].

$$2CH_3OH \rightleftarrows CH_3OCH_3 + H_2O \qquad \Delta H_{298\ K} = -23.5kJ/mol \qquad (12.5)$$

In the direct method, which is a new method, methanol dehydration is performed simultaneously with the hydrogenation of CO and CO_2 in a reactor [51]. The direct technology has been developed by companies like Haldor Topsoe, Japan JFE Holdings Company, and Korea Gas Corporation. In the direct conversion, two reactions perform simultaneously in the presence of a dual-function catalyst (hybrid catalyst), first the methanol synthesis and second the methanol dehydration. The hybrid catalysts were initially used to convert synthetic gases directly to DME [50].

$$CO_2 + 3H_2 \rightleftarrows CH_3OH + H_2O \quad \Delta H_{298\ K} = -49.4kJ/mol \tag{12.6}$$

$$CO_2 + H_2 \rightleftarrows CO + H_2O \quad \Delta H_{298\ K} = -41kJ/mol \tag{12.7}$$

$$CO + 2H_2 \rightleftarrows CH_3OH \quad \Delta H_{298\ K} = -90.4kJ/mol \tag{12.8}$$

$$2CH_3OH \rightleftarrows CH_3OCH_3 + H_2O \quad \Delta H_{298\ K} = -23kJ/mol \tag{12.9}$$

DME production technology from CO_2 hydrogenation is in development and is expected to be commercialized in the next 10 years [52]. The polygeneration plant for power and DME production is suggested by some other researchers like Catizzone et al. [53], Khojasteh Salkuyeh and Adams [18], Farooqui et al. [36], and Salman et al. [22].

– **Environmental Evaluation.**

For each ton of DME, 1.9 tons of CO_2 are consumed, and considering the market of 11.4 million tons of DME, the annual CO_2 consumption is 21.78 million tons [40]. Figure 12.12 shows the life cycle of DME production from methanol and its use as fuel. If renewable energy is used in the production of H_2 and the biological system to absorb CO_2, since DME is a clean fuel, the process is a comprehensive green process, and its emissions are very low. In the scheme below, the goal is to generate DME using flue gas according to the cycle. In this process, by separating CO_2 from the chimneys of power plants, synthetic gas is produced, and then it is used to produce DME.

– **Economic Studies.**

According to Marketsandmarkets, the DME market in 2014 was based on a volume of 3.7 million tons and grew by 15.67% by 2020. The raw materials used to produce DME are divided into natural gas, coal, methanol, and other sources such as biological materials [54]. According to Transparency Market Research in 2016, methanol accounted for 90% of total DME market demand. This is due to the ease of synthesis and the cheap method of converting methanol to DME. After that, the

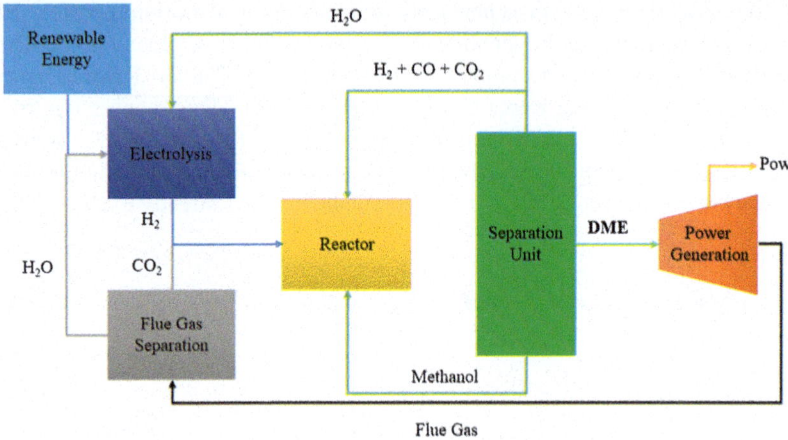

Fig. 12.12 General scheme of the DME production process from the flue gas

Table 12.5 Power generation costs: comparison of gasoil, DME, and mazut

Feed	Power generation potential	FOB cost (March 2019)	Power generation cost ($/kWh)
Gasoil	0.017 lit/kWh	$/lit 0.49	0.008
DME	0.026 lit/kWh	$/lit 0.43	0.011
Mazut	0.013 lit/kWh	$/lit 0.34	0.004

Table 12.6 Power generation cost: comparison of direct and indirect synthesis

Feed	Production cost ($/ton)	Power generation cost ($/kWh)
Flue gas (indirect)	99	0.0017
Flue gas (direct)	150	0.0026

largest share related to coal, which has decreased its use due to pollution. Production of DME with natural gas is also costly [55].

Based on the announced prices in the Persian Gulf for oil and petrochemical products, the cost per kilowatt hour of electricity produced is as shown in Table 12.5.

The cost of domestic production of DME from flue gas (based on the product price of FOB March 2019) is given in Table 12.6. The production cost is taken from the Khalili-Garakani et al. [1].

As can be seen, if the DME is produced domestically, due to the reduction in the cost of fuel produced, the resulting electricity cost is reduced and can compete with other liquid fuels. Also, the cost of saved diesel can be provided to the investor in line with part of the investment costs, which can increase the implementation of this plan for the private sector.

12.4 Status Quo, Challenges, and Outlook

In this work presented here, different CO_2 mitigation techniques have been explored. The most important categories are energy and carbon intensity reduction. Another discussed method is carbon sequestration, which must be developed in order to have more flexibility in options for suppressing the GHG emissions. The last discussed option is captured and conversion of CO_2 to the value-added materials. In this regard, significant advances have been achieved, from both academic and industrial research, in order to reduce as much as possible CO_2 emissions and conversion to the valuable materials. By using polygeneration approaches, the multiple feeds can be converted to multiple products such as power, fuels, chemical products, etc. In this review some of the applicable polygeneration systems for production of DME, methanol, synthesis gas, etc. from CO_2 emissions feed have been studied. Furthermore, some polygeneration systems based on power-to-X have been separately reviewed regarding its importance.

Finally, at the last part, some polygeneration case studies on methanol, urea, and DME production with use of renewable-based electricity supply have been evaluated in aspects of technical, emissions, and economy.

Of course there are still some barriers to the commercialization of these systems, which are discussed in the following:

- The most important challenge is the high activation energy of the CO_2 molecule which is required to prepare it to react with protons or other molecules. As the CO_2 molecule is relatively stable.
- Utilization of renewable energy is another concern as the application of fossil energy would eliminate the benefit of CO_2 mitigation. But, according to intermittent behavior of renewable energy resources, if put the hydroelectric energy apart, CO_2 mitigation process based on these resources is expensive. Polygeneration process in power plant can solve this problem.
- A high-concentration CO_2 source is required for the CO_2 mitigation processes to be economical.
- Regarding the conditions, the distributed or centralized production should be optimized.
- New and consistent policies are required not only for helping these green technologies to develop and grow but also to become established.

References

1. Khalili-Garakani A, Ghaseminezhad M., Davoudi, S. Applied processes to achieve optimal methods of manufacturing value added products from power plant flue gas. 2021.
2. Nazari, S., Shahhoseini, O., Davari, S., & Paydar, R. (2010). Experimental determination and analysis of CO2, SO2 and NOx emission factors in Iran' s thermal power plants. *Energy, 35*, 2992–2998. https://doi.org/10.1016/j.energy.2010.03.035

3. Yamasaki, A. (2003). An overview of CO2 mitigation options for global warming-emphasizing CO2 sequestration options. *Journal of Chemical Engineering of Japan, 36*, 361–375.
4. Steinberg, M. (1978). Synthetic carbonaceous fuels and feedstocks from oxides of carbon and nuclear power. *Fuel, 57*, 460–468. https://doi.org/10.1016/0016-2361(78)90154-0
5. Olah, G. A., Prakash, G. K. S., & Goeppert, A. (2011). Anthropogenic chemical carbon cycle for a sustainable future. *Journal of the American Chemical Society, 133*, 12881–12898. https://doi.org/10.1021/ja202642y
6. Olah, G. A., Goeppert, A., & Prakash, G. K. S. (2009). *Beyond oil and gas: The methanol economy* (2nd ed.). Wiley. https://doi.org/10.1002/9783527627806
7. Aresta, M., Dibenedetto, A., & Angelini, A. (2014). Catalysis for the valorization of exhaust carbon: From CO2 to chemicals, materials, and fuels. Technological use of CO2. *Chemical Reviews, 114*, 1709–1742. https://doi.org/10.1021/cr4002758
8. Thomas, J. M., & Harris, K. D. M. (2016). Some of tomorrow's catalysts for processing renewable and non-renewable feedstocks, diminishing anthropogenic carbon dioxide and increasing the production of energy. *Energy & Environmental Science, 9*, 687–708. https://doi.org/10.1039/C5EE03461B
9. Lefebvre, J., Friedemann, M., Manuel, G., Graf, F., Bajohr, S., Reimert, R., et al. (2016). Renewable power-to-gas : A technological and economic review. *Renewable Energy, 85*, 1371–1390. https://doi.org/10.1016/j.renene.2015.07.066
10. Aresta, M., Dibenedetto, A., & Quaranta, E. (2016). State of the art and perspectives in catalytic processes for CO2 conversion into chemicals and fuels: The distinctive contribution of chemical catalysis and biotechnology. *Journal of Catalysis, 343*, 2–45. https://doi.org/10.1016/j.jcat.2016.04.003
11. Abanadesa, J. C., Rubinb, E. S., Mazzottic, M., & Herzog, H. J. (2017). On the climate change mitigation potential of CO2 conversion to fuels. *Energy & Environmental Science, 10*, 2491–2499. https://doi.org/10.1039/C7EE02819A
12. Agarwal, A. S., Rode, E., Sridhar, N., & Hill, D. (2015). Conversion of CO2 to value-added chemicals: Opportunities and challenges. *Handbook of Climate Change Mitigation and. Adaption*, 1–40.
13. Jana, K., Ray, A., Mansouri, M., Assadi, M., & De, S. (2017). Polygeneration as a future sustainable energy solution – A comprehensive review. *Applied Energy, 202*, 88–111. https://doi.org/10.1016/j.apenergy.2017.05.129
14. Baliban, R. C., Elia, J. A., Weekman, V., & Floudas, C. A. (2012). Process synthesis of hybrid coal, biomass, and natural gas to liquids via Fischer – Tropsch synthesis, ZSM-5 catalytic conversion, methanol synthesis, methanol-to-gasoline, and methanol-to-olefins / distillate technologies. *Computers and Chemical Engineering, 47*, 29–56. https://doi.org/10.1016/j.compchemeng.2012.06.032
15. Adams, T. A., & Barton, P. I. (2011). Combining coal gasification and natural gas reforming for efficient polygeneration. *Fuel Processing Technology, 92*, 639–655. https://doi.org/10.1016/j.fuproc.2010.11.023
16. Chen, Y., Adams, T. A., & Barton, P. I. (2011). Optimal design and operation of static energy Polygeneration systems. *Industrial and Engineering Chemistry Research, 50*, 5099–5113. https://doi.org/10.1021/ie101568v
17. Chen, Y., Adams, T. A., & Barton, P. I. (2011). Optimal design and operation of flexible energy polygeneration systems. *Industrial and Engineering Chemistry Research, 50*, 4553–4566. https://doi.org/10.1021/ie1021267
18. Khojasteh Salkuyeh, Y., & Adams, T. A. (2014). A new power, methanol, and DME polygeneration process using integrated chemical looping systems. *Energy Conversion and Management, 88*, 411–425. https://doi.org/10.1016/j.enconman.2014.08.039
19. Naqvi, M., Dahlquist, E., & Yan, J. (2017). Complementing existing CHP plants using biomass for production of hydrogen and burning the residual gas in a CHP boiler. *Biofuels, 8*, 675–683. https://doi.org/10.1080/17597269.2016.1153362

20. Galanti, L., Franzoni, A., Traverso, A., & Massardo, A. F. (2011). Existing large steam power plant upgraded for hydrogen production. *Applied Energy, 88*, 1510–1518. https://doi.org/10.1016/j.apenergy.2010.11.033
21. Gustavsson, C., & Hulteberg, C. (2016). Co-production of gasification based biofuels in existing combined heat and power plants – Analysis of production capacity and integration potential. *Energy, 111*, 830–840. https://doi.org/10.1016/j.energy.2016.06.027
22. Salman, C. A., Naqvi, M., Thorin, E., & Yan, J. (2017). A polygeneration process for heat, power and DME production by integrating gasification with CHP plant: Modelling and simulation study. *Energy Procedia, 142*, 1749–1758. https://doi.org/10.1016/j.egypro.2017.12.559
23. Li, S., Gao, L., & Jin, H. (2017). Realizing low life cycle energy use and GHG emissions in coal based polygeneration with CO2 capture. *Applied Energy, 194*, 161–171. https://doi.org/10.1016/j.apenergy.2017.03.021
24. Jana, K., & De, S. (2017). Environmental impact of biomass based polygeneration – A case study through life cycle assessment. *Bioresource Technology, 227*, 256–265. https://doi.org/10.1016/j.biortech.2016.12.067
25. Huang, H., & Yang, S. (2017). Design concept for coal-based polygeneration processes of chemicals and power with the lowest energy consumption for CO2 capture. *Computer Aided Chemical Engineering, 44*, 1381–1386. https://doi.org/10.1016/B978-0-444-64241-7.50225-1
26. Farooqui, A., Bose, A., Ferrero, D., Llorca, J., & Santarelli, M. (2018). Techno-economic and exergetic assessment of an oxy-fuel power plant fueled by syngas produced by chemical looping CO2 and H2O dissociation. *Journal of CO2 Utilization, 27*, 500–517. https://doi.org/10.1016/j.jcou.2018.09.001
27. Stanger, R., Wall, T., Spörl, R., Paneru, M., Grathwohl, S., Weidmann, M., et al. (2015). International journal of greenhouse gas control Oxyfuel combustion for CO2 capture in power plants. *International Journal of Greenhouse Gas Control, 40*, 55–125. https://doi.org/10.1016/j.ijggc.2015.06.010
28. Nguyen, V. N., & Blum, L. (2015). Syngas and synfuels from H2O and CO2: Current status. *Chemie Ingenieur Technik, 87*, 354–375. https://doi.org/10.1002/cite.201400090
29. Krenzke, P. T., Fosheim, J. R., & Davidson, J. H. (2017). Solar fuels via chemical-looping reforming. *Solar Energy, 156*, 48–72. https://doi.org/10.1016/j.solener.2017.05.095
30. Yadav, D., & Banerjee, R. (2016). A review of solar thermochemical processes. *Renewable and Sustainable Energy Reviews, 54*, 497–532. https://doi.org/10.1016/j.rser.2015.10.026
31. Han, S., & Sik, C. (2014). Applicability of dimethyl ether (DME) in a compression ignition engine as an alternative fuel. *Energy Conversion and Management, 86*, 848–863. https://doi.org/10.1016/j.enconman.2014.06.051
32. Bakhtyari, A., & Rahimpour, M. R. (2018). Methanol to dimethyl ether. In D. ABF (Ed.), *Methanol* (1st ed.). Elsevier.
33. Gądek, M., Kubica, R., & Jędrysik, E. (2013). Production of methanol and dimethyl ether from biomass derived syngas – A comparison of the different synthesis pathways by means of flowsheet simulation. *Computer Aided Chemical Engineering, 32*, 55–60. https://doi.org/10.1016/B978-0-444-63234-0.50010-5
34. Ren, F., Wang, J., & Li, H. (2006). Direct mass production technique of dimethyl ether from synthesis gas in a circulating slurry bed reactor. *Studies in Surface Science and Catalysis, 159*, 489–492. https://doi.org/10.1016/S0167-2991(06)81640-X
35. Hankin, A., & Shah, N. (2017). Process exploration and assessment for the production of methanol and dimethyl ether from carbon dioxide and water. *Sustain Energy Fuels, 1*, 1541–1556. https://doi.org/10.1039/C7SE00206H
36. Farooqui, A., Di, F., Bose, A., Ferrero, D., Llorca, J., & Santarelli, M. (2019). Techno-economic and exergy analysis of polygeneration plant for power and DME production with the integration of chemical looping CO2/H2O splitting. *Energy Conversion and Management, 186*, 200–219. https://doi.org/10.1016/j.enconman.2019.02.043
37. Wolfersdorf, C., Forman, C., Keller, F., & Gootz, M. (2017). CO2-to-X and coal-to-X concepts in pulverized coal combustion power plants. *Energy Procedia, 114*, 7171–7185. https://doi.org/10.1016/j.egypro.2017.03.1821

38. SacitHerdem, M., Farhad, S., Dincer, I., & Hamdullahpur, F. (2013). Thermodynamic modeling and assessment of a combined coal gasification and alkaline water electrolysis system for hydrogen production. *International Journal of Hydrogen Energy, 39*, 3061–3071. https://doi.org/10.1016/j.ijhydene.2013.12.068
39. Samiee, L., & Gandzha, S. (2019). Power to methanol technologies via CO2 recovery: CO2 hydrogenation and electrocatalytic routes. *Reviews in Chemical Engineering*. https://doi.org/10.1515/revce-2019-0012
40. Chauvy, R., Meunier, N., Thomas, D., & De Weireld, G. (2019). Selecting emerging CO2 utilization products for short- to mid-term deployment. *Applied Energy, 236*, 662–680. https://doi.org/10.1016/j.apenergy.2018.11.096
41. Yara Fertilizer Industry Handbook (2017). Oslo - Norway.
42. Dawson, C. J., & Hilton, J. (2011). Fertiliser availability in a resource-limited world: Production and recycling of nitrogen and phosphorus q. *Food Policy, 36*, S14–S22. https://doi.org/10.1016/j.foodpol.2010.11.012
43. World Bank. Population Estimates and Predictions n.d.. https://datacatalog.worldbank.org/dataset/population-estimates-andprojections. Accessed 27 Feb 2019.
44. Stewart, W. M., & Roberts, T. L. (2012). Food security and the role of fertilizer in supporting it. *Procedia Eng, 46*, 76–82. https://doi.org/10.1016/j.proeng.2012.09.448
45. Meessen, J. H., Petersen, H. (2010). Urea. Uhlmann's Encycl. Ind. Chem., Wiley-VCH Verlag CmbH & Co. KGaA; https://doi.org/10.1002/14356007.a27_333.
46. Barzagli, F., Mani, F., & Peruzzini, M. (2011). From greenhouse gas to feedstock: Formation of ammonium carbamate from CO2 and NH3 in organic solvents and its catalytic conversion into urea under mild conditions. *Green Chemistry, 13*, 1267–1274. https://doi.org/10.1039/c0gc00674b
47. Driver, J. G., Owen, R. E., Makanyire, T., Lake, J. A., Mcgregor, J., & Styring, P. (2019). Blue urea: Fertilizer with reduced environmental impact. *Front Energy Research, 7*, 1–15. https://doi.org/10.3389/fenrg.2019.00088
48. Technical and economic feasibility studies of Assaluyeh carbon dioxide collection project and construction of urea production unit. 2019.
49. Basu, A., Gradassi, M., Sills, R., Technology, U., & Exploration, B. P. (2001). *Use of DME as a gas turbine fuel* (pp. 1–6). ASME Turbo-Expo, New Orleans.
50. Matzen, M., & Demirel, Y. (2016). Methanol and dimethyl ether from renewable hydrogen and carbon dioxide: Alternative fuels production and life-cycle assessment. *Journal of Cleaner Production, 139*, 1068–1077. https://doi.org/10.1016/j.jclepro.2016.08.163
51. Mevawala, C., Jiang, Y., & Bhattacharyya, D. (2019). Techno-economic optimization of shale gas to dimethyl ether production processes via direct and indirect synthesis routes. *Applied Energy, 238*, 119–134. https://doi.org/10.1016/j.apenergy.2019.01.044
52. Álvarez, A., Bansode, A., Urakawa, A., Bavykina, A. V., Wezendonk, T. A., & Makkee, J. G. (2017). Challenges in the greener production of Formates/formic acid, methanol, and DME by heterogeneously catalyzed CO2 hydrogenation processes. *Chemical Reviews, 117*, 9804–9838.
53. Catizzone, E., Bonura, G., Migliori, M., Frusteri, F., & Giordano, G. (2018). CO2 recycling to dimethyl ether : State-of-the-art and perspectives. *Molecules, 23*, 1–31. https://doi.org/10.3390/molecules23010031
54. Marketsandmarkets. Dimethyl Ether Market by Raw Materials (Coal, Methanol, Natural gas and Bio-based feedstock), by Applications (Aerosol Propellant, LPG Blending, Transportation fuel and Others), and by Region - Trends & forecasts to 2020 n.d.. https://www.marketsandmarkets.com/Market-Reports/dimethyl-ether-market-1120.html.

55. Transparency Market Research. Dimethyl Ether Market (Raw Material - Coal, Methanol, and Natural Gas; Application - Aerosol Propellants, LPG Blending, Transportation Fuel, Power Plant Fuel, and Chemical Feedstock) - Global Industry Analysis, Size, Share, Growth, Trends, and Forecast 20 n.d.. https://www.transparencymarketresearch.com/dimethyl-ether-market.html.

Chapter 13
The Role of Distributed Multi-vector Energy Assets in Economic Decarbonisation: Early Findings of a UK Demonstrator

Mohammad Royapoor, Kunpeng Wang, Robin Wardle, and Vahid Vahidinasab

13.1 Introduction

Several market and regulatory forces have brought about the prospect of a new age of small-scale, distributed low-carbon assets across wide geographical areas working uniformly to provide multiple energy services. Given their geographically distributed nature, the aggregation and optimum management of these distributed assets are increasingly examined with virtual power plant (VPP) describing the overarching nature of such heterogeneous but controllable energy system assets. Several features are forming an axiomatic foundation for these multi-site technologies to be united under a VPP umbrella, first the flow of power and information between stakeholders and energy entities [1], second the existence and exploitation of real (i.e. batteries) or virtual (i.e. building inertia and deferable loads) energy storage, third technologies that can add virtual inertia to the system and/or couple multiple sectors and finally a single or a cascade of cloud-based VPP platforms/controllers that can extract operational features from such active yet distributed energy assets and guide the overall system towards its objectives (minimising cost and carbon, maximising comfort, enhancing demand response (DR) capabilities and/or peer-to-peer trading, etc.) [2]. In the following five sections and against a backdrop of historical practices, techno-economic considerations for the design of VPP are outlined before greater focus are brought on the specific attributes (and initial learnings) of the case study Smart Local Energy System examined in this chapter.

M. Royapoor (✉) · K. Wang · R. Wardle
School of Engineering, Newcastle University, Newcastle upon Tyne, UK
e-mail: mohammad.royapoor@ncl.ac.uk

V. Vahidinasab
Department of Engineering, School of Science and Technology, Nottingham Trent University, Nottingham, UK

© The Author(s), under exclusive license to Springer Nature Switzerland AG 2022

V. Vahidinasab, B. Mohammadi-Ivatloo (eds.), *Whole Energy Systems*, Power Systems, https://doi.org/10.1007/978-3-030-87653-1_13

13.1.1 Status Quo, Challenges and Outlooks

The age of petroleum experienced its zenith in the twentieth century, where a boom in industrialisation and spread of internal combustion engines created vast demands for fossil fuels. Energy systems dominated government infrastructure programmes that were planned centrally and, before the second and third generation of nuclear power stations, were mostly reliant on oil and its derivatives [3]. The environmental and safety legacy of the twentieth century energy systems is one of the prevailing challenges of the twenty-first century, where emerging visions of the future are dominated by decentralised yet interlinked energy systems enabled by renewable generation and distributed storage that connect multiple sectors (transport, heat, power, etc.). This approach has the benefit of exploiting all user flexibilities to create virtual inertia, facilitate greater use of data to enable predictive controls and notably transfer prosperity from energy giants to local communities [4]. Such 'smart energy systems' (SES) will not make legacy utility infrastructures redundant but will utilise them to interlink sectors and regions (even transnationally) and facilitate a shift from single-sector thinking to coherent and integrated systems that achieves greater resilience by exploiting synergies between sub-sectors. Importantly SES are planned, designed and managed with host communities at heart. In addition to their economic and social aspects, full realisation of SES is also heavily reliant on integrated energy management that optimises the dispatch of a number of services while taking heed of carbon, cost and human comfort constraints. Figure 13.1

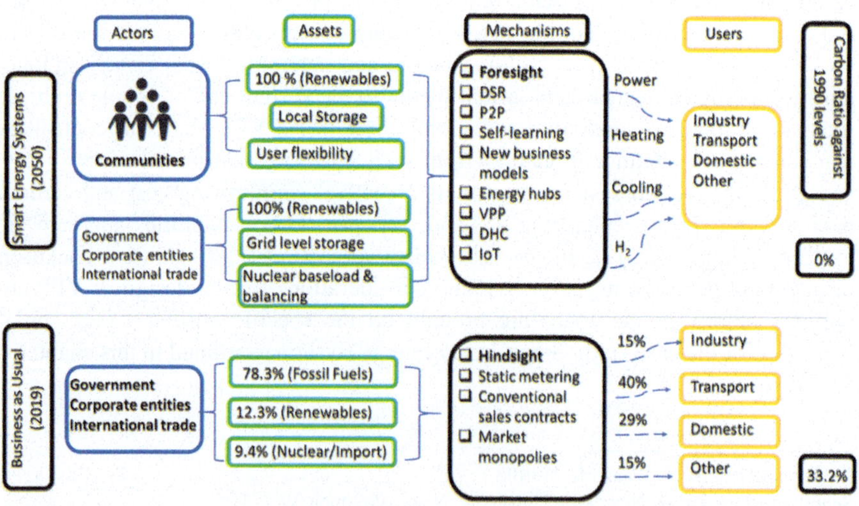

Fig. 13.1 UK energy system outline comparing a 2019 baseline against an envisaged 2050 SES (*VPP* virtual power plant, *DHC* district heating and cooling, *DSR* demand-side response, *IoT* Internet of Things)

utilises 2019 UK energy mix statistics [5] and system architecture to provide a simplified comparison between business as usual and a 2050 SES model.

It should be noted that Fig. 13.1 attempts to simplify an extremely complex landscape with multiple actors engaged over a cascade of interactive layers; however this illustration focuses on the emerging importance of interconnecting sectors and exploiting system-level flexibilities to create a new platform for business models, asset sharing and virtual inertia. The baseline (year 2019) involved total UK energy consumption of 197.6 million tonnes of oil equivalent that resulted in 351.5 million tonnes of equivalent CO_2 (which is reported in this figure as a 'carbon ratio' of 33.2% against a 1990 baseline).

13.1.2 Smart Local Energy Systems, Rationale and Potentials

While the definitions of the term SES are evolving, a sharp increase in the use of this term was pointed out by Lund et al. [4]. As noted, the fundamental feature of SES is a transition from single-sector thinking with centralised assets to a more integrated and holistic approach involving distributed assets. The crucial difference between the term 'smart energy systems' and other similar terms such as 'smart grid' is the holistic and cross-sectoral aspects where multiple services (tri- or quad-generation, i.e. cooling, heating, power, hydrogen) are provided to the hosting community. The transition to distributed local energy systems (involving generation, storage and flexibility of control) provides a greater opportunity for community engagement, and the centrality of the host community has led to the term integrated community energy systems (ICESs) gaining traction in literature [6]. 'Self-provision' and 'system-support' have also been viewed as axiomatic for an ICES which offers the possibility of giving the power of change back to the consumers. Note that system support refers to the ability of ICES to provide services to the wider energy network beyond its own boundary. While technological, socio-economic, environmental and institutional issues act as principal barriers to the wider adaption of ICES, rising electricity prices and falling costs of generation and storage are reported to make grid defection a widespread future reality in Australia and the USA, where rich solar resources exist [7, 8]. However, under current market mechanisms, the economics and system reliability of grid-defected ICES were found to be poor in the Netherlands as a result of much larger capital expenditure (CapEx) required for oversized DERs [9]. Being able to market surplus generation to neighbouring communities might improve economic prospects of an off-grid ICES, and the importance of existing grids in enabling efficient exchanges of surplus resources between ICESs has been highlighted by several authors [10]. Nationwide and international driving forces behind the ambition to achieve a 100% renewable energy system are examined by Young, J. et al. [11] who argue that in addition to favourable national and local green policies, government guarantees are also required to minimise investment risks and to steer energy systems devolution. These clearly point to an evolving technical landscape that will lead to the formation of highly individual SLES platforms as a function of local regulatory regimes, site-specific market mechanisms, availability of RES and forms of demand and level of engagement from the hosting community.

13.1.3 Consumer Attitude Towards Shared Assets and Monetising Flexibility

A study by Richter and Pollitt [12] discusses the findings from a 2015 survey of UK energy consumers about what services are most important to them and what terms of contracts they would be willing to accept. The authors break down the sample into four main clusters of consumers, sorted by their risk aversion and openness to having their data monitored. Across all four clusters, consumers highly valued technical support and were willing to pay for it. They also find that on average, consumers are willing to pay 34% of the gained energy bill savings from ICES on platform providers that enable the realisation of energy savings. High levels of fixed compensations do not usually entice consumers; rather it is the ongoing energy savings that were the most important factor in their decisions. Richter and Pollitt suggest that platform companies can add fees for technical support and data privacy protection to recoup CapEx and operational costs. The biggest difference in the customer clusters was found to be in the extent of control and monitoring by the energy service/platform companies that customers would welcome at a certain price point. A recent optimisation approach [13] suggested a deterministic/stochastic model for transforming passive consumers with storage and/or PV into active prosumers to participate in the day-ahead market. A two-step process was designed for the aggregator to use clusters of prosumers and predict flexibility capacity to then transform that flexibility into supply and demand bids. This centroid-based clustering approach was found to save on average an additional 20% in energy costs for prosumers and up to 40% in transaction and trading costs for the aggregator. The model proved centroid-based clustering can limit uncertainty in prediction while maintaining a high quality of energy bids. This predictive modelling can be almost entirely automated to reduce costs further. Gissey et al. [14] use prisoners' dilemma (a scenario used to illustrate a key tenant of game theory, where rational individuals will not cooperate, even though they would be better off in doing so) to support their findings that most domestic PV and storage owners will set up their usage patterns to their own advantage, regardless of system-wide priorities (and despite system-wide efficiency leading to greater overall savings for individuals too). While capital costs for storage are inversely related to size, 'spikier' residential loads offer a larger opportunity for storage to provide efficiencies. This also means that a greater number of players in the residential system are harder to control (as opposed to a few large commercial participants). Authors note that if an aggregator were to leave the control of energy assets entirely up to residential users, the volatility in the system would be on average 2–3% higher, and energy prices would be 4–7% higher than if the assets were centrally controlled. They call for regulators to alter pricing structures and retail tariffs to follow wholesale costs more closely, hence tightening the gap between system and individual priorities. They estimate the total cost to the UK energy system to be £407 m per year if the flexibility resources are not centrally controlled. Therefore, a wide range of regulatory and managerial space is left to be occupied by either a new generation of enterprises or supportive policy to enable niche technologies, business models and a completely new way of engaging the users. Learning and adaption, reiteration and improvement is also an integral part of the process for SLES as a novel and new approach that can be designed to deliver against a wide range of outcomes, many of which may not be compatible [15].

13.1.4 The Role of District Heating

Even if the Northern European building portfolio can be all built to Passivhaus levels, there is still a very strong case for district heating (DH) within smart energy systems. This is reflected in the UK where a recent publication outlines heat networks as a key part of UK plans to provide low cost and carbon heating [16]. DH's role in SLES is due to their ability to absorb surplus power generation and waste industrial heat to provide domestic hot water (and space heating when needed). Lund, H. et al. conducted an EnergyPro modelling examination of a 100% renewable energy system in 2060 to identify the best future heating scenarios [17]. Danish district heating (currently serving 46% of Danish net heat demand) and its expansion to neighbouring areas (to cover 53%, 63% and 70% of net heat demand) was found to be the most cost- and carbon-effective solution, followed by individual air-sourced heat pumps (ASHPs), while hydrogen-powered micro-CHPs were found to suffer from low overall efficiencies and high costs, and natural gas-powered micro-CHPs offer carbon effectiveness in the short run while becoming more expensive than DH in the long run. In order to better plan for consumer integration and expansion of DHs, creating and maintaining municipal heat Atlases has been advocated by Karl and Möller who demonstrated via EnergyPLAN software simulations that DH expansion in 2012 offered little improved efficiency of the overall energy system against a business as usual scenario but will be most cost and carbon efficient for a future scenario where increased power/heat/transport energy exchanges will exist and are complimented by end-user demand reduction [18]. The SLES case study examined in this chapter involved a 240 kW_{th} marine-source heat pump that would supply a base thermal demand via a fifth-generation ambient temperature DH system to a number of commercial sites. Given the seasonal variations of the heat source (seawater at the south coast of England), the flow temperature was investigated to be between 8 and 15 °C. This base thermal energy would then be boosted at site level via water-to-water heat pumps to deliver heating and domestic hot water (DHW) to reflect the requirements of the users at each site.

13.1.5 Hydrogen as a Vector Coupling Solution

Despite low round-trip efficiencies, hydrogen is unique in facilitating multi-vector energy interchanges and short- and long-term storage capabilities. The power sector can interact successfully with both heating and transport (via CHPs or EVs). But heating and transport (both major primary energy consumers) are less able to share capacity or load except with hydrogen deployment. Hydrogen is also capable of delivering a power, heat or transport service with zero emissions at the point of use and has been proposed as a means of creating additional resilience and inertia in future power systems to counteract the potential instabilities that renewables and their associated controls may cause [19]. Even at the single household level, hydrogen has been investigated as a plausible hybrid solution with solar power where the excess output of a 0.5 kWp PV array was used to feed a 0.1 kW proton exchange membrane (PEM) fuel cells to charge a storage tank that contributes to the continuous power demand at domestic level [20]. However, it is at much larger scales that hydrogen can begin to gain greater economic advantage as both short- and long-term storage mechanism but also a multi-vector element within a SLES. Despite the established nature of fuel cells and electrolysers, it is chiefly the economics of hydrogen and its overall efficiencies that constitute the main barriers to its wider adoption [21]. Additionally most of the hydrogen produced today is heavily reliant on fossil fuels, with a staggering 96% derived from technologies that reform fossil fuel feedstock [22]. This leaves only 4% of hydrogen production that is a product of electrolysis technologies which currently forms the only segment of H_2 production that can be considered low or zero carbon if wholly produced, processed and stored by renewable technologies.

Current generation efficiencies of hydrogen range from 1.6% (photosynthesis) [21] to reported values of between 43% and 53% (thermolysis in membrane reactors [23]) with the highest figures reported for steam methane reforming (generally around 85% with a theoretical limit of 93.8% also reported [24]). By the time energy expenditures in the liquification or pressurisation of hydrogen are accounted for and added to efficiency penalties in the final energy conversion mechanisms (i.e. a hydrogen boiler, fuel cell, etc.), the overall round-trip efficiency of hydrogen as an energy source remains notably low. Despite this, several justifications exist that make hydrogen a convincing candidate in the future energy systems. These include zero-emission credentials at point of use; extreme versatility across heat, power, transport, chemical and potentially future aviation sectors [25]; facilitating excess renewable diffusion across power and gas [26]; and being regarded as the most efficient inter-seasonal energy storage mechanism [27]. Aug 2019 blackout in the UK was noted as a unique stress test that exposed electrical grid fault lines ensuing from excessive penetration of new equipment and controls associated with renewables and smart grid technologies [19]. It is against this challenging background that hydrogen has emerged as both a unique energy source and a versatile storage mechanism that can bring much greater 'virtual inertia' and stability across the entire energy system and enable its phased decarbonisation. The demonstrator concerned in this chapter included a 36 kg/h electrolyser that utilised proton exchange membrane (PEM) technology and was intended to be deployed close to emergency services to facilitate an initial trial of hydrogen vehicle adaption within public sector service providers.

13.2 The Case-Study Demonstrator

Referred to as SmartHubs SLES, this demonstrator was one of four UK energy system demonstrator projects aimed at encouraging innovative low- or zero-carbon energy solutions that could serve its host community with cheap and clean energy. Additionally, these demonstrators (backed by over £100 m of government finance) also aimed to encourage a new wave of entrepreneurial activities that would see greater adoption of niche technologies, greater engagement of end users in energy management and a broader emphasis of creating prosperity at local level through an energy revolution. SmartHubs SLES demonstrator had a wide range of assets that in combined form offered 1.96 MW_p of renewable generation, 24 MWh_p of household and grid-level electrical storage, 3.87 MW_p of EV charging, 3.49 MW_{th} of air and marine source heat pump thermal capacity and 36 kg/h of H_2 production using proton exchange membrane (PEM) technology. The platform for deploying these assets was 250 homes and 40 commercial sites spread across a wide geographical area in the South of England. Figure 13.2 attempts to offer an outline of the magnitude of assets and their spread across heat, power and transport vectors.

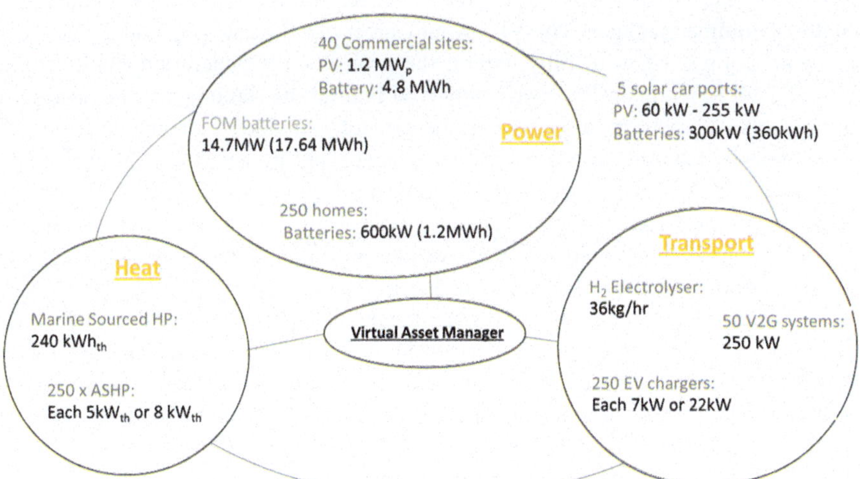

Fig. 13.2 Overview of SmartHubs SLES assets where local and global optimisation of services was pursued through a virtual asset manager (*V2G* vehicle to grid, *FOM* front of meter)

Within the transport vector, a major global car manufacturer was to install 250 EV chargers, and 50 additional chargers with vehicle-to-grid (V2G) technology incorporated. A 2 MW proton exchange membrane was also intended to produce 36 kg/h of green hydrogen and stationed close to emergency services to support a first generation of hydrogen vehicles commissioned for ambulance and police service. The electrolyser had the ability to be turned on/off instantly with five cycles per hour feasible; therefore it could participate in several ancillary services as outlined in Table 13.1. Hydrogen would be produced at 20 bar and compressed to pressures of 700–900 bars to support vehicle filling duties.

The heating vector involved the installation of 250 ASHP in domestic properties that depending on the size would be fitted with either 5 kW_{th} or 8 kW_{th} units. These units where all equipped with a machine learning algorithm that would extract building thermal response and occupant heating preferences across its initial trial period to then use the natural inertia of buildings to respond to grid signals and provide several demand-side response services. A total thermal output of 240_{th} was to be met by one or several modularised marine-sourced heat pumps (MSHPs) to feed an ambient temperature (8–15 °C) district heating network supporting commercial sites and schools. The baseload provided by the MSHP(s) would then be boosted at site level via water-to-water heat pumps to deliver heating and domestic hot water (DHW). The MSHP unit(s) were expected to be able to respond to a grid signal immediately and therefore could engage in a number of auxiliary services with the only constraint being a maximum of six starts per hour.

A more diverse range of assets were intended to be deployed on the power vector, with both generation and storage at building and district levels (Fig. 13.2). This included 600 kW_p (1.2 MWh) of new Li-ion batteries deployed behind the meters at 250 homes and a total of 22.44 MWh of second-life lithium-ion batteries deployed at multiple commercial sites. The VPP would supervise the charging and discharging of these storage assets to fully exploit the project's PV generation (with a peak rate of 1.96 MW_p from PV arrays deployed both at 40 commercial sites and also 5 car ports). Clearly in combination, the project contained a diverse portfolio of generation and storage assets across heat, power and transport vectors that could be configured to deliver against a multiplicity of outcomes. Figure 13.3 outlines how the physical assets, the power/energy flow and data streams had to support a static contractual but dynamic aggregation and market engagement layers. The project data was also to be made available to Newcastle University to pursue two overall research questions. First how the technical optimisation of the VPP could be improved over a long project horizon as increasing high-frequency asset data became available. The second was to quantify (against a business-as-usual (BaU) scenario) the level of carbon mitigation that all assets in unity can realise, and how VPP could enhance asset portfolio carbon mitigation even further.

Table 13.1 A list of local vs. centralised (VPP) optimisation actions that project asset portfolio had the potential to offer

	Control action	Data required	Optimisation	Services	
FOM batteries	Charge/discharge	AS details BCS	VPP	TA – CM – FR	VPP
BTM batteries	Charge/discharge	AS details BCS	Local/VPP	TA – CM – FR	
Car port PV and battery	Charge/discharge battery – charge EV	AS details BCS EV status Solar irradiation	Local/VPP	TA – CM – PA	
Distributed ASHPs[a]	Zone heating On/off	AS details Zone thermal requirement Building thermal response	Local/VPP	CM – TA	
MSHP	Zone heating On/off	Thermal demand Thermal system inertia	Local/VPP	CM – TA	
H_2 electrolyser	On/off	Electrolyser status	VPP	CM – TA – PA	

Notes
(i) *TA* Triad avoidance
(ii) *CM* Constraint management
(iii) *FR* Frequency response
(iv) *PA* Price arbitrage
(v) *AS* Ancillary services (that include i to iv above)
(vi) *BCS* Battery charge status
(vii) *BTM* Behind the meter (battery)
[a]ASHPs have a long ramp time of 6 min and therefore are not suitable for FR

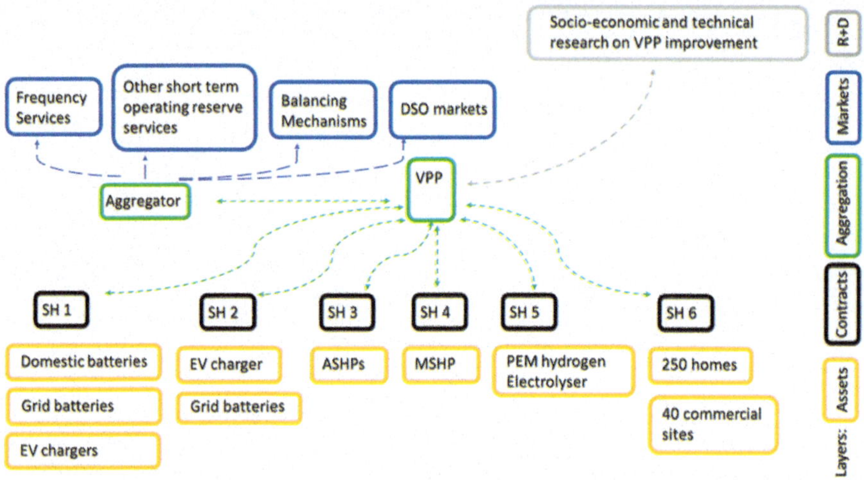

Fig. 13.3 Overview of SLES assets where local and global optimisation of services was pursued through a virtual asset manager performing VPP duties (*SH* stakeholder, *DSO* distribution system operators)

13.3 Virtual Power Plant: Benefits and Challenges

Quite clearly a portfolio of distributed assets operating in or across interlinked heat, power and transport vectors can be managed to satisfy a variety of different objectives. Initially the funding of four UK demonstrators (of which SmartHubs SLES was one) intended to deliver low cost and carbon energy to hosting communities, showcase ancillary services and demand response from a heterogenous RES asset base, encourage energy enterprise and green job creation and enhance the level of engagement and interest of the host communities. The broader socio-economic objectives of the project however are beyond the scope of this chapter. VPP architecture is examined under commercial (CVPP) and technical (TVPP) frameworks [28], and this chapter focuses primarily on the TVPP framework of SmartHubs SLES demonstrator. Table 13.1 outlines the control actions that a TVPP can undertake to execute on each class of assets and the data required from any asset to enable a revenue-generating or carbon-mitigating flexibility service. In addition to services provided by each individual asset class, the VPP design seeks to further optimise the overall generation of value from the combined asset portfolio; however the prioritisation of these services could only be finalised in consultation with the 250 domestic property occupants that was an original objective of this demonstrator.

A notable difficulty is that the performance of some asset classes can be optimised locally (e.g. car port PVs can charge local battery or the EV), while others can only be addressed by the VPP (H_2 electrolyser or front of meter (FOM) batteries). This remained a challenge within the project since a successful TVPP design had to find ways of reconciling onboard control philosophies of each individual asset entity with a system-level priority. While multi-objective optimisation and prioritising actions using weight coefficients are a tried and tested control theory subject, in the case of SmartHubs SLES, control design necessitated a completely fresh approach that wasn't addressed in the conventional global optimum of cost, comfort and carbon solutions. A specific example of this was the onboard (and hence distributed) machine learning (ML) algorithm that controlled ASHPs by extracting building physics features from operational data to then dictate optimum starts, DR services and avoidance of times of high grid carbon intensity while still delivering occupant comfort. ML-derived control actions of 250 distributed ASHP that each had developed a unique operational regime to reflect property thermal response and occupant heating patterns therefore needed to be united with a centralised TVPP seeking to perform an ancillary service. These aggregated vs. local control signals required careful safeguards to avoid control conflicts, excessive plant cycling and occupant discomfort.

13.4 Unlocking the Value of Data

Successful management of SmartHubs SLES assets through a VPP requires communication, automation, analysis of historical data and dynamic engagement of individual components that inevitably lead into the production of a substantial amount of data. This requires a robust solution for data management and analysis as well as cyber-security. As outlined in [29], one of the earliest steps taken within the project was a Smart Energy Platform (SEP) architecture that was designed to conduct data management and analysis for this demonstrator in order to provide optimal planning and technical operation as well as socio-environmental and financial research.

Figure 13.4 illustrates the data processing structure of the SEP which collects, backhauls and stores data and enables near-real-time as well as retrospective experimental data analysis. The SEP data processing program retrieves recorded operational data and stores it in the database by calling Python REST APIs. These include raw power network data, static asset and dynamic weather data and socio-economic and policy data. Depending on the requirements, different datasets could be selected to feed any specific analysis. For instance, SEP retrieves relevant data to facilitate assessment of balancing energy supply and demand, explores the added value that the flexibility of assets can provide and delivers cost and carbon benefits for stakeholders, including businesses (on commercial sites) and residents of 250 homes as the asset portfolio continues generating data in a low-carbon future. Given that the project asset portfolio was initially planned for installation in 2020, SmartHubs SLES would have been operational from third to beyond fifth UK carbon budgets (CB), where reductions of 8.6% (third CB), 23.3% (fourth CB) and 11.5% (fifth CB) are enshrined in UK law as it moves towards a legislated 2050 net zero carbon deadline [30].

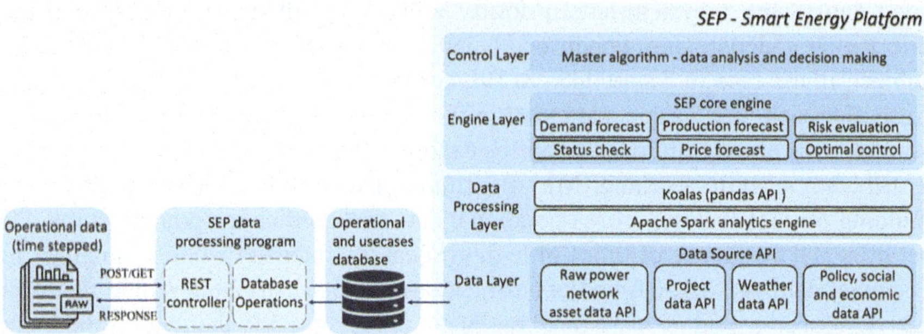

Fig. 13.4 Smart Energy Platform data analysis

The modularised multiple-layer design of the SEP also makes the interoperability and reuse of existing implementations possible [31]. It can be easily maintained, customised or extended in the future for different applications and research purposes.

A greater perspective of data management needed to be taken by the project partners to safeguard the communication infrastructure, VPP management, user identity and information and supply of services against cyber incidents. Initial stages of cyber-security arrangement required encryption of data, firewall protection for devices and authentication of users via secure digital apps. Cyber-security of energy systems however is a new and evolving discipline, and against a number of global energy-related incidents, additional protection and ongoing learnings are required to adequately protect energy system integrity and service offering [32].

13.5 Discussion, Challenges and Findings

Over an initial project timespan of 24 months, the team experienced four notable challenges, first the procurement process (which remains beyond the scope of this chapter), second design and specification of individual assets and their onboard control, third the design of a TVPP while avoiding control conflict and finally the challenge of honouring autonomy of choice in particular for residential buildings while retaining full flexibility of the asset portfolio to deliver auxiliary and balancing services. The Covid-19 pandemic presented substantial logistical limitations since 250 homes required energy system retrofits with PV, behind the meter Li-ion battery and ASHPs. This has meant a temporary halt of project activities; however the design stage socio-economic insights of the project team can be summarised as follows:

(i) There is little existing guidance on the contractual format and extent of asset control that could maximise consumer adoption of data collection and remote asset management while simultaneously maximising financial returns for VPP platform providers.

(ii) The UK has a clear decarbonisation pathway through five carbon budgets that are designed to provide forward guidance to businesses and communities. Yet the shape and speed of decarbonisation trajectory (and future cost) of different primary fuels carry substantial uncertainty. For instance, annual rate of decarbonisation of grid electricity or the penetration (and economics) of hydrogen in transport or heating vector (via blending with natural gas) can only be studied across a broad range of plausible scenarios. The wide spread of these scenarios propagates into large uncertainties that can hinder conclusive results. This appears to suggest that TVPP design involving a wide range of assets across multiple energy vectors will have to incorporate an iterative aspect whereby an initial design aimed at system robustness is improved and adjusted periodically to reflect learnings accumulated through operational data and socio-technical realities that emerge with advancing time.

(iii) The UK has attempted to ease the participation in electricity markets via its New Electricity Trading Agreement. This includes (i) future, (ii) day ahead, (iii) reserve and (iv) real-time balancing markets. Despite having an aggregate capacity to dispatch services at magnitudes of multiples MWs, SmartHubs SLES asset availability could not be confidently guaranteed at its maximum capacity given the multiplicity of stakeholders, operational regimes and asset types. This limited TVPP participation in mostly intraday, ancillary and real-time balancing markets, which may limit the full realisation of asset profitability and the formation of contracts that are in place prior to asset deployment.

(iv) Existing data to benchmark the current carbon intensity of UK power, heating and transport vectors are limited and sporadic. This creates difficulties in establishing a business-as-usual baseline which in turn allows quantifying the amount of carbon saving achieved via the deployment of assets in isolation and finally the added value of VPP. This can underestimate the value of a VPP managing low carbon and DR-ready assets given the large CapEx required to displace legacy alternatives. This is despite future carbon risks, for instance, a VPP platform managing a SLES may emerge as more profitable in a future scenario where increasing levels of carbon taxations are levied, which then justifies upfront SLES CapEx.

From a technical perspective, the main project findings and challenges of VPP design and characterisation are as follows:

(i) Most thermal conversion energy assets (heat pumps, hybrid boilers, etc.) are underutilised over their working life. For the existing generation of thermal conversion assets, their lack of adaptability to being remotely controlled via a VPP platform (and yet delivering their duties in a more dynamic manner) presents technical challenges in unlocking their additional value. A case may exist for regulatory interventions to mandate thermal conversion asset designs to support active involvement with power (and energy) systems management to replace existing manufacturing practice of design as siloed entities with static onboard control solutions.

(ii) Actual levels of flexibility available to a VPP operator from real-world assets (i.e. marine or air source heat pumps, electrolysers) are still not very well defined by plant manufacturers. For instance, the number of plausible cycling (on/off) that a piece of plant can be subjected to each hour without jeopardising its integrity and the duration that the plant needs to remain operational after a DR event before the asset is ready to be deployed to perform another DR service are little understood in particular for heating and hydrogen plants.
(iii) Modelling the interactions between assets from multiple energy vectors (heat/power/transport/hydrogen) is quite challenging as these energy vectors are represented by different parameters, for instance, voltage, current and frequency (power), mass flow and temperature (DH and thermal plants) and Kg of H_2 production (H_2 electrolyser). This (when combined with the two previous points) reduces the ability to extract combined flexibility of a multi-vector energy system and added value of a VPP at design stage.
(iv) Taking heed of uncertainties outlined in (i)–(iii) in the form of a bandwidth poses limitations to VPP design and its global optimum. This is particularly challenging as future VPPs will have to reconcile cost, carbon and comfort optimums with occupant autonomy and energy asset characteristics.
(v) Prior to the availability of operational data, existing parametric modelling solutions have a limited scope to extract system features particularly of a multi-vector, multi-site energy system and at low temporal resolutions. This hinders determination of what emergent properties may follow from aggregating the control of all assets. Therefore, the confidence in the design of a VPP for a SLES system that fully exploits dynamic interaction between participants and energy assets remains limited and speculative during SLES design and prior to the availability of operational data.

13.6 Conclusion

This chapter reported on the initial findings of a smart local energy system (SLES) demonstrator part funded by UK government and referred to as SmartHubs SLES. While a solitary low-carbon energy asset can deliver carbon savings against comparative conventional technologies, this demonstrator sought to deliver a much broader range of cost and carbon benefits from aggregating heterogeneous RES assets using a VPP platform. These additional services relied on low-loss fifth-generation district energy, distributed generation and storage, EV charging and hydrogen electrolyser assets to offer ancillary services, increase self-consumption via BTM batteries and VPP signals, increase system virtual inertia through exploiting deferable loads in domestic and commercial buildings and encourage prosumer activity, niche technology deployment and energy entrepreneurship. Except for power system assets, technical knowledge on how to deploy hydrogen and thermal plants for DR events were found to be in early stages. This resulted in a conservative magnitude of services that the demonstrator could offer in response to grid signals, which in turn limited project ability to arrive at conclusive remarks on the economics of a VPP and reduced the confidence of quantifiable revenue streams (from future energy markets).

Most SLES asset entities were manufactured with onboard controls tasked with a local optimum. This local control could be terminated by the VPP in preference for a global optimal service that needed to observe stringent cycling thresholds particularly with thermal and hydrogen assets containing liquid refrigerants and pressurised fluids. While the number of cycles to failure and operational robustness of power system components are well characterised, more manufacturer data is required to allow future VPP design to fully exploit non-power system assets. This is instrumental for SLES design as the load magnitude of heating and transport class of assets are often substantial.

At SLES design stage, arriving at clarity on the combined value of distributed assets is hindered by the order of priority of services and weighting of objectives (cost, comfort, carbon, system integrity), levels of prosumer engagement and full knowledge of asset performance under a DR event together with the location- and case-specific formation of SLES portfolios. This will mean that an optimal design for the governing VPP is more likely to be mapped post inauguration of energy entities and through data-driven approaches that can extract unique system features from actual analytics. Therefore, next-generation SLESs need to be commissioned for operational robustness and are only likely to realise their full potential as VPP entities from observed operational flexibilities, asset responsiveness, data-driven machine learning insights and participant activity levels. It is only with advancing time that such SLES portfolios can mature into dynamic agents capable of delivering seasonally and diurnally different outcomes against a changing demand and supply landscape.

References

1. Adu-Kankam, K. O., & Camarinha-Matos, L. M. (2018). Towards collaborative virtual power plants: Trends and convergence. *Sustainable Energy, Grids and Networks, 16*, 217–230.
2. Rouzbahani, H. M., Karimipour, H., & Lei, L. (2021). A review on virtual power plant for energy management. *Sustainable Energy Technologies and Assessments, 47*, 101370.
3. Cleveland, C. J. (2004). *Encyclopedia of energy*. Boston University.
4. Lund, H., et al. (2017). Smart energy and smart energy systems. *Energy, 137*, 556–565.
5. Department for Business, E.I.S (2020). *UK energy in brief* (BEIS, Ed.). Crown.
6. Koirala, B. P., et al. (2016). Energetic communities for community energy: A review of key issues and trends shaping integrated community energy systems. *Renewable and Sustainable Energy Reviews, 56*, 722–744.
7. Bronski, P., et al. (2014). *The economics of grid defection: When and where distributed solar generation plus storage competes with traditional utility service* (pp. 1–73). Rocky Mountain Institute.
8. CSIRO. (2013). *Change and choice: The future grid forums analysis of Australia's potential electricity pathways to 2050*. CSIRO.

9. Koirala, B. P., et al. (2016). Local alternative for energy supply: Performance assessment of integrated community energy systems. *Energies, 9*(12), 981.
10. Bačeković, I., & Østergaard, P. A. (2018). Local smart energy systems and cross-system integration. *Energy, 151*, 812–825.
11. Young, J., & Brans, M. (2017). Analysis of factors affecting a shift in a local energy system towards 100% renewable energy community. *Journal of Cleaner Production, 169*, 117–124.
12. Richter, L.-L., & Pollitt, M. G. (2018). Which smart electricity service contracts will consumers accept? The demand for compensation in a platform market. *Energy Economics, 72*, 436–450.
13. Iria, J., & Soares, F. (2019). A cluster-based optimization approach to support the participation of an aggregator of a larger number of prosumers in the day-ahead energy market. *Electric Power Systems Research, 168*, 324–335.
14. Castagneto Gissey, G., et al. (2019). Value of energy storage aggregation to the electricity system. *Energy Policy, 128*, 685–696.
15. Ford, R., et al. (2021). Smart local energy systems (SLES): A framework for exploring transition, context, and impacts. *Technological Forecasting and Social Change, 166*, 120612.
16. Department for Business, E.I.S. (2020). Guidance on heat networks. Available from: https://www.gov.uk/guidance/heat-networks-overview
17. Lund, H., et al. (2010). The role of district heating in future renewable energy systems. *Energy, 35*(3), 1381–1390.
18. Sperling, K., & Möller, B. (2012). End-use energy savings and district heating expansion in a local renewable energy system – A short-term perspective. *Applied Energy, 92*, 831–842.
19. Bialek, J. (2020). What does the GB power outage on 9 August 2019 tell us about the current state of decarbonised power systems? *Energy Policy, 146*, 111821.
20. Bayrak, Z. U., et al. (2016). A low-cost power management system design for residential hydrogen & solar energy based power plants. *International Journal of Hydrogen Energy, 41*(29), 12569–12581.
21. Dawood, F., Anda, M., & Shafiullah, G. M. (2020). Hydrogen production for energy: An overview. *International Journal of Hydrogen Energy, 45*(7), 3847–3869.
22. Fakeeha, A., Ibrahim, A. A., Khan, W. U., Seshan, K., Al Otaibi, R. L., & Al-Fatesh, A. S. (2018). Hydrogen production via catalytic methane decomposition over alumina supported iron catalyst. *Arabian Journal of Chemistry, 11*, 405–414.
23. Chiesa, P., Romano, M. C., & Kreutz, T. G. (2013). 10 – Use of membranes in systems for electric energy and hydrogen production from fossil fuels. In A. Basile (Ed.), *Handbook of membrane reactors* (pp. 416–455). Woodhead Publishing.
24. Peng, X. D. (2012). Analysis of the thermal efficiency limit of the steam methane reforming process. *Industrial & Engineering Chemistry Research, 51*(50), 16385–16392.
25. Schmidt, O., et al. (2017). Future cost and performance of water electrolysis: An expert elicitation study. *International Journal of Hydrogen Energy, 42*(52), 30470–30492.
26. Simonis, B., & Newborough, M. (2017). Sizing and operating power-to-gas systems to absorb excess renewable electricity. *International Journal of Hydrogen Energy, 42*(34), 21635–21647.
27. International Energy Agency. (2014). *Energy technology perspectives 2014*. International Energy Agency.
28. Naval, N., & Yusta, J. M. (2021). Virtual power plant models and electricity markets – A review. *Renewable and Sustainable Energy Reviews, 149*, 111393.
29. Wang, K., & Wade, N. (2021). An integration platform for optimised design and real-time control of smart local energy systems. In *2021 12th international renewable energy congress (IREC)*.
30. Royapoor, M., et al. (2019). Carbon mitigation unit costs of building retrofits and the scope for carbon tax, a case study. *Energy and Buildings, 203*, 109415.
31. Wang, K., Siebers, P.-O., & Robinson, D. (2017). Towards generalized co-simulation of urban energy systems. *Procedia Engineering, 198*, 366–374.
32. Bhuiyan, E. A., et al. (2021). Towards next generation virtual power plant: Technology review and frameworks. *Renewable and Sustainable Energy Reviews, 150*, 111358.

Index

© The Author(s), under exclusive license to Springer Nature Switzerland AG 2022

V. Vahidinasab, B. Mohammadi-Ivatloo (eds.), *Whole Energy Systems*, Power Systems, https://doi.org/10.1007/978-3-030-87653-1

The manufacturer's authorised representative in the EU is Springer
Nature Customer Service Centre GmbH, Europaplatz 3, 69115 Heidelberg,
Germany. If you have any concerns regarding our products, please
contact ProductSafety@springernature.com

Printed and bound by CPI Group (UK) Ltd, Croydon, CR0 4YY
30/09/2026
02225491-0011